International Symposium of the Society of Core Analysts (SCA 2022)

E3S Web of Conferences Volume 367 (2023)

Austin, Texas, USA
19-22 September 2022

Editor:

Jin Hong Chen

ISBN: 978-1-7138-6643-5

Printed from e-media with permission by:

Curran Associates, Inc.
57 Morehouse Lane
Red Hook, NY 12571

Some format issues inherent in the e-media version may also appear in this print version.

For additional information, please contact EDP Sciences – Web of Conferences at the address below.

EDP Sciences – Web of Conferences
17, Avenue du Hoggar
Parc d'Activité de Courtabœuf
BP 112
F-91944 Les Ulis Cedex A
France

Phone: +33 (0) 1 69 18 75 75
Fax: +33 (0) 1 69 28 84 91

contact-edps@webofconferences.org

Additional copies of this publication are available from:

Curran Associates, Inc.
57 Morehouse Lane
Red Hook, NY 12571 USA
Phone: 845-758-0400
Fax: 845-758-2633
Email: curran@proceedings.com
Web: www.proceedings.com

TABLE OF CONTENTS

Austin, Texas USA

Sept 19-23, 2022

Final Program

Core Analysis in Carbon Geosequestration, Geothermal, and Nuclear-Waste Disposal

From the 2022 President

Dear Colleagues and Friends

On behalf of the Society and the Board of Directors, I am pleased to welcome you to the Society of Core Analysts (SCA) International Symposium 2022. Happy to see you back in person in Austin, TX USA after two years of online formats!

This year's symposium theme is "Augment Core Analysis to Embrace Energy Transition". The world is changing rapidly and we are in the middle of important debates about energy security, energy transition and climate change. As the world changes, so do applications for SCAL. It can be stated that oil and gas will continue to play a major role as energy sources. However, there is consensus that emissions from fossil fuels must be avoided and the energetic geological potential must be used as sustainably as possible. This opens up great opportunities for our society to expand the SCA portfolio to include alternative energies and climate protection. Even if topics such as geological carbon and hydrogen storage and the use of geothermal energy have been subject for several years, we are making them explicit with this year's symposium – let's move on to the future!

Organizing an SCA symposium takes a lot of effort and requires passionate people. I would like to thank everyone who actively participated in the organization to make this symposium possible. The backbone of the symposium is the VP Technology, JinHong Chen, and his Technical Committee, who are responsible for all aspects of the technical program. JinHong and his team put together an excellent program with the high technical standard that we are used to from SCA. VP Arrangement, Tracie Walker organized the venue and the social program of the symposium. In this way, she takes care that everything runs smoothly, of our well-being and that we can enjoy the meeting – she is the key person. Last but not least, I would like to thank Melanie Young, our SCA Executive Director, for her invaluable support in all matters – Thank you all!

I am sure that we will have an excellent conference with interesting presentations, fruitful technical discussions and - finally again - lots of socializing and that we will move forward as a community. I look forward to seeing you in Austin!

Holger Ott

SCA President

From the 2022 VP Arrangements

Howdy and Welcome Ya'll!

On behalf of the Organizing Committee, I would like to welcome the attendees of the 2022 SCA Annual Symposium in Lakeway Texas. We hope that you will enjoy the beautiful setting here in the Texas Hill Country with the incredible views of Lake Travis. Welcome to this year's Symposium and we hope each of you enjoy our Symposium theme "Augment Core Analysis to Embrace Energy Transition".

During our conference, in the heart of the Hill Country, you will hopefully discover and enjoy the beauty and charm of the area and the city of Austin with all the many different aspects.

Austin, Texas is geographically and politically the center of the Lone Star State. Austin residents are made up of a diverse group of individuals, including musicians, politicians, university professors, students, and so on. The city is home to enough large sites of major technology corporations to have earned it the nickname "Silicon Hills." The slogan that many residents have picked up is Keep Austin Weird. The slogan is also used for a campaign to preserve smaller local businesses and resist excessive commercialization.

Austin, Texas is known for being the "Live Music Capital of the World" and is home to more than 100 live music venues. Austin is home to the Limits Music festival (ACL) which features 130 bands in over 3 days. Austin is not only known for their music, but is also known for the Texas Longhorns.

This gorgeous city averages around 300 days of sunshine a year. All that sunshine can be enjoyed through various outdoor activities and attractions. Austin has it all, whether you enjoy biking, hiking, mountain climbing, fishing, swimming, or boating.

Our young professionals' event will be spent right here on Lake Travis as we venture out onto the Lake on a "Party Boat". Individuals joining in on this fun adventure will have an opportunity to see some beautiful outcrops while enjoying the "cool" evening air and hopefully everyone gets an opportunity to see our gorgeous Hill Country fireflies.

We will discover more of the Texas Hill Country when we visit Hamilton Poole and Reimer's Ranch to learn more about the geology of the area with the traditional Friday field trip, led by the Bureau of Economic Geology.

With a large variety of vendors, technical veterans, industry leaders, and rising young professionals, the SCA meeting is a great opportunity for professional interaction, to share ideas, innovations, knowledge, best practices, products, and services. In the exhibition halls the vendors showcase their latest technical innovations and offerings in equipment and services.

I would like to thank the sponsors and vendors for their support, the authors and technical committee of the SCA for their dedication to make the SCA a passionate conference, and a thank you for all the volunteers who helped in the organization. A very special thanks to Melanie Young who is the backbone of this organization.

Welcome Ya'll!

Tracie Walker,

VP Arrangement

From the 2022 VP Technology

Dear Colleagues and Friends,

Welcome to the 2022 SCA International Symposium in Austin, Texas!

The theme of the 35th symposium is "Augment core analysis to embrace energy transition". People working in the energy and related sectors are facing a critical challenge as the world increasingly transitions towards clean energy. SCA, with our special expertise, can embrace the challenge and contribute to the transition. In consideration of this, I have invited four experts to give an opening workshop "Core Analysis in Carbon Geosequestration, Geothermal, and Nuclear-Waste Disposal". I would like to thank the four final presenters: Martin J. Blunt, Jon Burger, Igor Faoro, and Garrard Rodney. I am sure these presentations will be interesting and informative to all attendees.

The 2022 SCA technical program include 32 oral presentations in 11 sessions and 31 posters in two sessions. There are 46 manuscripts and 17 abstracts accepted from the 94 submitted abstracts.

Although I have attended the SCA multiple times and have organized other conference as chairperson, I was pleasantly surprised by the dedication, time, and efforts of the authors and technical committee to improve the manuscript quality. Quite a few manuscripts went through multiple rounds of review and revision. I would like to thank all the authors and the Technical Committee who represent the best of our society and are responsible for the high technical standard of this symposium.

My special thanks to Melanie Young, our SCA Executive Director, for her invaluable support in all matters and for her patience with my frequent impromptu phone calls.

Finally, I am looking forward to seeing you all in Austin, Texas for a successful gathering and technical communication after a two-year break!

JinHong Chen

SCA VP of Technology

Current SCA Board

EXECUTIVE DIRECTOR	Melanie Young Society of Core Analysts
PRESIDENT	Holger Ott Montanuniversität Leoben
VP TECHNOLOGY	Jinhong Chen Aramco Research Center - Houston
VP MEMBERSHIP	Matthias Appel Shell
VP PUBLICATIONS	Josephina Schembre-McCabe Chevron
VP ARRANGEMENTS	Tracie Walker Diversified Well Logging
SECRETARY/TREASURER	Christa Vindum Vindum Engineering
PAST PRESIDENT	Benjamin Nicot TOTALEnergies
PRESIDENT ELECT	Carl Fredrik Berg NTNU
YOUNG PROFESSIONALS DIRECTOR	Alvaro Munoz-Beltran Weatherford Labs
INDUSTRIAL RELATIONS LIAISON	Craig Lindsay Core Specialist Services Limited
MARKETING DIRECTOR	Stefano Pruno Weatherford Labs
AISA DIRECTOR	Hu Dong iRock Technologies
AUSTRALIA/ASIA DIRECTOR	Christoph Arns University of New South Wales

2022 Technical Committee

The SCA Technical Committee is the backbone of the Society of Core Analysts. Its members rank and evaluate the submitted abstracts and the technical quality of the manuscripts. In the review of each individual manuscript and in-depth discussions with the authors, many hours are spent until a manuscript is accepted. In many cases, the interaction between authors and TC members leads to a higher quality of individual manuscripts. As a result, the annual SCA symposia are packed with high-quality presentations. For this, sincere acknowledgement go to this year's Technical Committee members:

Adam Moss

Benjamin Nicot

Bo Gao

Carl Fredrik Berg

Carl Sondergeld

Cyril CAUBIT

David Potter

Einar Ebeltoft

Eric Withjack

Fabrice Pairoys

Harry Xie

Hassan Mahani

Hendrik Rohler

Holger Ott

Issa Abu Shiekah

Jim Funk

James Howard

John Mills

Jos Maas

Kory Holmes

Matthias Halisch

Mike Dick

Olga Vizika

Patrick Egermann

Rudolf Held

Ryan Armstrong

Shannon Eichmann

Stacey Althaus

Stephanie Perry

Stefano Pruno

Steffen Berg

Subhash Ayirala

Will Richardson

Technical Achievement Darcy Award Recipient
James Howard

James J. Howard spent much of his 40+ year career in the laboratory making measurements on a wide range of geological and petrophysical properties on reservoir rocks. He has co-authored 80+ relevant papers that cover a wide range of topics in core analysis and has several patents. He has B.S. and Ph.D. degrees in the geosciences, with his graduate work focused on geochemistry and clay mineralogy. Much of his career was in the oil and gas industry, with several short interludes at academic institutions along the way, and has ended up as an advisor to a software company involved with the digital rock world. His work has included the early development of NMR interpretation based on multi-exponential distributions, in-situ monitoring of multiphase flow experiments with NMR and then MRI, wettability characterization, high resolution imaging of rocks that included mineral mapping from SEM images, the CO2-CH4 exchange process in natural gas hydrates and a range of measurements on very low permeability rocks at reservoir conditions. He is most proud of the accomplishments of the dozens of students and young professionals who have worked in his labs over the years.

He has belonged to a wide range of technical societies throughout his career, with the SCA being his favorite since he first heard about it in the late 1980s. He has served on the SCA Technical Review committee for a number of years. Several of his 19 co-authored SCA papers were selected as “Best of SCA” and published in Petrophysics.

The Venue

The hotel is a beautiful setting for our conference and is admittedly a bit spread out. Below is a map of the facility and a listing of where you can find important meeting spaces. All coffee breaks will take place in the Exhibition Hall. There will also be signs throughout the Hotel to help guide you to your destination and volunteers to show you the way!

Opening Reception located in Sunset Room = G

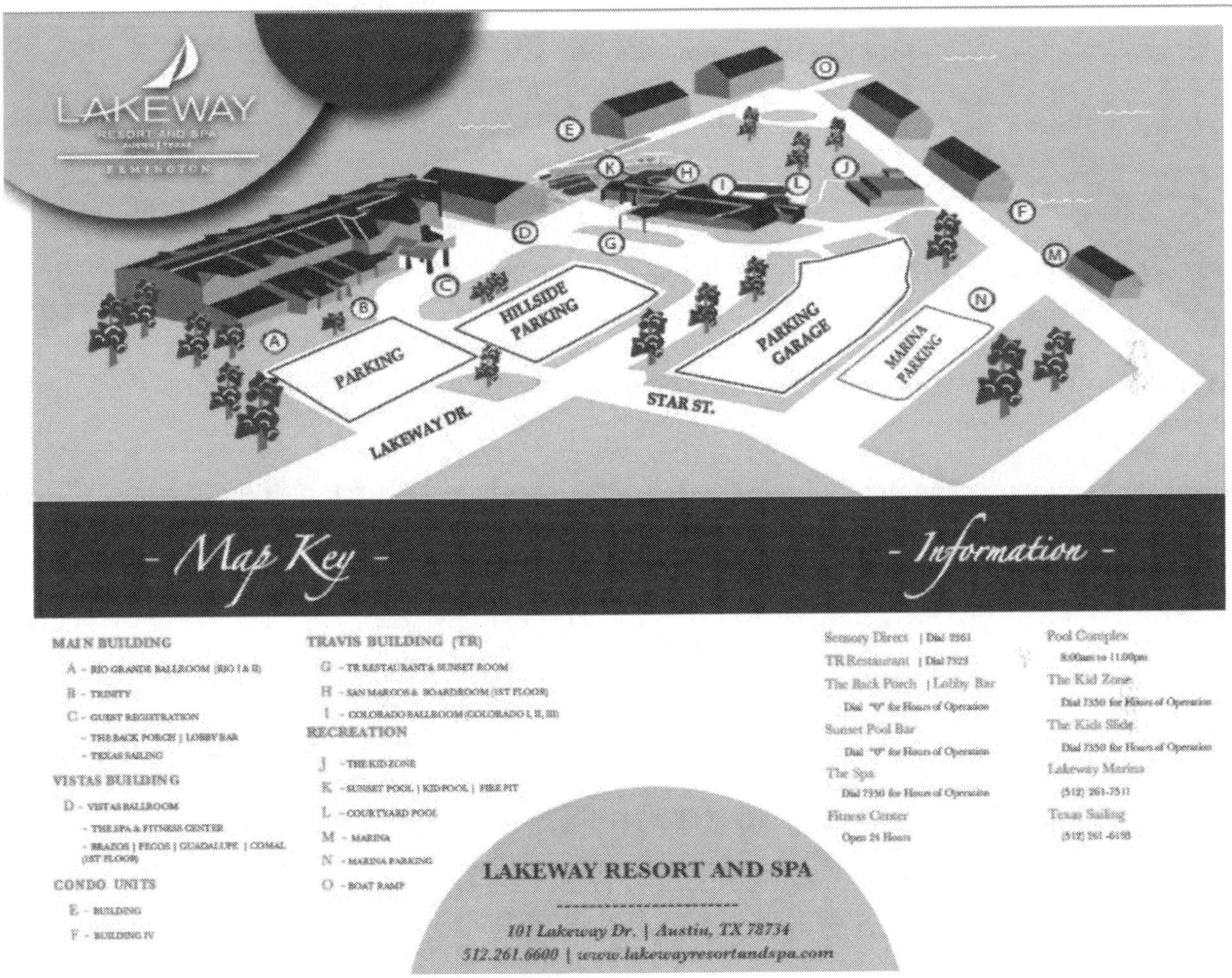

Presentations in Rio Grande Ballroom = A

Exhibitions/Coffee Breaks in Vistas Ballroom = D

Lunch in Colorado Ballroom = I

Young Professional Event at the Marina = M

Golf – Sunday 18th, 8:00 a.m. – 2:30 p.m.

Course Name: The Hills of Lakeway, Live Oak Course

Prizes
1st, 2nd, and 3rd Place
Closest to the Pin
Long Drive

Format will be individual Gross Scoring….no need to provide a handicap.

Tee Time: 9:00 am. With 8-minute intervals between foursomes.

Cap at 16 players

Short Courses – Monday 26th, 8:50 a.m. – 12:30 p.m.

Short Workshop: Core Analysis in Carbon Geosequestration, Geothermal, and Nuclear-Waste Disposal

A special thanks to the presenters who helped put this short workshop together; Martin Blunt, Jon Burger, Igor Faoro and Garrard Rodney.

Coffee and tea will be available during breaks. Lunch will be provided to attendees.

Short Course is Kindly Sponsored by:

Opening Reception – Monday 19th, 5:30p.m. -8:30 p.m.

The Opening "Icebreaker" Reception will be held in the Sunset Room. Snacks and drinks will be served to foster a relaxed atmosphere to meet and greet both old and new colleagues.

Dress Code: Business casual

Technical Sessions – Monday 19th, Tuesday 20th, Wednesday 21st and Thursday 22nd, 8:15 a.m. - 5 p.m.

(With the exception for start at 1:30 p.m. on Monday and end at 3:30 p.m. on Thursday).

Oral presentations: The Symposium will offer 32 oral presentations, 25 minute presentations followed by 5 minutes for discussion.

Poster sessions: The symposium will offer 31 posters, distributed in two poster sessions. Tuesday (Poster session Odd) and Wednesday (Poster Session Even).

Young Professionals Event – Tuesday 20th, 6:00-9:00 p.m.

Lakeway Marina – Party Boat Tour

Lakeway Marina 6:00 – Easy 2 minute walk to the marina from the hotel

Party Boat – 3 Hour Tour

Cook Out – Hot Dogs, Hamburgers, Vegan Burgers – Assorted Chips, etc.

Drinks – Beer, Whiteclaws, Canned Sodas, Water

Kindly Sponsored by: Shell

Awards Gala Dinner – Wednesday 21st, 6:30 - 9:30 pm

The gala dinner will be held at the Star Hill Ranch, located at 15000 Hamilton Pool Rd. Austin, TX 78738. Buses will be picking everyone up at the Hotel at 6pm for transportation to the event. We will enjoy some great BBQ and Whiskey from the region.

Dress code: (Super Casual) Casual dress code or go all out Cowboy and Cowgirl Style. The evenings may still be warm in Texas and the venue may be a bit dusty so dress comfortably.

Gala Dinner is Kindly Sponsored by:

Optional Field Trip – Friday Sept 23rd, 9:00 a.m.- 3:15 p.m.

Optional Friday Field Trip - Hamilton Pool Preserve and Reimers Ranch

Leave Lakeway at 9:00 am on Friday, September 23rd

- Transportation and lunch provided.
- Hamilton Pool: 9:45 to 11:00 am
- Reimers Ranch – 11:30 am
- Lunch and Safety Briefing
- Geological Tour Outcrops 12:30 – 2:30 pm
- Arrive Back at Lakeway Hotel 3:15 pm

Note: Reimers ranch cores will be at the SCA conference for viewing throughout the week.

Trip Leaders: Charlie Kerans, Brian Hunt, Charlotte Sullivan, and Toti Larsen.

Publication of Proceedings

The proceedings are prepared in USB format and will be given at the Symposium to all registered participants. Additional USBs may be ordered the SCA web site:

www.SCAweb.org

The SCA has decided to no longer carry printed copies of the proceedings.

Exhibition Hours

Monday: 8:00 a.m. – 5:00 p.m.	*Exhibition Build-up starts at 8am
Tuesday: 8:15 a.m. – 5:00 p.m.	
Wednesday: 8:15 a.m. – 5:00 p.m.	
Thursday: 8:15 a.m. – 3:30 p.m	*Exhibition Break-down can start after the afternoon break and complete by

Exhibitors

AMETEK Chandler Engineering – AMETEK Chandler Engineering produces the highest quality instruments and measurement systems for the Oil and Gas Industry. Our portfolio includes Quizix precision pumps used in core flooding (NEW models available!), SCAL, EOR, and other fluid delivery applications, the 6100 Formation Response (formation damage) systems, Core flood/EOR flow systems, custom core flow systems for Steady State/Unsteady State permeability measurements, and the Chandler 3000 Series PVT systems for phase behavior studies.

Core Laboratories – Core Laboratories is a leading provider of proprietary and patented Reservoir Description and Production Enhancement services.
Core Laboratories remains dedicated to providing the technology you need to enhance your production. We continue to develop and acquire technologies that complement our existing products and services, and we disseminate these technologies throughout our global network. Core Laboratories' reservoir optimization technologies are used to increase total recovery from existing fields. Our services enable our clients to optimize reservoir performance and maximize hydrocarbon recovery from their producing fields. Core Laboratories has taken extensive measures to ensure the services and data provided by all of our worldwide companies are of the highest quality and integrity. Our commitment to applying and developing new technologies to optimize reservoir performance is unsurpassed in the oilfield service industry. This commitment to technology and to your bottom line makes Core Laboratories, The Reservoir Optimization Company™.

Core Specialist Services - Core Specialist Services are a consultancy that can support every type of core-based project from planning to core, through coring & wellsite, routine, SCAL, formation damage, and all kinds of specialized studies. Our strength lies not only within our internal team with 72 years of experience, but also with our extensive network of associate subject matter experts (SMEs) in a wide range of disciplines, including reservoir engineering, geomechanics, coring, well planning, and much more. If you need a core-related skill we don't have in-house, we can source it

Core Specialist Services support projects related to oil and gas, geothermal, carbon capture & storage, hydrogen storage, and radioactive waste management – principally from a core-related perspective but also with a wider scope in many cases. We can create a virtual team to support your project no matter where it is located globally.

Core Specialist Services has no bias toward a specific provider but seeks only to source the best services and solutions for our clients.

An exception is our working relationship with Craytive Technologies. We are showcasing their BaselineZ platform at this Symposium. BaselineZ is where core analysis and the sub-surface meet the Metaverse. The power of the platform has many facets but includes the ability to integrate and manage all types of core and sub-surface data and imagery in the Virtual Core Shed and Virtual Data Room - from nano-CT, thin section, to the whole core, well, field, basin, or region. It maximizes the value of collaboration as team members can join virtual meetings from anywhere there is an internet connection worldwide – all appearing in the same virtual space as avatars.

At Core Specialist Services our enthusiasm for core is second to none – please do come past of booth or talk to us anytime. We endorse all and every service provider exhibiting at the SCA and will promote any service & service provider we believe provides innovative and cost-effective solutions to our mutual clients in this rapidly changing world.

DCI Corporation – DCI offers innovative solutions to your core testing requirements. From turnkey systems to system components we can help you with your laboratory needs. DCI designs and manufactures both custom and standard core holders, accumulators, syringe pumps, acoustic separators, core flood systems, electrical resistivity systems, rock mechanics systems and much more. Stop by our booth to see how DCI can help you make better measurements on core properties.

Diversified Well Logging – Diversified Well Logging, LLC. is a forward looking geological service company and leader in Surface Measurement While Drilling (SMWD) operations. In our 70 year history, we have been a leader both offshore and onshore with our conventional and unconventional surface evaluation services. Our focus has been, and always will be, the delivery of valuable data and solutions to our clients and partners, helping drive their capital efficiency and return on investment through our innovation and expertise.

Diversified employs a competent staff of over 150 field geologists and field engineers along with experienced field and support personnel. In addition. we maintain an experienced support staff in our operations centers of Reserve, Louisiana (Gulf Coast), Pennsylvania (Northeast Office), Cannonsburgh Pennsylvania (Geo steering Center), Corpus Christi, Texas (South/Central Texas Office), Midland, Texas (West Texas Office and Geo steering Center) and Conroe, Texas (Corporate Office). Our international branches include offices in Villahermosa, Mexico and the Asia Pacific region. All operations centers follow a comprehensive quality control & Safety program.

Our classic surface logging services are enhanced by rock and gas geochemistry performed in realtime. This geochemical/chemostratigraphic data in turn allows our geo steerers to geo position each well with more accuracy and precision than by using gamma ray alone. We also deliver innovation with sample collection automation (RoboLogger™) which improves rock characterization whilst lowering HSE risk. We are now incorporating drilling data and offset well data utilizing A.I. Artificial intelligence and machine learning to generate petrophysical quality logs at lower cost and lower risk. The new deliverables improve our customer's capital efficiency and lower their finding and development costs.

H2 Laboratories - H2 Laboratories, a division of Green Imaging Technologies (GIT), is a rock core, NMR-based laboratory located in New Brunswick, Canada. By utilizing our long-term collaboration with Oxford Instruments, H2's team of NMR experts are able to create and develop client-specific NMR workflows with the objective to provide tangible solutions to our client's unique and ever-changing challenges.

GeoTek - Geotek is a group of companies specialising in the non-destructive analysis of geological cores. We supply our range of Multi-Sensor Core Logger (MSCL), hyperspectral imaging, and X-ray CT systems that use multiple geophysical and geochemical sensors to rapidly and automatically gather measurements on sediment or rock cores. The rugged nature of the equipment makes it suitable for use in either an onshore laboratory/repository environment or onboard survey and drilling vessels. Stop by our booth and discuss our dedicated core plug and core flood CT systems, and find out about our brand-new systems: BoxScan and Hyperspectral Core Imaging System integrating SpectraMap's Infrared Spectrometer technology

Green Imaging Technologies, Inc – Green Imaging Technologies is the industry leader in NMR rock core analysis. Our software is the backbone of the Oxford GeoSpec line of NMR Rock Core Analyzers, which are used by all major oil producers, oil service companies and the most active research institutes worldwide. Our customers have access to exclusive, patented measurements including NMR capillary pressure and quantitative saturation profiles. Our team of experts are focused on rock core analysis, and as such have developed relationships with the most respected researchers and experts in the NMR rock core analysis field. At this years' SCA we will be showcasing new applications such as our latest work on the Oxford Instruments 10,000 PSI overburden cell. We will also be talking about how NMR can be utilized for other applications of interest, such as analyzing potential carbon storage reservoirs. Stop by the Green Imaging/Oxford Instruments booth in the exhibit hall to find out what is new in NMR rock core analysis.

HOT Microfluidics GmbH - HOT Microfluidics is the leading provider of turnkey microfluidic solutions & services for IOR/EOR, conformance control, hydrogen storage and CCS/CCUS. Our InspIOR® is the industry-leading reservoir-condition microfluidics technology platform – ready to target your fluidics challenges.

Math2Market GmbH - Math2Market develops the GeoDict software to provide a complete solution to our clients comprising software, support, project work, user training, and customized software development. The Digital Rock Physics and Digital Core Analysis (DRP-DCA) suite of GeoDict, in combination with imaging capabilities, is a purpose-built tool that enables its users to perform the entire workflow digitally and in-house with unmatched fast runtimes, handling of extremely large data sets, and low hardware requirements. Latest applications of GeoDict are the simulation of two-phase flow with consideration of the entire hysteresis cycle or the digital analysis of rocks for CO2 sequestration techniques and storage mechanisms.

MetaRock Laboratories - A unique, diversely skilled company, MetaRock Laboratories has been providing a range of Automated Integration Solutions, Testing Systems and Services since 1996. Our custom-designed products, built to simulate and withstand very high temperatures and pressures, service a high-value segment in the Oil & Gas, Mining, Geotechnical & Medical Industry.
While our initial drive was to be known at the forefront of the Geomechanics testing field, our strong commitment to staying ahead of the technology and innovation curve–as well as meeting clients' needs has allowed us to extend our portfolio to a wide array of technology and automation solutions.
On the strength of our cross industry experience, MetaRock Laboratories provides a wide range of consulting services in areas of Rock Mechanics and Core Analysis, Custom Software Development and Automation.

Object Research Systems - All things scientific imaging. Server and workstation solutions for visualization, deep learning image enhancement and automated segmetation, image analysis and numerical modeling.

Oxford Instruments – Oxford Instruments will display the renowned GeoSpec line of NMR rock core analysers, including GeoSpec12 which offers ten times greater sensitivity and 100 times faster measurements on tight rocks and low porosity samples. The GeoSpec range measures standard core parameters such as pore size distributions, BVI, FFI, porosity, and T2 cut-off on a single instrument, and can perform advanced measurements such as capillary pressure and spatially resolved T2 distributions with the exclusive use of

Green Imaging Technologies' software. We will also be available to discuss new applications such as gas isotherms, wettability and relative permeability.

PanTerra – PanTerra is an integrated laboratory, geosciences, and engineering consultancy serving the international energy industry for more than 30 years. Our services include conventional and special core analysis, PVT, production chemistry, Enhanced Oil Recovery, subsurface evaluation and modelling, field development, souring studies, engineering and project management services. Capitalizing on our in-house expertise, PanTerra also specializes in recruitment and secondment of subsurface professionals and additionally offers a unique blend of E&P learning customized to individual needs. For more information please visit www.panterra.nl or connect via our LinkedIn page.

Prores As– Prores is a Norwegian employee-owned corporation built on knowledge and with an ambition to bridge the gap from ideas to solutions. Prores consists of an integrated team of researchers and engineers with an extensive network offering petroleum expertise, technology, and software development. We offer high quality solutions for petroleum asset operations. All our products are developed in-house in close cooperation with our industrial partners. The latest spin-offs from Prores are WellGuard AS and WellStarter AS, where Prores remains a majority owner. WellGuard technology is a wireline logging tool for precision measurements of cement barrier integrity through multiple consecutive casings to detect debonding, micro-annuli, cracks, and fluid channels prior to P&A operations. Wellstarter HIPlog is a wireless downhole flow monitoring solution based on heat pulses released into the well stream providing the inflow profile of oil, gas, and water in producing wells. At-the-Bit Mud Loss Control is a new initiative currently being developed in Prores AS. Our Sendra software is the market leading software for experimental history matching and SCAL analysis. For more information on our solutions and services, please visit our website at www.prores.no.

Rotunda Scientific Technologies LLC - Rotunda Scientific Technologies LLC provides innovative radiation measurement and protection products to the energy industry. Our offering includes several gamma spectrum core loggers, multichannel analyzers, and a positive displacement pump manufactured for use with corrosive solvents such as brine. The GMS310 and GMS312 Gamma Spectrum Core Loggers are designed for core sample analysis (Spectral Gamma, API and Percent Concentration) in the rugged environment of the exploration site and will be available at our booth for you to evaluate. In addition to the GMS310 and GMS312, we offer many other state of the art radiation detection and protection products for use during exploration or nondestructive testing. You are invited to see these products at our booth #13 during the SCA Annual Meeting. We look forward to meeting you for the first time or seeing you again and catching up!

Qmineral – Qmineral is an independent material test laboratory and one of the most important mineralogy/XRD labs worldwide. We were the first lab to win the "Reynolds cup" (aka the World Championship of quantitative mineralogy) for two consecutive times. We are specialized in the analysis of clay bearing samples and the detailed structural analysis of a rock's clay minerals.

Qmineral has just launched its new service for quantitative mineralogy, named HSMQ (High Speed Mineralogical Analysis). HSMQ is a tool to quantify the mineralogy of a sample very accurately and quickly - about 40 times faster than with XRD, allowing you to characterize a virtually unlimited amount of samples in a very short time

Vinci Technologies – Vinci Technologies' origins began with the manufacturing of highly specialized laboratory and field instruments for the oil & gas industry. Throughout this history, Vinci has continuously developed new instruments to address emerging challenges such as carbon footprint, environment and global awareness.

Vindum Engineering Inc. - MANUFACTURER OF PRECISION HIGH PRESSURE PULSE-FREE PUMPS, VALVES AND FLUID FLOW EQUIPMENT

Vindum VP Series Pumps: High Precision, Continuous Pulse-Free Metering Pumps
- Latest generation design: lower cost, higher performance
- Models up to 25,000 psi
- Ambient Temperature or High Temperature (up to 160C) Options
- Gas or Liquids, including CO2

New Vindum VPL Series Pumps: Large Volume Cylinder Pumps
- Easily upgrade to continuous pulse-free flow with 2 pumps
- Syringe Piston Design with cylinder wash area
- CV Valves Standard
- VPware Software Included
- Gas or Liquids, including CO2

CV Valves: Constant-Volume, High-Pressure Valves
- Air activation
- High Temperature applications (up to 300C)
- Gas or Liquids, 2-Way and 3-Way Models

MV Hastelloy Valves & Fittings: Modular Manual Valve System
- Mountable, compact design saves space
- O-ring seals: replaceable, various seal materials
- Lower Cost Hastelloy Design
- In-stock for immediate delivery

Booth Layout

1. Diversified Well Logging
2. Core Laboratories
3. PanTerra Laboratory Services BV
4. Core Specialist Services
5. Qmineral
6. Vindum Engineering
7. Vindum Engineering
8. HOT Microfluids
9. Geotek Ltd
10. Green Imaging Technologies, Inc
11. Oxford Instruments
12. H2 Laboratories
13. Rotunda Scientific Technologies
14. AMETEK Chandler Engineering
15. Math2Market GmbH
16. DCI Corporation
17. Vinci Technologies
18. MetaRock Laboratories
19. Object Research Systems
20. Prores AS

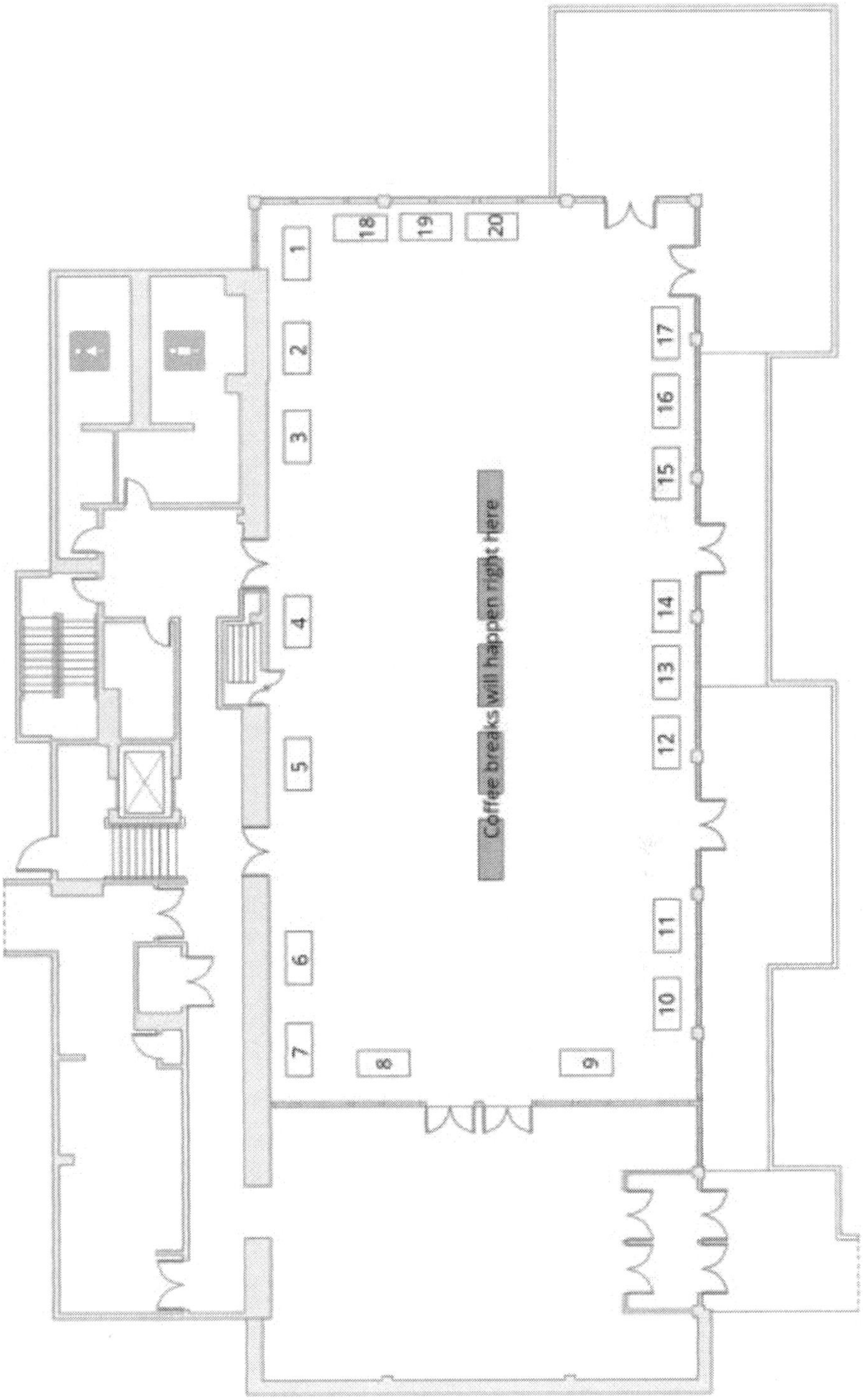

Sunday, Sep 18	
8:00 – 2:30	Optional Golf Event - The Hills of Lakeway, Live Oak Course Please meet in Hotel Lobby for 8:00am

Monday, Sep 19	
8:00 – 5:00	**Registration Desk Open**

Short Course Program

Short Workshop: Core Analysis in Carbon Geosequestration, Geothermal, and Nuclear-Waste Disposal

MONDAY, September 19, 2022

Short Workshop: Core Analysis in Carbon Geosequestration, Geothermal, and Nuclear-Waste Disposal	
Moderator: JinHong Chen	
8:30 – 8:50	Welcome and safety moment
8:50 – 9:40	Flow in Porous Media in the Energy Transition Presenter: Martin Blunt
9:40 – 10:30	What Do We Need to Know for Designing a Reliable Carbon Storage Project? Presenter: Jon Burger
10:30 – 11:00	Coffee Break **Kindly Sponsored by: AMETEK Chandler Engineering**
11:00 – 11:45	Core Analysis in Geothermal: Supporting the Energy Transition Presenter: Igor Faoro
11:45 – 12:30	Core and Petrophysics Analysis as Complementary Datasets for Geological Disposal of Nuclear Waste at NAGRA Presenter: Garrard Rodney
12:30 – 1:30	Lunch **Kindly Sponsored by: TOTALEnergies**

Augment Core Analysis to Embrace Energy Transition

Technical Program

Monday, September 19, 2022

1:30 – 3:00	**Session 1: Core Analysis Embracing Energy Transition**
	Chairs: C. Berg and H. Xie
SCA01	Displacement Stability Revisited – A New Criterion for the Onset of Viscous Fingering
	Jos G. Maas, Niels Springer, Albert Hebing, and Steffen Berg
SCA02	Advanced Digital-SCAL Measurements of Gas Trapping in Sandstone
	Ying Gao, Tibi Sorop, Niels Brussee, Hilbert van der Linde, Ab Coorn, Matthias Appel, and Steffen Berg
SCA03	Integrated Thermo-Poro-Mechanical Characterization for CO2 Sequestration at Deep Aquifer Conditions
	Sudarshan Govindarajan, Munir Aldin, Akshay Thombare, Omar Abdulbaki, Deepak Gokaraju, Abhijit Mitra, and Robert Patterson
3:00 - 3:30	**Coffee Break**
	Kindly Sponsored by: AMETEK Chandler Engineering
3:30- 5:00	**Session 2: Improved SCAL Techniques and Interpretation 1**
	Chairs: E. Ebeltoft and M. Dick
SCA04	Wireless Acquisition for Resistivity Index in Centrifuge – Wiri: A Comparative Study of Three Pc-RI Methods
	Quentin Danielczick, Ata Nepesov, Laurent Rochereau, Sandrine Lescoulie, Victor De Oliveira Fernandes, and Benjamin Nicot
SCA05	Unraveling Electrokinetics – A Brand New and Innovative Workflow for the Quantification of Electrokinetic Properties of Siliciclastic Rocks
	Matthias Halisch, Stephan Kaufhold, and Christian Weber

SCA06	Analytical Models for Predicting the Formation Resistivity Factor and Resistivity Index at Overburden Conditions
	Meysam Nourani, Stefano Pruno, Mohammad Ghasemi, Muhamet Meti Fazlija, Byron Gonzalez, and Hans-Erik Rodvelt

5:30 - 8:30 **Opening Reception "Icebreaker Reception" – Sunset Room**

Tuesday, September 20, 2022

7:30 – 5:00 **Registration Desk Open**

8:15 - 9:45 **Session 3: Laboratory Core Analysis 1**

Chairs: B. Gao and S. Althaus

SCA07 THz Imaging to Map the Microporosity Distribution in Carbonate Rocks

Shannon L. Eichmann, Jacob Bouchard, Hooisweng Ow, Doug Petkie, and Martin Poitzsch

SCA08 Innovations in Low UCS Core Acquisition and Quality Assessment Using Digital Rock Physics

Dmitry Lakshtanov, Jennie Cook, Yuliana Zapata, Dave Saucier, Robin Eve, Mark Lancaster, Nathan Lane, Glen Gettemy, Kevan Sincock, Elizabeth Liu, Rosemarie Geetan

SCA09 Angle-Dependent Ultrasonic Wave Propagation in Rocks for Estimating High-Resolution Elastic Properties of Complex Core Samples

Daria Olszowska, Gabriel Gallardo-Giozza, Domenico Crisafulli, and Carlos Torres-Verdín

9:45 - 10:15 **Coffee Break**

Kindly Sponsored by: Math2Market

10:15 – 11:45 **Session 4: Pore Scale Imaging and Modeling 1**

Chairs: M. Halisch and S. Pruno

SCA10 Pore Network Simulations Coupled with Innovative Wettability Anchoring Experiment to Predict Relative Permeability of a Mixed-Wet Rock

Mohamed Regaieg, Franck Nono, Titly Farhana Faisal, Clément Varloteaux, and Richard Rivenq

SCA11 ElRock-Net: Assessing the Utility of Machine Learning to Initialize 3D Electric Potential Simulations

Bernard C. Chang, Javier E.Santos, Rodolfo Victor, and Maša Prodanović

SCA12 A Bayesian Optimization Approach to the Extraction of Intrinsic Physical Parameters from T2 Relaxation Responses

Rupeng Li, Igor Shikhov, and Christoph H. Arns

11:45 - 1:00	**Lunch** **Kindly Sponsored by: Chevron**
1:00 – 3:00	**Poster Session (Odd Numbers)**
3:00 - 3:30	**Coffee Break** **Kindly Sponsored by: Math2Market**
3:30 – 4:00	**Exhibitor presentations** H2 Laboratories, Green Imaging Technologies, Inc., Oxford Instruments, Vindum Engineering, Inc., AMETEK Chandler Engineering, Qmineral, Core Laboratories.
4:00 – 5:00	**Session 5: Unconventionals and Source Rocks** **Chairs: J. Howard and E. Eichmann**
SCA13	Shale Characterization Using Magnetic Resonance Mohammad Sadegh Zamiri, Jiangfeng Guo, Florea Marica, Laura Romero-Zerón, and Bruce J. Balcom
SCA14	The Effect of Nanoconfinement on the Phase Behavior of Ethane/N-Propane Binary Mixture: An Experimental Study at Varying Pore Sizes and Compositions Keerti Vardhan Sharma, Rami M. Alloush, Karem Al-Garadi, and Mohammad Piri
6:00 – 9:00 p.m.	**Young Professional Event – Lakeway Marina – Party Boat Tour** **Kindly Sponsored by: Shell**

Wednesday, September 21, 2022

7:30 – 5:00	**Registration Desk Open**
8:15 – 9:45	**Session 6: Pore Scale Imaging and Modeling 2**
	Chairs: J. Maas and S. Eichmann
SCA15	Forced Imbibition and Uncertainty Modelling Using the Morphological Method
	Pit Arnold, Mario Dragovits, Sven Linden, Fatime Zekiri, and Holger Ott
SCA16	Initial States of Core Flooding Techniques Evaluation: A Global Pore-Scale Investigation
	Franck Nono, Cyril Caubit, and Richard Rivenq
SCA17	Numerical Study of NMR Relaxation Responses in Synthetic Clayey Sandstone by Dual-Scale Modeling
	Yingzhi Cui, Igor Shikhov, and Christoph Arns
9:45 - 10:15	**Coffee Break**
10:15 – 11:45	**Session 7: Application of Artificial Intelligence/Machine Learning**
	Chairs: H. Ott and J. Schembre-McCabe
SCA18	Artificial Intelligence Assisted Quantitative Petrophysical Properties Analysis using Core Images and Well Logs
	Tao Lin, Mokhles Mezghani, Chicheng Xu, and Weichang Li
SCA19	Combining High-Resolution Core Data and Machine Learning Schemes to Develop Sustainable Core Analysis Practices
	Christophe Germay, Tanguy Lhomme, Luc Perneder, and John Cummings
SCA20	Prediction of Centrifuge Capillary Pressure Using Machine Learning Techniques
	Brandon Jeremy Bursey, Erfan Mohagheghian, Edison Sripal, and Lesley Anne James

11:45 - 1:00	**Lunch**

1:00 – 3:00 **Poster Session 2 (Even Numbers)**

3:00 – 3:30	**Coffee Break**

3:30 – 4:00 **Exhibitor Presentations**

Core Specialist Services, Rotunda Scientific Technologies, PanTerra Laboratory Services BV, Geotek Ltd, HOT Microfluidics, Diversified Well Logging, Math2Market GmbH, Vinci Technologies, DCI Corporation

4:00 – 5:00 **Session 8: Wettability & others**

Chairs: J. Maas and S. Althaus

SCA21 Digital Rock Workflow to Calculate Wettability Distribution in A Reservoir Rock

Ashraful Islam, Rafael Tio Salazar, and Bernd Crouse

SCA22 Fast Wettability Assessment on Small Rock Samples Using A 3D, High-Resolution, Image-Based Amott-Like Test

Maria Repina, Regis Brugidou, Alexandre Dufour, and Richard Rivenq

6:30 – 9:30	**Gala Dinner – Star Hill Ranch**

Thursday, September 22, 2022

7:30 – 3:00	**Registration Desk Open**
8:15 - 9:45	**Session 9: Improved SCAL Techniques and Interpretation 2**
	Chairs: S. Pruno and W. Richardson
SCA23	Hybrid Technique for Setting Initial Water Saturation on Core Samples
	Victor Fernandes, Cyril Caubit, Benjamin Nicot, Fabrice Pairoys, Henri Bertin, and Jean Lachaud
SCA24	Water-Gas Imbibition Relative Permeability: Literature Review, Direct versus Indirect Methods and Experimental Recommendations
	Fabrice Pairoys, Cyril Caubit
SCA25	Geomechanical Deformation of Saturated Porous Media under Various Wettability Conditions: A Pore-scale Investigation
	Ahmed Zankoor, Rui Wang, Maziar Arshadi1 and Mohammad Piri
9:45 - 10:15	**Coffee Break**
10:15 – 11:45	**Session 10: Laboratory Core Analysis 2**
	Chairs: E. Ebeltoft and H. Xie
SCA26	A Combinational NMR and Dielectric Technique Using Spectral NMR Mapped Distributions of Dielectric Relaxation
	James J. Funk, Michael Myers, and Lori Hathon
SCA27	Experimental Time-Lapse Visualization of Mud-Filtrate Invasion and Mudcake Deposition in Complex Rocks Using X-Ray Radiography
	Pierre Aérens, Carlos Torres-Verdín, and Nicolas Espinoza
SCA28	Causal Protocols to Assess the Viability of Native State or Restored State Preparation
	Jules Reed, Stefano Pruno, Izaskun Zubizarreta, and Rolf Johansen

11:45 – 12:15	**Business Meeting and Lunch**
12:15 - 1:30	**Lunch**

1:30 – 3:30 **Session 11: Displacement Mechanisms/EOR/IOR**

Chairs: J. Funk and M. Halisch

SCA29 Carbonated Water Injection for Heavy Oil Recovery

Jinxun Wang, Abdulkarim M. AlSofi, Hassan Behairy, Abdullah M. Boqmi, and Sinan Caliskan

SCA30 Carbonated Smart Water Injection for Optimized Oil Recovery in Chalk at High Temperature

Md Ashraful Islam Khan, Sander Haaland Kleiberg, Ivan Dario Pinerez Torrijos, Tina Puntervold, and Skule Strand

SCA31 Nano-Colloid Based Suspensions and Emulsions Used as Means for Enhanced Oil Recovery

Anastasia Strekla, Christina Ntente, Maria Theodoropoulou, and Christos Tsakiroglou

SCA33 Simultaneous Interpretation and Uncertainty Analysis of SCAL Data from Complex Rocks

Omideza Amrollahinasab, Siroos Azizmohammadi, and Holger Ott

3:30 **Closing Remarks**

Posters Presentations

	** Posters with odd numbers present on Tuesday and with even numbers present on Wednesday
Posters with Manuscript	
SCA34	Capturing the Wetting State of An Aged-Carbonate Core Through Pore-Scale Multiphase Flow Simulations Tingting Wang, Ying Da Wang, Chenhao Sun, James E. McClure, Peyman Mostaghimi, and Ryan T. Armstrong
SCA35	Comparison of Three-Dimensional Permeability Inversion from Positron Emission Tomography Experimental Data Using Convolutional Neural Networks and Ensemble Kalman Filter Zitong Huang and Christopher Zahasky
SCA36	Digital Rocks Portal (Digital Porous Media): Connecting Data, Simulation and Community Maša Prodanović, Maria Esteva, James McClure, Bernard C. Chang, Javier E. Santos, Anuradha Radhakrishnan, Ankita Singh, and Hasan Khan
SCA37	Pore-scale Analysis of CO2-brine Displacement in Berea Sandstone and Its Implications to CO2 Injectivity Guangyuan Sun, Zhuang Sun, Andrew Fager, and Bernd Crouse
SCA38	Core Characterization of Patterson #5-25 Well for Carbon Capture and Storage in Western Kansas Thomas Paronish, Rhiannon Schmitt, Dustin Crandall, Franek Hasiuk, Eugene Holubnyak, and Jingyao (Jenny) Meng
SCA39	An Approach for Image-Based Quantification of Fines Migration in Geologic Columns and Core Samples Collin R. Sutton and Christopher Zahasky
SCA40	An Integrated Petrophysical Analysis Based on NMR, Organic Geochemistry and Mineralogy. The Vaca Muerta Source Rock-Unconventional Play at Different Thermal Maturities Diana Masiero, Marcos Comerio, Esteban Domené, Gabriela Vila, Bernarda Epele, Mariano Cipollone, Mariela Silka, Carlos Camacho, Lourdes Vera López, and Silvina Chiappero

SCA41	Some Useful Guidelines for Whole Core CT-Scanning for Petrophysical Applications Shameem Siddiqui
SCA42	Investigation Pore Geometry Wettability Preference in Oolitic Oil Reservoir: Pore Scale Imaging and Modelling Study Hussien Al-Ajaj, Ralph Flori, Saleh Alsayegh, Haidar AlMubarak, and Waleed Al-Bazzaz
SCA43	Rapid, High Resolution Probe Screening Techniques for Core Analysis and their Potential Usefulness for Hydrocarbon or Energy Transition Applications Emmanuel Okwoli and David K. Potter
SCA44	Applications of Temperature Dependent Paramagnetic Properties for Quantifying Mineral Content and Extending the Use of Paramagnetic Dopants for Laboratory or Borehole Analysis of NMR Data Cody W. Good and David K. Potter
SCA45	Modification of the SDR Equation for Permeability Prediction Andreas Weller and Zeyu Zhang
SCA46	Towards Multiscale Digital Rocks: Application of a Sub-Resolution Production Model to a multiscale Sandstone Rafael Salazar-Tio, Andrew Fager, Guangyuan Sun, Bernd Crouse, Rui Xu, Brett Wendt, and Adam Lewis
SCA47	Chemostratigraphic Analysis as A Powerful Tool for the Lateral Continuity of Structurally Complex Reservoirs: A Case Study Liborius-Parada Andreina, Medina-Macedo Marlen, Tonner Dave, Hughes Simon, and McCulley Meri
Posters without Manuscript	
SCA48	Carbon Capture and Storage (CCS), Evaluation of Carbon Dioxide Storage Efficiency at the Western Siberia Field Pavel Golub, Andrei Cheban, and Evgenii Romanov
SCA49	A Joint Workflow Towards a Reliable Quantification and Understanding of NMR Surface Relaxivity Matthias Halisch, Raphael Dlugosch, Zeyu Zhang, and Andreas Weller

SCA50	Comparison of Geophysics- and Core-Based Wettability Assessment Methods: An Experimental Study Using Artificial Grain Packs Zulkuf Azizoglu and Zoya Heidari
SCA52	Multi-Scale 3D Carbonate Digital Rock Reconstruction: Traditional or Machine Learning Approaches? Yiteng Li, Xupeng He, Marwa AlSinan, Hyung Kwak and Hussein Hoteit
SCA53	NMR T2 Response versus Roughness: A Numerical and Analytical Study Yiteng Li, Xupeng He, Marwa AlSinan, Hyung Kwak, and Hussein Hoteit
SCA54	Nuclear Magnetic Resonance Laboratory Study of A Tight Sandstone for Robust Permeability Prediction Jun Gao, Hyung Kwak, Abdullah Alkhaldi, and Gabor Hursan
SCA55	NMR Spin-spin Relaxation in Unconventional Source Rocks Z. Harry Xie
SCA56	SEM Image-Constrained Process-Based Modeling for Relative Permeability Estimation of Carbonate-Rich Mudrock Christopher J. Landry and Masa Prodanovic
SCA57	Multi-Phase Flow in Fractured Rocks: From Pore-Scale Processes to Field-Scale Responses Xupeng He, Marwa AlSinan, Hyung Kwak, and Hussein Hoteit
SCA58	Direct Measurement of In-Situ Hydrogen-Water-Quartz System Relative Permeability for Underground Hydrogen Storage in A Depleted Gas Reservoir Scott Higgs, Ying Da Wang, Jonathan Ennis-King, Samuel J. Jackson, Ryan T. Armstrong, and Peyman Mostaghimi
SCA59	Characterization of Surface Conductivity of Clays Viacheslav Emelianov, Zeyu Zhang, Konstantin Titov, Matthias Halisch3, and Andreas Weller
SCA60	Manganese-Ion Based Tailored Waterflooding Processes for Carbonates Amani Alghamdi, Saleh Salah, Mohammed Otaibi, Subhash Ayirala, and Ali Yousef

Friday, September 23, 2022	
9:00 a.m. – 3:15 p.m.	Optional Friday Field Trip Hamilton Pool Preserve and Reimers Ranch Leave Lakeway at 9:00 am on Friday, September 23rd • Transportation and lunch provided. • Hamilton Pool: 9:45 to 11:00 am • Reimers Ranch – 11:30 am • Lunch and Safety Briefing • Geological Tour Outcrops 12:30 – 2:30 pm • Arrive Back at Lakeway Hotel 3:15 pm Note: Reimers ranch cores will be at the SCA conference for viewing throughout the week. Trip Leaders: Charlie Kerans, Brian Hunt, Charlotte Sullivan, and Toti Larsen.

- Delegate Bags kindly sponsored by Vindum Engineering
- Lanyards kindly sponsored by H2 Laboratories
- Pens and Pads kindly sponsored by Strata Technology
- Speaker Gifts kindly sponsored by Stratum Reservoir
- USB Technical Proceedings kindly sponsored by Petricore

Notes

Thanks to all Symposium Sponsors

SCA 2022 Sponsors - THANK YOU!

TotalEnergies

Chevron

MATH
2 MARKET

AMETEK
CHANDLER
ENGINEERING

Laboratories
A division of Green Imaging Technologies

VINDUM
ENGINEERING, INC.
Performance – Quality – Value

petricore

STRATA
TECHNOLOGY LTD

STRATUM
RESERVOIR

SOCIETY OF CORE ANALYSTS

139 Whittaker Rd • Rusagonis, NB E3B 7X3, Canada

E-Mail: office@scaweb.org • www.SCAWeb.org

edp sciences

Web of Conferences

Statement of Peer review

In submitting conference proceedings to *Web of Conferences*, the editors of the proceedings certify to the Publisher that

1. They adhere to its **Policy on Publishing Integrity** in order to safeguard good scientific practice in publishing.
2. All articles have been subjected to peer review administered by the proceedings editors.
3. Reviews have been conducted by expert referees, who have been requested to provide unbiased and constructive comments aimed, whenever possible, at improving the work.
4. Proceedings editors have taken all reasonable steps to ensure the quality of the materials they publish and their decision to accept or reject a paper for publication has been based only on the merits of the work and the relevance to the journal.

Title, date and place of the conference

Title: The 35th International Symposium of the Society of Core Analysts
Date: September 19-22, 2022
Place: The Lakeway Resort and Spa, Austin, TX USA

Proceedings editor(s): **JinHong Chen**

Date and editor's signature

 https://doi.org/10.1051/e3sconf/202336700001

PREFACE

The 2022 International Symposium of the Society of Core Analysts (SCA), founded in 1986, is a chapter - at -large of the Society of Petrophysicists and Well Log Analysts (SPWLA) . In 2022, the symposium was held in Austin, TX, between September 19 and 23, featuring a workshop on "Core Analysis in Carbon Geosequestration, Geothermal, and Nuclear-Waste disposal" and various technical sessions showcasing 33 oral presentations, 14 posters with manuscripts and 17 posters. The symposium theme was "Augment Core Analysis to Embrace Energy Transition"

The collection of articles published in the SHS Web of conferences consists of fifteen (15) articles prepared for oral presentations. Thirty-one (31) proceedings presented in this symposium, published in other journals, and seventeen (17) posters are not included in this collection. Here is the list of papers no included in this collection.

Papers presented at the 2022 SCA Symposium published in other journals

Displacement Stability Revisited – A New Criterion for the Onset of Viscous Fingering
J. G. Maas, N. Springer, A. Hebing, and S. Berg

Advanced Digital-SCAL Measurements of Gas Trapping in Sandstone
Y. Gao, T. Sorop, N. Brussee, H. van der Linde, A. Coorn, M. Appel, and S. Berg

Wireless Acquisition for Resistivity Index in Centrifuge – Wiri: A Comparative Study of Three Pc-RI Methods
Q. Danielczick, A. Nepesov, L. Rochereau, S. Lescoulie, V. De Oliveira Fernandes, and B. Nicot

Unraveling Electrokinetics – A Brand New and Innovative Workflow for the Quantification of Electrokinetic Properties of Siliciclastic Rocks
Matthias Halisch, Stephan Kaufhold, and Christian Weber

Analytical Models for Predicting the Formation Resistivity Factor and Resistivity Index at Overburden Conditions
M. Nourani, S. Pruno, M. Ghasemi, M. M. Fazlija, B. Gonzalez, and H-E Rodvelt

THz Imaging to Map the Microporosity Distribution in Carbonate Rocks
Shannon L. Eichmann, Jacob Bouchard, Hooisweng Ow, Doug Petkie, and Martin Poitzsch

Innovations in Low UCS Core Acquisition and Quality Assessment Using Digital Rock Physics
D. Lakshtanov, J. Cook, Y. Zapata, D. Saucier, R. Eve, M. Lancaster, N. Lane, G. Gettemy, K. Sincock, E. Liu, and R. Geetan

Angle-Dependent Ultrasonic Wave Propagation in Rocks for Estimating High-Resolution Elastic Properties of Complex Core Samples
D. Olszowska, G. Gallardo-Giozza, D. Crisafulli, and C. Torres-Verdín

Pore Network Simulations Coupled with Innovative Wettability Anchoring Experiment to Predict Relative Permeability of a Mixed-Wet Rock
M. Regaieg, F. Nono, T. F. Faisal, C. Varloteaux, and R. Rivenq

Capturing the Wetting State of An Aged-Carbonate Core Through Pore-Scale Multiphase Flow Simulations
T. Wang, Y. Da Wang, C. Sun, J. E. McClure, P. Mostaghimi, and R. T. Armstrong

Comparison of Three-Dimensional Permeability Inversion from Positron Emission Tomography Experimental Data Using Convolutional Neural Networks and Ensemble Kalman Filter
Z. Huang and C. Zahasky

Core Characterization of Patterson #5-25 Well for Carbon Capture and Storage in Western Kansas
T. Paronish, R. Schmitt, D. Crandall, F. Hasiuk, E. Holubnyak, and J. Meng

An Approach for Image-Based Quantification of Fines Migration in Geologic Columns and Core Samples
C. R. Sutton and C. Zahasky

Investigation Pore Geometry Wettability Preference in Oolitic Oil Reservoir: Pore Scale Imaging and Modelling Study
H. Al-Ajaj, R. Flori, S. Alsayegh, H. AlMubarak, and W. Al-Bazzaz

Rapid, High Resolution Probe Screening Techniques for Core Analysis and their Potential Usefulness for Hydrocarbon or Energy Transition Applications
E. Okwoli and D. K. Potter

Applications of Temperature Dependent Paramagnetic Properties for Quantifying Mineral Content and Extending the Use of Paramagnetic Dopants for Laboratory or Borehole Analysis of NMR Data
C. W. Good and D. K. Potter

Chemostratigraphic Analysis as A Powerful Tool for the Lateral Continuity of Structurally Complex Reservoirs: A Case Study
A. Liborius-Parada, M. Medina-Macedo, D. Tonner, S. Hughes, and M. McCulley

Posters presented at the 2022 SCA Symposium

Carbon Capture and Storage (CCS), Evaluation of Carbon Dioxide Storage Efficiency at the Western Siberia Field
P. Golub, A. Cheban, and E. Romanov

A Joint Workflow Towards a Reliable Quantification and Understanding of NMR Surface Relaxivity
M. Halisch, R. Dlugosch, Z. Zhang, and A. Weller

Comparison of Geophysics- and Core-Based Wettability Assessment Methods: An Experimental Study Using Artificial Grain Packs
Z. Azizoglu and Z. Heidari

Multi-Scale 3D Carbonate Digital Rock Reconstruction: Traditional or Machine Learning Approaches?
Y. Li, X. He, M. AlSinan, H. Kwak and H.Hoteit

Integrated thermo-poro-mechanical characterization for CO_2 Sequestration at deep aquifer conditions

Sudarshan Govindarajan[1, *], *Munir* Aldin[1], *Akshay* Thombare[1], *Omar* Abdulbaki[1], *Deepak* Gokaraju[1], *Abhijit* Mitra[1] and *Robert* Patterson[1].

[1]MetaRock Laboratories, 2703 Highway 6 S, Houston, Texas, U.S.A.

Abstract. Modelling and forecasting of injected CO_2 plume behaviour is an essential step in the baseline, monitoring, and verification [BMV] process in the CO_2 sequestration lifecycle. The goal of reduction of uncertainty through forecasting models, can be better realized by accounting for the thermo-poro-mechanical nature of the deep subsurface reservoir systems. The current study focusses on developing and refining a laboratory workflow which will help in generating representative static and dynamic datasets at ambient and deep aquifer conditions. The workflow involves characterizing the poroelastic Biot coefficient and mechanical properties at ambient, high temperatures and at reservoir representative stress conditions. This information will be combined with the dataset from a CO_2 flood experiment which replicates the displacement of brine by super critical CO_2 at ambient and high temperatures and at reservoir representative stresses. Resistivity and acoustic signals will be monitored throughout the flood experiment. Existing analytical models for fluid substitution such as the Biot-Gassmann-Brie populated with representative data will be evaluated for finding the best description of the experimental observations. The integrated results of the workflow are meant to help develop better informed static and dynamic models improving the confidence in the BMV process of CCUS.

1 Introduction

The SR15 report [1] released by IPCC in 2018 states the assumption of large-scale deployment of CO_2 removal measures to limit warming to 1.5°C by year 2100. Carbon capture, utilization, and storage (CCUS), as a CO_2 removal measure, has to implemented on a massive scale to meet the necessitated ambitious net zero emission goals. Geological storage or sequestration has the highest uncertainty through the CCUS lifecycle. Baseline, monitoring and verification programmes have been acknowledged as a fundamental and critical requirement for reducing the uncertainty in long term containment of CO_2 in the subsurface [2]. Rules promulgated by both the EPA in the US and the equivalent CO2CARE program in the EU require post injection site care satisfying strict qualification criteria, to be part of the project scope [3]. Monitorability of the storage complex is dependent on both the subsurface and the surface environment [4]. The method chosen for monitoring whether geophysical (seismic, resistivity), borehole/well based, geochemical tracing, satellite or micro-seismic based geomechanical monitoring is entirely dependent on the purpose (leak detection vs. volumetrics) as well as the geography [4, 5]. Lateral (areal) coverage for plume monitoring is enabled by surface measurements while higher resolution data can be acquired through borehole-based measurements. [4]. An integrated monitoring approach is the ideal recommendation but requires heavy investment [3, 4]. The critical reasoning and potential benefits behind investing in long term post injection monitoring as compared to higher investment in the initial characterization and preventative predictions have been compared in [3]. The inefficiency of long-term monitoring in lower certainty storage complexes due to lack of established remedial measures was pointed out in [3]. This was contrasted against the advantages of increasing certainty in the storage complexes by investing in accelerated laboratory studies, utilization of natural analogues, scaled field experiments and enhanced understanding of reservoir heterogeneity. Core analysis being the guiding standard has been the norm in the oil and gas industry. The importance of actual core-based laboratory investigations, to be used in an integrated approach with log analysis, field scale geophysical characterization and associated static-dynamic reservoir modelling is well recognized in the historical subsurface industries. This integrated approach is now being applied with greater emphasis, for increasing the certainty in the CCUS lifecycle, from the initial confidence building predictive exercises (site characterization and initial models) to establishing predictive post injection monitoring targets [3, 4, 6, 7, 8, 9, 10].

Core analysis performed at representative conditions on representative rock directly contributes to decreasing uncertainties in subsurface engineering. The comprehensive picture of the subsurface required for the numerical modelling [11] can be obtained with higher certainty when rock properties are measured at representative in-situ conditions. Deep saline aquifers are the most promising storage complexes for CCUS [8]. Characterising the displacement behaviour of the CO_2-brine systems is essential for pore space management as well as risk assessment [8, 11] The ideal integrated site characterization exercise would incorporate the high stress, pressure and temperature conditions encountered in the deep saline aquifers in the core analysis exercise. This study is a laboratory investigation of

* Corresponding author: sudarshan@metarocklab.com

the influence of in-situ conditions on CO_2 induced brine displacement and the applicable petrophysical models which help estimate the observed phenomenon.

2 Experimental Approach

The uncertainty associated with lack of understanding of rock heterogeneity at the sub seismic resolution scale leads to surprises at the field scale [12]. Geophysical monitoring data and reservoir characterization data, together, help obtain quantitative estimates necessary for verification of field behaviour against modelling estimates within uncertainty bounds [13]. Geophysical parameter behaviour is dependent on the rock minerology, porosity, pore fluid content, fluid saturations, in-situ temperatures, and pressures [14, 15]. The ability of seismic (acoustic) methods to monitor trapped CO_2 has been successfully demonstrated in the field however detecting the migrating plume front requires further investigation [16]. Prior uncertainty in petrophysical parameter estimations have a significant effect on geophysical monitorability and there is a need for accurate benchmark measurements and reservoir characterization [14]. Injecting CO_2 under supercritical conditions which can be sustained in the deep aquifers has inherent injectivity benefits [17]. The major challenge for laboratory experimental characterization for CCUS is performing the test at in-situ representative pressure and temperatures while accounting for the solubility between the supercritical CO_2 (SC-CO_2) and water and the properties of the SC-CO_2 [12]. There have been multiple studies aimed at characterizing the fluid substitution phenomenon in the lab space [12, 14, 15, 16, 17, 18, 19, 20, 21]. To replicate CO_2 injection into the deep aquifers, most approaches involved injecting dry CO_2 gas or CO_2-brine solutions into brine saturated rock samples at varied conditions of temperature, pressure, and usually low effective stresses. Experimental observations were obtained through either isolated or combined acoustics, CT scanning and resistivity-based monitoring methods. The current investigation employs the porous plate technique and incorporates injection of dry SC-CO_2 into a brine saturated rock at reservoir representative stress, temperature, and pressure conditions while measuring the resistivity during the entire duration of the experiment. The porous plate technique is a reliable method for replication of the typical distribution of fluids in reservoir rocks [22,23]. Electrical/electromagnetic based surveys have been found to be a good complement to seismic measurements in the field as resistivity is more sensitive to saturation changes than compressional velocities [13]. Resistivity measurements in the lab have been found to be more reliable using the porous plate technique when performed at very slow injection rates following steady state equilibration protocols and a coherent brine saturation – resistivity index measurement is critical for validating logs by developing higher accuracy in fluid saturation definitions [22].The current study adopts a modification of the established I-S_W protocol by injecting at slow flow rates and using the porous plate for preferential effluent expulsion. The current study also incorporates typical reservoir representative high effective stress conditions for greater reduction in uncertainty.

3 Theoretical Background

Supercritical fluids have the advantage of having low viscosity and relatively low compressibility (Appendix A) [24]. Injecting a material in the supercritical state enables maximising the pore space available in the subsurface. The study of the displacement of brine by non-wetting fluid using the single sample-water wet membrane, porous plate technique following Archie's equations is well established in the industry [22]. The porous plate acts as a barrier to the non-wetting fluid and allows for gradual development towards irreducible water saturations. The resistivity being monitored during this gradual displacement can be plotted in a log-log plot against the water saturation (I-S_W plot) for obtaining the saturation exponent, "n". The saturation exponent has been historically estimated for typical lithologies and reliable ranges for associated saturation exponents have been made. The I-S_W plot is a reliable indicator of the saturation state of the sample. In the present study the brine in the samples is displaced by SC-CO_2 while monitoring the resistivity and the deviation from linearity in the I-S_W plot is used as an indicator of breakthrough or end of experiment. Fig. 1 shows an example of a typical I-S_W plot.

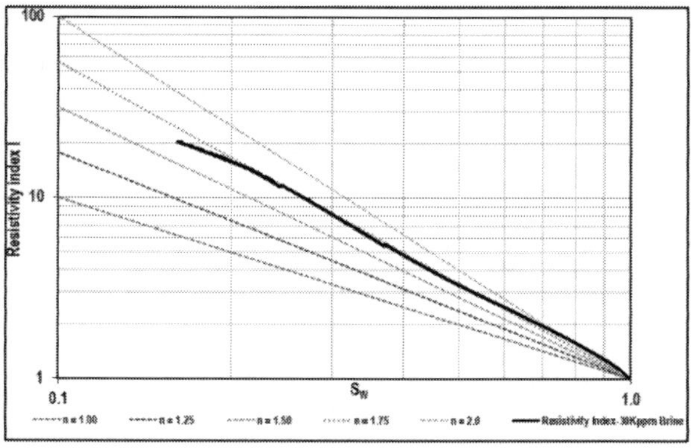

Fig. 1. Resistivity Index Vs. water saturation on log-log plot.

Fluid substitution as a phenomenon occurring during subsurface engineering has been well addressed by the hydrocarbon industry [25]. The general formula for the saturated porous system bulk modulus can be represented as under:

$$K_{SAT} = K_{dry} + \frac{(1-(K_{dry}/K_0))^2}{(\varphi/K_F)+((1-\varphi)/K_0)-(K_{dry}/K_0^2)} \quad (1)$$

where,

K_{SAT}: Bulk modulus of saturated porous medium
K_{dry}: Drained Bulk modulus
K_0: Grain modulus = Skeletal modulus = Mineral bulk modulus
K_F: Fluid Bulk modulus
Φ: Porosity as a fraction

The velocity can then be calculated as:

$$V_{PSAT} = \sqrt{\frac{1}{\rho_{sat}} * (K_P + K_{dry} + \mu\frac{4}{3})} \quad (2)$$

where,

$$K_P = \frac{(1-(K_{dry}/K_0))^2}{(\varphi/K_F)+((1-\varphi)/K_0)-(K_{dry}/K_0^2)} \quad (3)$$

V_{PSAT}: Saturated compressional velocity
ρ_{SAT}: Density of saturated rock
μ: G = Shear modulus of rock
Theoretically, the shear modulus for a rock is the same irrespective of saturated or dry condition as the fluids have zero shear velocity. The above equations have been utilised previously such as in reference [18] for interpreting laboratory measurements. However, the critical points mentioned in reference [25] are, that the above formulation was defined for seismic frequencies and may not work as well at the laboratory ultrasonic (1MHz) scale and that the K_{dry} has to be the drained bulk modulus i.e., either at constant pore pressure with free flow of fluid or with pore pressure drained to atmospheric condition in the absence of pore fluid. Reference [17] illustrated another expression for the undrained bulk modulus for a fluid saturated rock as:

$$K_U = K_{dry} + \frac{\alpha^2}{(\varphi/K_F)+((\alpha-\varphi)/K_0)} \quad (4)$$

where,
K_U: undrained bulk modulus of saturated porous medium
α: Biot poroelastic coefficient which is given by:

$$\alpha = 1 - (K_{dry}/K_0) \quad (5)$$

The fluid bulk modulus may be defined for a finely mixed fluid system by the Reuss average (parallel formulation):

$$\frac{1}{K_F} = \sum \frac{S_i}{K_i} \quad (6)$$

The fluid bulk modulus for a patchy saturation can be expressed by the Voigt average (series formulation) as:

$$K_F = \sum S_i * K_i \quad (7)$$

It was also shown in [17] that the Biot-Gassmann-Brie formulation was able to best describe the experimental observations.

$$K_F = S_B^N * K_B + (1 - S_B^N) * K_{CO_2} \quad (8)$$

where:
S_i: saturation of ith component
K_i: Bulk modulus of ith component
K_B: Brine modulus
S_B: Brine saturation
K_{CO2}: Bulk modulus of CO_2
The combined saturated density of the porous solid filled with the mix of fluids can be estimated as

$$\rho_{SAT} = \rho_{dry} + (\varphi) * \rho_F \quad (9)$$

where:
ρ_{dry}: Dry Bulk density of sample

$$\rho_F = S_B * \rho_B + (1 - S_B) * \rho_{CO_2} \quad (10)$$

where:
ρ_B: Density of brine at temperature, pressure
ρ_{CO2}: Density of CO_2 at temperature, pressure

4 Experimental System Overview

Prolonged experiments with SC-CO_2 at high temperatures and pressure require specialised equipment. The rock mechanics testing system, shown in Fig. 2 has been modified to contain the SC-CO_2 without any leaks. The system consists of a servomechanical frame press capable of reaching up to 300,000 lbs within an isothermal enclosure. The confining pressure pump has multivariable control capability and can apply up to 30000 psi (2068 MPa) with µ-inch resolution. The pore pressure pumps are 30 cc pumps capable of reaching 15000 psi (103 MPa) with nano-inch resolution. The pore pressure pumps can handle SC-CO_2 and have the necessary higher resolution to address the volume monitoring for saturation estimations. The system has also been modified with requisite channels for making the resistivity measurements. The system has also been modified with a high-resolution acoustics monitoring system, shown in Fig. 3 for live acquisition and display of the compressional velocity data as noted in [26].

Fig. 2. Rock Mechanics Test Equipment.

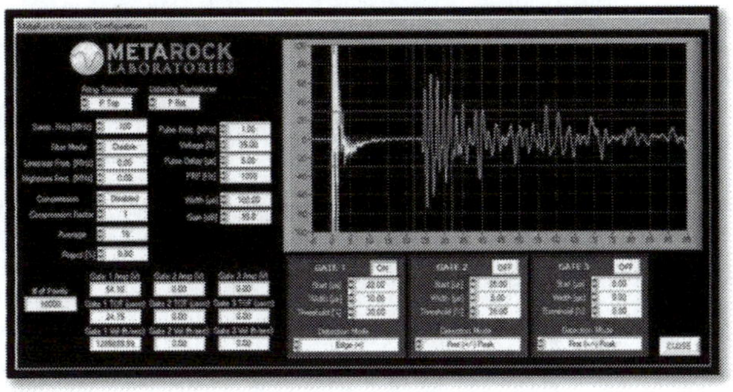

Fig. 3. Data acquisition and monitoring system for live high resolution compressional wave acoustics data.

Ultrasonic (1MHz) acoustic transducers are isolated from pressure and housed within the endcaps. The system is also rated for achieving high temperatures up to 450°C (842F) with acoustic measurement achievable up to 200°C (392F).

5 Experimental Suite, Results and Discussion

5.1. Testing Suite Design and Sample Statistics

The testing suite was defined to address the input requirements for the equations that govern the fluid substitution challenge. The poroelastic coefficient, the dry bulk modulus and the grain modulus are all required inputs. Additional triaxial testing at saturated condition was performed to be able to establish the range of expected velocity variation. The SC-CO_2 flow testing was performed on Buff Berea sandstone at ambient 35°C (95F) and higher temperature of 65°C (149F). Samples belonged to the same outcrop block and were prepared to be approximately 1" in diameter and 2" in length meeting all recommended ASTM/ISRM standards. The permeability range of the sample as obtained from the outcrop database was between 150-350 mD brine (KCl) permeability and 400-500 mD Klinkenberg permeability with N_2 gas. Grain volume and bulk volume of the samples were also measured to ensure uniformity. Table 1 shows the detailed statistics for the sample set.

Table 1.Sample Statistics

Sample Detail	Mean	Std Dev	Coefficient Of Variation
Length (in)	1.997	0.004	0.182
Diameter (in)	0.994	0.001	0.088
Weight (g)	51.074	0.177	0.347
Bulk Volume (cc)	25.507	0.048	0.186
Grain Volume (cc)	19.261	0.066	0.341
Pore Volume (cc)	6.246	0.085	1.361
Grain Density (g/cc)	2.651	0.001	0.056
Bulk Density (g/cc)	2.002	0.008	0.385
Porosity (%)	24.489	0.307	1.253

5.2. Triaxial testing with static and dynamic characterization:

All testing was performed at a high effective stress of 2000 psi (13.78 MPa). Triaxial tests were carried out for the characterization of the dry bulk modulus at both temperature conditions, at confining pressure of 2000 psi (13.78 MPa), under drained conditions, at slow loading rates. The differential stress cycles were unloaded at approximately 7000 psi (48.26 MPa) without taking the sample to failure. Compressional and shear velocities were acquired throughout the duration of the test. Results from the base triaxial testing are displayed from Fig. 4 to Fig. 7. The comparison between the axial and radial strain responses at 35°C (95F) and 65°C (149F) indicate that Buff Berea (being a quartz-rich rock) does not display a marked change in deformation behaviour within that temperature range. However, the compressional and shear velocities are both slightly lowered with the increase in temperature. The choice of 65°C (149F) was done as sandstones do not exhibit significant changes in elastic properties between 25°C (77F) to 45°C (113F) [27], however even at 65°C (149F) we did not observe significant change in the mechanical and acoustic properties.

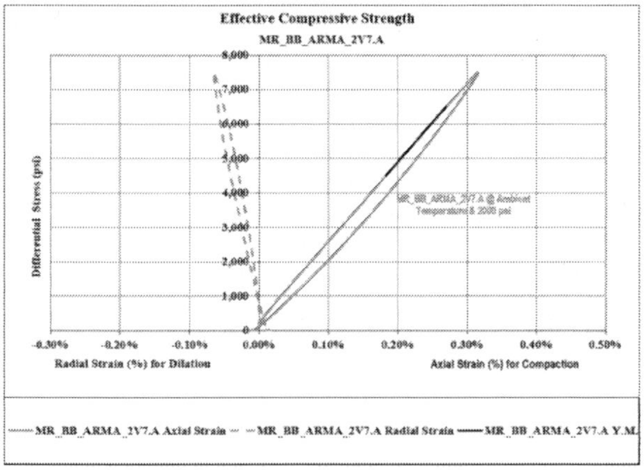

Fig. 4. Stress-strain plot for the ambient (35°C) temperature triaxial test and confining pressure of 2000 psi.

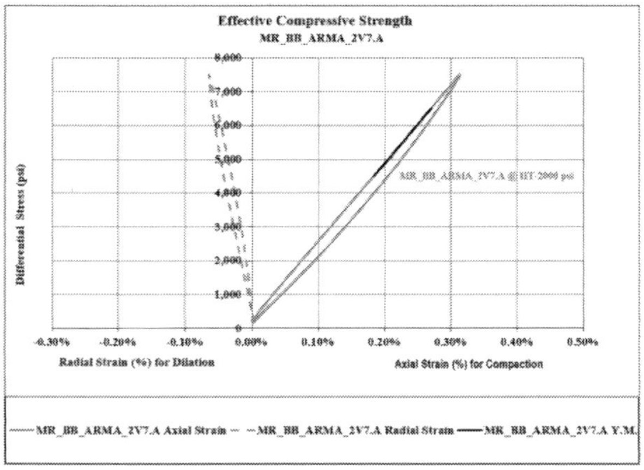

Fig. 5. Stress-strain plot for the high (65°C) temperature triaxial test and confining pressure of 2000 psi.

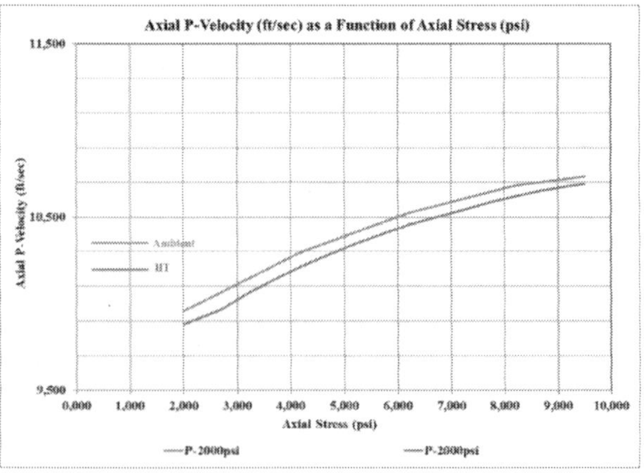

Fig. 6. Compressional velocity comparison at ambient and high temperature.

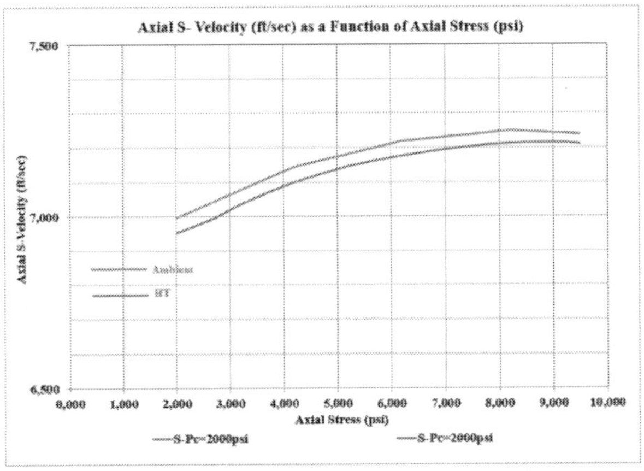

Fig. 7. Shear velocity comparison at ambient and high temperature.

5.3. Poroelastic characterization:

Poroelastic coefficient characterization as per reference [28], was carried out by cycling confining and pore pressures while maintaining the effective stress of 2000 psi (13.78 MPa). Fig. 8, Fig. 9 display the summary and results of the poroelastic coefficient characterization.

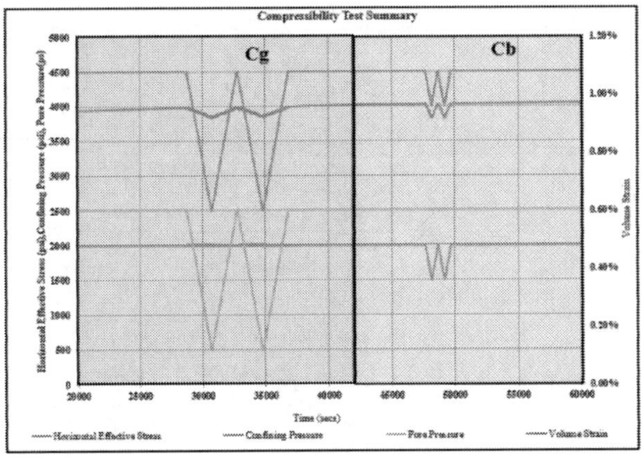

Fig. 8. Poroelastic coefficient measurement experimental summary.

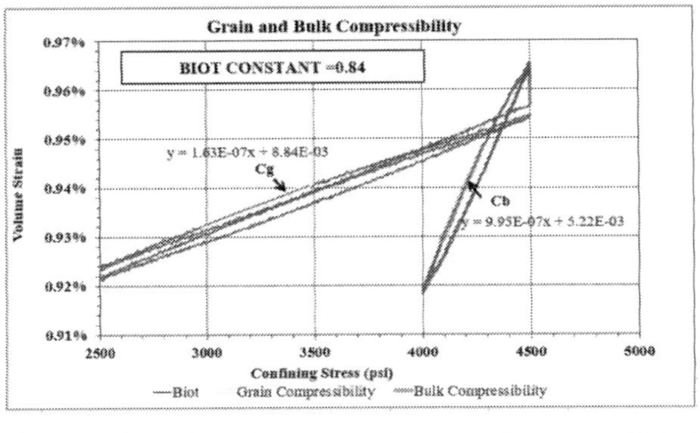

Fig. 9. Experimental results for poroelastic coefficient measurement.

5.4. SC-CO_2 flow study:

5.4.1 Set-up and experimental conditions

The fluid substitution experiment was carried out on two different plugs at ambient and high temperature. The protocol for both involved saturating the sample with brine at stress and flowing SC-CO_2 from the top to the bottom with the porous plate positioned at the bottom end. Fig. 10 shows the simplified flow schematic for the fluid substitution experiment.

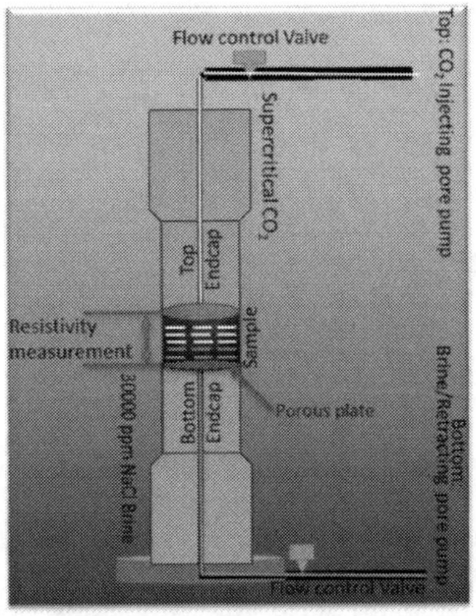

Fig. 10. Simplified flow schematic for the fluid substitution experiment.

Resistance and compressional velocity were monitored throughout the duration of the experiment. Experiments were performed maintaining an effective stress of 2000 psi (13.78 MPa) with confining pressure at 4500 psi (31.02 MPa) and pore pressure at 2500 psi (17.236 MPa). Samples were initially saturated with 30000 ppm NaCl brine at 100 psi pore pressure and confining of 2000 psi (13.78 MPa). The conditions were then progressed to desired experimental conditions of 4500 psi (31.02 MPa) confining and 2500 psi (17.236 MPa) pore pressure at stabilized temperature. Once stabilization was achieved (strain and pore pressure), injection of SC-CO_2 was begun from the top. The total experiment ran for 150 hours approximately. The ideal porous plate I-SW experiment is meant to be performed at very low flow rates of one pore volume over 10-30 days. The injection of the SC-CO_2 results in the creation of differential pressure across the sample. It is recommended to displace the fluid/brine saturating the sample at a constant injection rate with super critical fluids.

The injection is to be done while monitoring resistivity change across the sample along with ultrasonic velocity and the differential pressure across the sample. This investigation studied the use of the differential pressure as a feedback channel for driving the injection in the interest of expediency.

As the injection was begun, initial differential pressure created was around 50 psi (2550 at upstream and 2500 psi at downstream). Once flow stabilized at this differential pressure the system was programmed to respond with flow rates necessary to maintain a differential pressure of 150 psi. The limiting differential pressure for the experiment, is the differential pressure sustainable across the porous plate which was approximately 200 psi. When injection is started, some time is consumed in establishing an interface between the SC-CO_2 and the brine and there is a delay in the downstream pump response to the upstream pump. The dead volumes within the system must be thoroughly defined to enable higher accuracy in the pump volume-based saturation definitions.

5.4.2 Saturation based on volume and resistivity monitoring:

The volume of effluent brine on the downstream side was deemed as indicative of saturation conditions within the sample. The results of the ambient temperature fluid substitution experiment are shown in Fig. 11 to Fig. 13**Fig. *13***. The injection was carried out by gradually increasing the desired differential pressure setpoint across the sample. The resistivity was measured across the whole sample in the axial plane. Brine is a conductive fluid. The replacement of the brine in the sample by SC-CO_2 results in the increase in the resistivity of the sample. The time at which the resistivity starts increasing is indicative of sample penetration by the SC-CO_2. The resistivity changed from an initial 5.357 (Ω-m) to 24.787 (Ω-m) which is an increase of 362.7%. The I-Sw plot Fig. 13 indicates that the saturation exponent for the Buff Berea sandstone is 2. This is a value typically expected for sandstones. The measured resistivity across the sample increases exponentially when no more brine is expelled from the bottom of the sample while SC-CO_2 injection is being continued. This could indicate either breakthrough by the SC-CO_2 or achieving residual saturations for that rock type. The ideal I-SW experiment implemented at very low injection rates would entail piston-like uniform displacement of the wetting fluid by the non-wetting fluid. However, since SC-CO_2-brine interactions could entail mixing of the two fluids to some degree, ideal piston like uniform displacement is not expected. The utilization of the differential pressure as a driving mechanism also resulted in relatively higher instantaneous flow rates. The I-SW for the ambient temperature experiment displays deviation from linearity at 48% brine saturation indicating rapid increase in resistivity. The phenomenon was also reflected in the reduced brine effluent volumes and the cessation of the retraction of the downstream pump. Based on the resistivity behaviour, the pump movement and effluent volumes, the residual saturation of water at the end of the experiment is deemed to be approximately 48%.

The current study utilized 30000ppm NaCl brine. The possibility of salt precipitates being left behind as the brine is displaced or due to interaction between brine and dry SC-CO_2 exists, however it was not observed in the current experiment based on visual inspection. Further investigation in the form of thin section analysis or XRD is needed to be performed for increasing confidence in the observation. The phenomenon of salt precipitation is of greater importance in high salinity brines and should be taken into consideration after formation water evaluation.

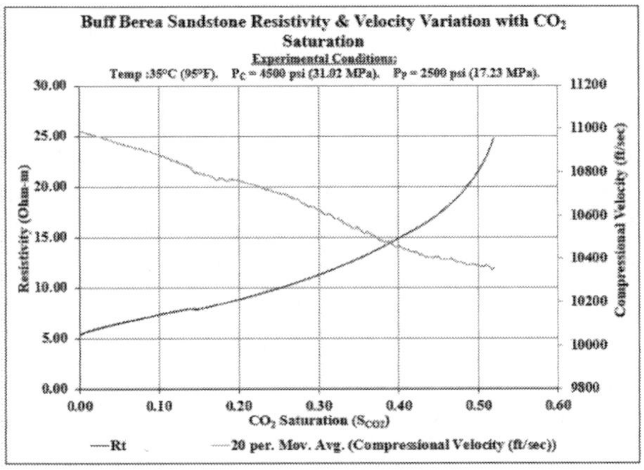

Fig. 11. Resistivity and velocity variation due to fluid displacement at ambient temperature.

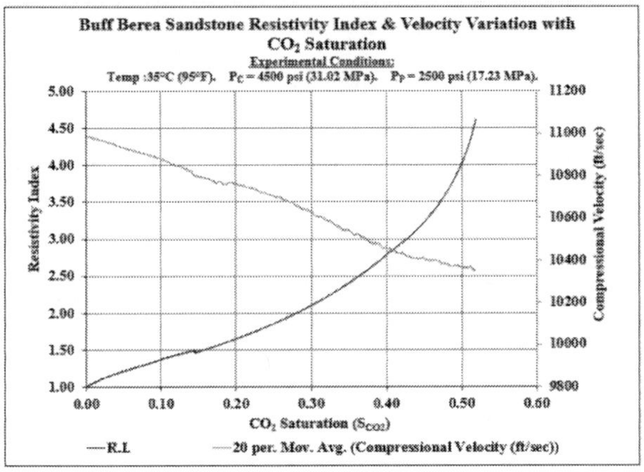

Fig. 12. Resistivity Index and velocity variation due to fluid displacement at ambient temperature.

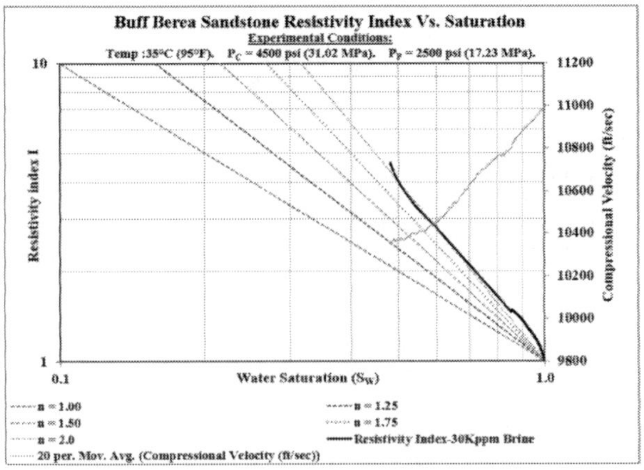

Fig. 13. I-S_W plot for brine displacement by SC-CO_2 at ambient temperature.

5.4.3 Acoustic monitoring:

Brine has greater density than SC-CO_2 under the established experimental conditions. As the brine gets displaced and volume of SC-CO_2 within the sample increases there is a decrease in the bulk density and modulus of the porous rock system. This decrease causes a decrease in the measured compressional velocity across the sample. The compressional velocity changed from an initial 10991 ft/sec to 10325 ft/sec which is a 6.44% change. This is similar to results obtained by [17], [18] and [27]. Greater decreases in velocities have been observed in carbonates but at lower effective stresses [16]. The sensitivity of acoustic measurements to change in fluid in the samples, were observed, at low effective stresses of 453 psi (3 MPa) and 290 psi (2 MPa) in [18] and [17] respectively. The acoustic data was found to be sensitive to change in water saturation for up to 10-20% S_W by [18] and up to 25% S_W by [17]. In the present study the effective stress utilised was 2000 psi (13.78 MPa). The high-resolution equipment enabled capture of acoustic data at the rate of a measurement every second. The sensitivity of the velocity seemed to be damped only at around 40 % saturation of SC-CO_2 i.e., after displacement of 40% of original brine volume. The different models for fluid substitution were compared against the measured experimental data. Fig. 14. shows the comparison of the different estimates with an averaged curve representing the high resolution, high volume acoustic dataset. The Reuss and Voigt based velocity predictions form the lower and upper bounds respectively for the experimental measurement. The velocity calculated using the modified Voigt i.e., the Biot-Gassmann-Brie expression matches very closely with the measured velocities. The absolute error when comparing the velocity calculated using the Reuss based parallel formulation for finely mixed fluids and the experimental measurements comes to a minimum of 0.01% and a maximum of 5.16%. The absolute error when comparing the velocity calculated using the Voigt based series formulation for patchy saturation and the experimental measurements comes to a minimum of 0.16% and a maximum of 1.7%. The absolute error when comparing the velocity calculated using the modified Voigt i.e., the Biot-Gassmann-Brie based series formulation for patchy saturation and the experimental measurements comes to a minimum of 0.49% and a maximum of 0.82%. The meandering nature of the experimental velocity curve indicates that the fluid distribution is more complicated than can be estimated by the Reuss and Voigt bounds, similar to [17]. The exponent value utilised for the Biot-Gassmann-Brie was 1.25. Even though the Biot-Gassmann-Brie curve matches well, it can still be seen that the experimental curve oscillates between indicating a mixed or patchy saturation. The overall trend in this case is more towards a series displacement which may be the result of the utilization of the porous plate technique. Since the maximum error was found to be between the Reuss or parallel formulation and the experimental measurement, a modification similar to the Biot-Gassmann-Brie is proposed as below:

$$K_F = \frac{K_B * K_{CO2}}{K_B * (1 - S_B^m) + K_{CO2} * S_B^m} \qquad (11)$$

where:
K_B: Brine modulus
S_B: Brine saturation
K_{CO2}: Bulk modulus of CO_2
m: modified Reuss exponent

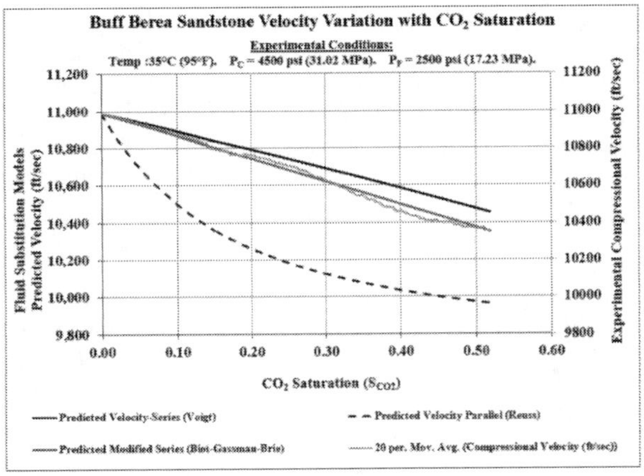

Fig. 14. Comparison of experimental measurement and velocity estimates using different models.

Fig. 15 shows the comparison of the velocity estimate based on the modified Reuss formulation with an "m" value of 0.3. The absolute error when comparing the velocity calculated using the modified Reuss based formulation and the experimental measurements comes to a minimum of 0.05% and a maximum of 1.48%. This enables a reduced uncertainty in the velocity estimate window.

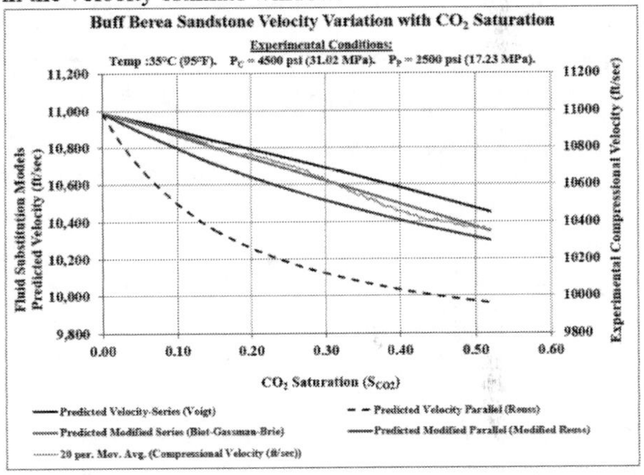

Fig. 15. Modified Reuss based velocity estimate comparison with experimental data.

5.4.4 Failure Envelope Characterization:

After completion of the fluid substitution testing, the same sample was utilised to conduct a multistage triaxial test as per [29]. The samples used for the current study were companion samples used for [29]. [29] had generated a Mohr Coulomb failure envelope based on multiple single stage triaxial testing. Fig. 16 shows the failure envelope and strength data for the fluid substitution sample and Fig. 17 displays the finalized failure envelope and strength data for the virgin Buff Berea sandstone samples. The unconfined compressive strength (U.C.S.) estimate for room temperature testing on virgin samples was 5929.7 psi and that on SC-CO_2 exposed sample was 4518 psi. This indicates a reduction of U.C.S. by 24%. The cohesion for the virgin samples was 1288.3 psi and the SC-CO_2 exposed sample exhibited 1058.4 psi. The reduction in cohesion was 18%. The friction angle changed from 42.33° for the virgin sample to 39.7°, a change of 6%. The fluid substitution samples did undergo a lot of processes such as saturation and desaturation along with exposure to CO_2 at high effective stress and pore pressure. However, most of those process steps are meant to reproduce the processes seen in the field. As such the difference in strength properties can be taken to be reasonably representative.

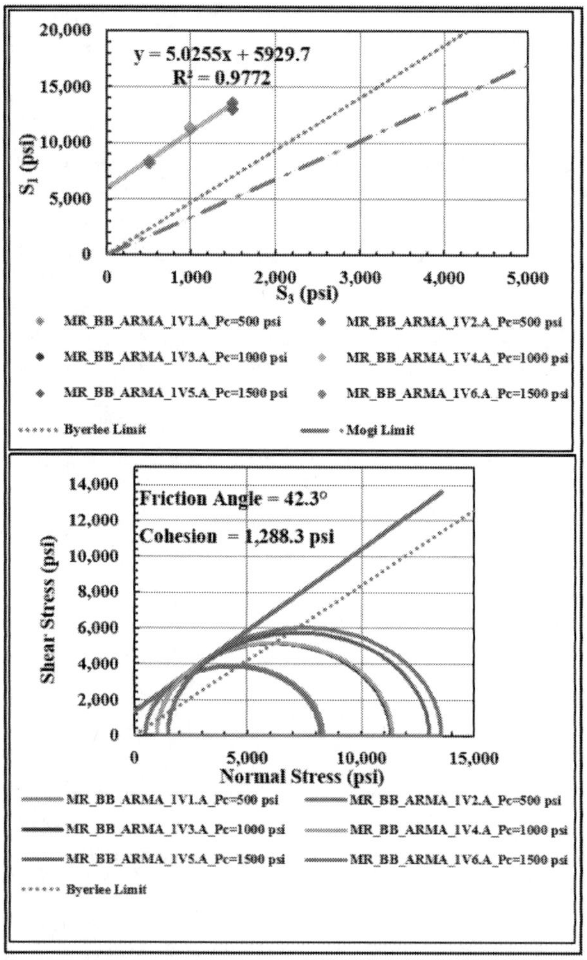

Fig. 16. Mohr Coulomb failure envelope and U.C.S. estimate for Buff Berea Sandstone.

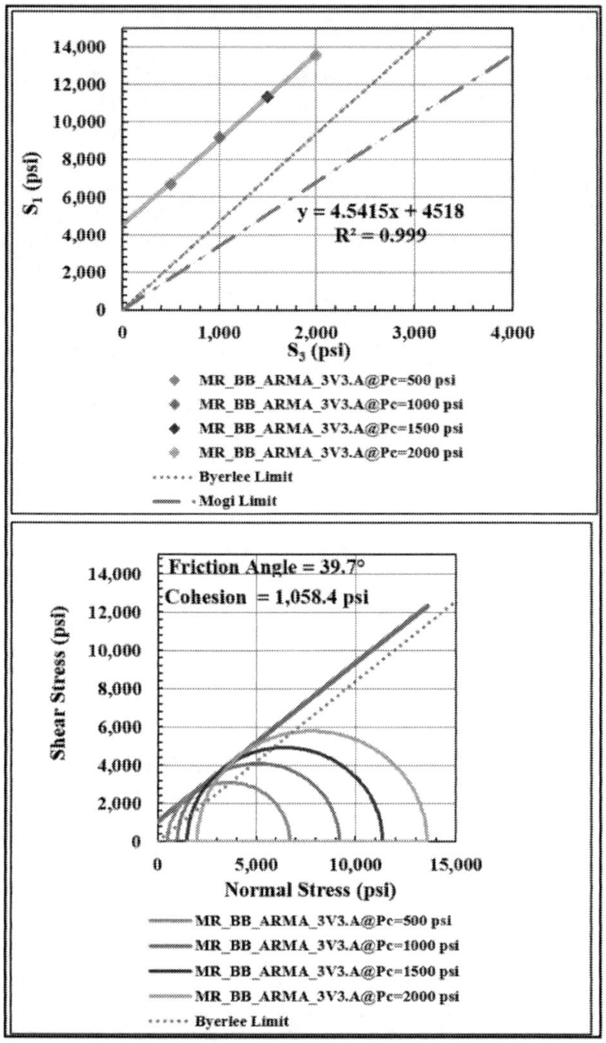

Fig. 17. Multistage triaxial test results on the fluid substitution sample tested at ambient temperature.

The high temperature fluid substitution experiment for the companion Buff Berea sample is summarised in Fig. 18.

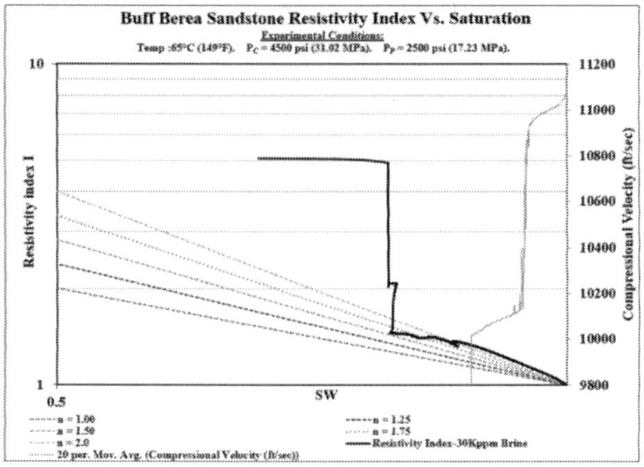

Fig. 18. 65°C (149F) fluid substitution -high injection rate induced breakthrough of SC-CO_2

 https://doi.org/10.1051/e3sconf/202336701001

The injection of the SC-CO_2 in the experiment was driven by the differential pressure across the sample. However, the selected setpoint for the differential pressure initiated a very high flow rate which caused a breakthrough of the SC-CO_2 through the sample. Fig. 19. illustrates an example of gradual sustained injection in a carbonate rock at high temperature of 120°C (248F) and high effective stress of 2500 psi (17.23 MPa). Resistivity index trend in the I-SW plot seems to indicate that the water saturation at the end of the experiment is 0.4 or 40%. The saturation exponent for the SC-CO_2 induced brine displacement through the rock is 1.75. The carbonate sample didn't have any companion samples for performing the triaxial and poroelastic coefficient characterizations required for the velocity predictions which were consequently not evaluated.

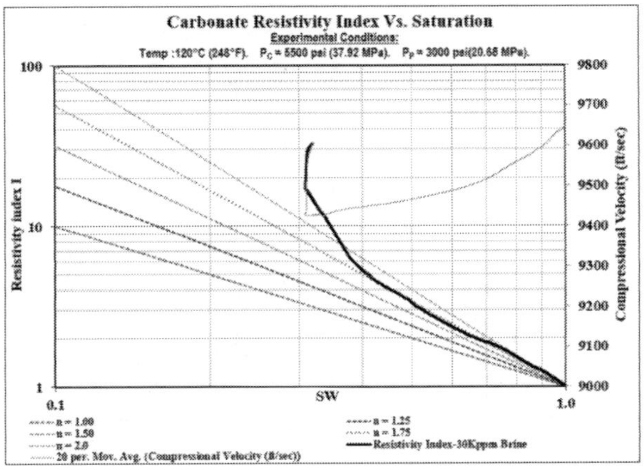

Fig. 19. High temperature, high pressure fluid substitution in carbonate sample.

6 Conclusions

Implementing an integrated, robust, laboratory program for addressing CCUS challenges helps reduce modelling and field scale uncertainty. The Biot-Gassmann-Brie formulation helps predict the acoustic behaviour during laboratory fluid substitution experiment reasonably well. The proposed modified Reuss formulation helps narrow the window for the model-experimental data match.

Appendix A: CO_2 Reference Data

All reference data has been sourced or calculated based on information from [32]. The critical temperature for CO_2 is 30.97°C, the critical pressure is 1070 psia and critical density is 0.467 g/cc. Fig. 20 to Fig. 22 show the variation of density, viscosity and compressibility of CO_2 with increase in pressure under different isothermal conditions of 30°C, 35°C and 65°C. At any temperature, with increase in pressure, phase change is observed. The critical point marks the transition to supercritical phase. Fig. 20 illustrates that in the gaseous phase, at lower pressures, the maximum magnitude of the density only reaches around 0.35 g/cc. In the supercritical phase, at all three temperatures, the densities are much higher. At 35°C, the density of the supercritical phase, is very close to the density of liquid CO_2 at 30°C. The region of interest (2500 psi) for the current study is marked in a red rectangle. The densities displayed by CO_2 in the supercritical phase are closer to liquid densities than gaseous densities especially with increase in pressure. The viscosity in the supercritical phase decreases with increase in temperature and remains lower than liquid phase viscosities at all pressures and temperatures. The compressibility shown in Fig. 22 has been calculated from density and velocity data as below:

$$C_b = \frac{1}{c * \rho * V_P^2} \tag{12}$$

Where: ρ: Density in kg/m^3

V_P: Compressional velocity in m/s

c: 1.4504E-04 (conversion to psi)

Fig. 22 illustrates that the compressibility of CO_2 at higher pressures in the supercritical phase is closer to liquid compressibility than gaseous compressibility.

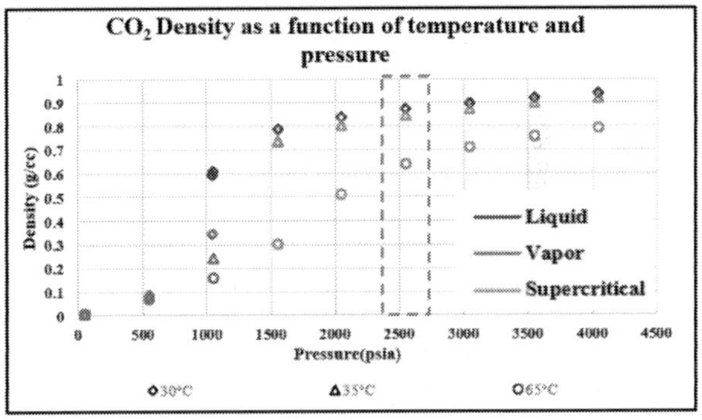

Fig. 20. Change in density of CO_2 as a function of pressure increase at three isothermal conditions of 30°C,35°C and 65°C. Phase change is mapped by colour coding.

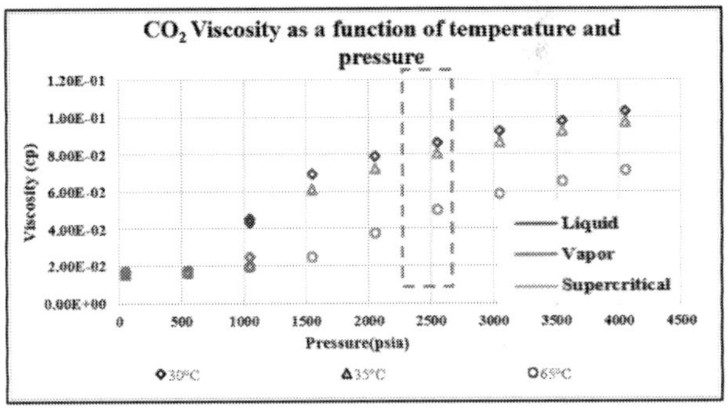

Fig. 21. Change in viscosity of CO_2 as a function of pressure increase at three isothermal conditions of 30°C,35°C and 65°C. Phase change is mapped by colour coding.

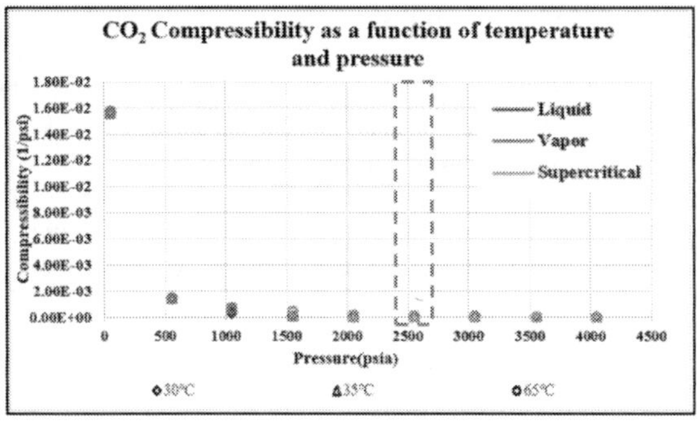

Fig. 22. Change in compressibility of CO_2 as a function of pressure increase at three isothermal conditions of 30°C,35°C and 65°C. Phase change is mapped by colour coding.

Authors would like to acknowledge the guidance of the late William Mickelson and the kind help of Hamad Aldin, Malik Al Salman, Hatim Aldin and Sneha Kalidindi in the execution of the experimental program.

References

1. Masson-Delmotte, V., P. Zhai, H.-O. Pörtner, D. Roberts, J. Skea, P.R. Shukla, A. Pirani, W. Moufouma-Okia, C. Péan, R. Pidcock, S. Connors, J.B.R. Matthews, Y. Chen, X. Zhou, M.I. Gomis, E. Lonnoy, T. Maycock, M. Tignor, and T. Waterfield. *IPCC, 2018: Global Warming of 1.5°C. An IPCC Special Report on the impacts of global warming of 1.5°C above pre-industrial levels and related global greenhouse gas emission pathways, in the context of strengthening the global response to the threat of climate change, sustainable development, and efforts to eradicate poverty.* In press.
2. A. Simone, E. Mackie and N. Jenvey, En. Proced. **1**, *GHGT-9*, 2219-2226, (2009).
3. S. D. Hovorka, En. Proced. **114**, *GHGT-13*, 3754-3756, (2017).
4. M. Smith, D. Campbell, E. Mckay, and D. Polson, *CO_2 aquifer storage site evaluation and monitoring*. Herriot-Watt University, Edinburgh (2011).
5. S. Carrol, K. Mansoor, X. Yang, T. A. Buscheck and Y. Sun. En. Proced. **114**, *GHGT-13*, 3628-3635, (2017).
6. K. Yamamoto. *CO2SC Symposium*.26-28, (2006).
7. J. Desroches, L. Jammes, and T. Berard. *CO2SC Symposium*.238-241, (2006).
8. S. Bachu and B. Bennion. *CO2SC Symposium*.193-195, (2006).
9. DOE-NETL, *BEST PRACTICES: Site Screening, Site Selection, and Site Characterization for Geologic Storage Projects* (2017). DOE/NETL-2017/1844.
10. J. Rutqvist, J. Birkholzer, and C.F. Tsang. *CO2SC Symposium*.230-232, (2006).
11. *DOE-NETL, BEST PRACTICES: Risk Management and Simulation for Geologic Storage Projects (2017). DOE/NETL-2017/1846*
12. R. Pini, S. Krevor. Chapter 7-Laboratory studies to understand controls on Flow and Transport for CO_2 storage. Editors, P. Newell, A. G. Ilgen. *Science of Carbon Storage in Deep Saline Formations*. Elsevier, pp 145-180, (2019).
13. B. Dupuy, A. Romdhane, P. Eliasson, H. Yan. Combined geophysical and rock physics workflow for quantitative CO2 monitoring. International Journal of Greenhouse Gas Control,106, 103217, (2021).
14. Z. Xue, T. Ohsumi, H. Koide.An experimental study on seismic monitoring of a CO2 flooding in two sandstones. Energy,30, 2352-2359, (2005).
15. A. JafarGandomi, A. Curtis. Assesing the monitorability of CO2 saturation in subsurface saline aquifers. International Journal of Greenhouse Gas Control.7. pp 244-260.(2012)
16. V. Vilarrasa, R.Y. Makhnenko, J. Rutqvist, Chapter 9 - Field and Laboratory Studies of Geomechanical Response to the Injection of CO2, Editor(s): P. Newell, A. G. Ilgen, *Science of Carbon Storage in Deep Saline Formations*, pp 209-236, Elsevier,(2019).
17. M. Gutierrez, D. Katsuki, A. Almarabat. Seismic velocity change in sandstone during CO2 injection. E3S Web of Conferences, 205, ICEGT (2020).
18. J. Kim, T. Matsuoka, Z. Xue. Monitoring and detecting CO_2 injected into water-satuated sandstone with joint seismic and resistivity measurements. *Exploration Geophysics*, 42, 58–68, (2011).
19. I. Falcon-Suarez, H. Marin-Moreno, F. Browning, A. Lichtschlag, K. Robert, L. J. North, A.I. Best. Experimental assesment of pore fluid distribution and geomechanical changes in saline sandstone réservoirs during and after CO2 injection.
20. T. Tsuji, T. Ikeda, F. Jiang. Hydrologic and elastic properties of CO_2 injected rock at various reservoir conditions : Insight into quantitative monitoring of injected CO_2. En. Proced. **114**, *GHGT-13*, 4047-4055, (2017).
21. H. Ott, S. Berg, S. Oedai. Displacement and mass transfer of CO_2/Brine in sandstone. *International Symposium of Society of Core Analysts*. SCA2011-05. (2011).
22. M.Z. Kalam, K.Al. Hammadi, O.B. Wilson, M. Dernaika and H. Samosir. Importance of porous plate measurements on carbonates at peseudo reservoir conditions. *International Symposium of Society of Core Analysts*. SCA2006-28, (2006).
23. D.G. Longeron, M.J. Argaud and L. Bouvier. Resistivity Index and capillary pressure measurements under reservoir conditions using crude oil. 64th Annual Technical Conference and Exhibition of the Society of Petroleum Engineers. (1989).
24. Richard Rosen, William Mickelson, Munir Sharf-Aldin, Basak Kurtoglu, Tobi Kosanke Meghana PaiAngle, Robert Patterson, Faraz Mir, Santhosh Narasimhan, and Amir Amini. Impact of Exp Studies on Unconventional Mechanisms.SPE-168965. SPE Unconventional Resources Conference – USA held in The Woodlands, Texas, USA, 1-3 April, (2014).
25. G. Mavko, T. Mukerji, J. Dvorkin. *The Rock Physics Handbook*. pp 266-339. (2009).

26. D. Gokaraju, M. Aldin, A. Thombare, A. Mitra, S. Govindarajan, R. Patterson. A novel method for experimental characterization of the Poroelastic Constants in Unconventional formations.URTeC :2902907, (2018).
27. V. Mikhaltsevitch, M. Lebedev, and Boris Gurevich. Measurements of the elastic and anelastic properties of sandstone flooded with supercritical CO_2. Geophysical Prospecting, **62**, pp 1266–1277, (2014).
28. J. W. Dudley, M. Brignoli, B. R. Crawford, R. T. Ewy, D. K. Love, J. D. McLennan, G. G. Ramos, J. L. Shafer, M. H. Sharf-Aldin, E. Siebrits, J. Boyer, M. A. Chertov. ISRM Suggested Method for Uniaxial-Strain Compressibility Testing for Reservoir Geomechanics. Rock Mech Rock Eng **49,** pp 4153-4178, (2016).
29. S. Govindarajan, M. Aldin, A. Guedez, A. Thombare, D. Gokaraju, A. Mitra, R. Patterson. Experimental Investigation for Selection of Unloading Criterion in Multistage Triaxial Testing. ARMA 21–1217, *55th US Rock Mechanics/Geomechanics Symposium* held in Houston, Texas, USA, 20-23 June 2021. (2021).
30. R. Span and W. Wagner. A new equation of state for CO_2 covering the fluid region from the triple point temperature to 1100K at pressures upto 800MPa. *J. Phys. Chem. Ref. Data* Vol. **25**, No 6 (1996).
31. M. Batzle and Z. Wang. Seismic properties of pore fluids. *GEOPHYSICS*, VOL. **57**, NO. 11 P. 1396-1408 (1992).
32. P.J. Linstrom and W.G. Mallard, Eds., NIST Chemistry WebBook, *NIST Standard Reference Database Number* **69**,National Institute of Standards and Technology, Gaithersburg MD, 20899,

    https://doi.org/10.18434/T4D303, (retrieved July 4, 2022)

A Bayesian optimization approach to the extraction of intrinsic physical parameters from T_2 relaxation responses

Rupeng Li[1], *Igor* Shikhov[1], and *Christoph H.* Arns[1,*]

[1] UNSW, School of Minerals & Energy Resources Engineering, Sydney 2052 NSW, Australia

Abstract. NMR transverse relaxation responses in porous media provide a sensitive probe of the micro-structure yet are influenced by a set of factors which are not easily detangled. Low-field T_2 transverse relaxation measurements can be carried out quickly and are frequently used to derive pore size distributions and determine derivate parameters like movable fluid volumes or permeability. Here we present an inverse solution workflow extracting related intrinsic physical parameters of the system by tightly fitting experiment and numerical simulation(s). We propose a Bayesian optimization approach that determines five T_2 related properties associated with two values of temperature simultaneously. This concurrent optimization (CO-OPT) utilizes Gaussian process regression to determine the intrinsic physical parameters leading to a match to experiment with a minimal number of function evaluations. A multi-modal search strategy is employed to identify non-unique solution sets of the problem. The workflow is demonstrated on Bentheimer sandstone, identifying five intrinsic physical parameters simultaneously, namely the surface relaxivity of quartz and the effective diffusion and relaxation times of the clay regions at 20°C and 60°C, providing the temperature-dependent quartz surface relaxivity and effective clay parameters. Given the generality of the method, it can easily be adapted to transverse relaxation experiments, or dynamic conditions where e.g., a change in wettability is monitored by intrinsic NMR parameters.

1 Introduction

NMR relaxometry has proven its high value both for laboratory and field petrophysical applications for estimation of pore size distributions [1], saturation of fluids [2], wetting conditions [3] and permeability [4].

The underlying assumption behind – relaxation occurs in fast diffusion regime (a surface governed process) in isolated or weakly coupled pores – must be valid, which is typically the case. This allows representing the observed relaxation rate in saturated rocks $1/T_2$ as a linear sum of three rate processes: bulk fluid relaxation rate $1/T_{2b}$, surface relaxation rate $1/T_{2S}$ and diffusion in the internal field $1/T_{2D}$ [5]

$$\frac{1}{T_2} = \frac{1}{T_{2b}} + \frac{1}{T_{2S}} + \frac{1}{T_{2D}}. \quad (1)$$

Here ρ_2 is transverse surface relaxivity, S and V are surface area and volume of an individual pore. In diamagnetic and weakly paramagnetic rocks, the surface relaxation term, i.e.,

$$\frac{1}{T_{2s}} = \rho_2 \frac{S}{V}, \quad (2)$$

is the primary relaxation mechanism, thus relating the T_2 distribution to pore-sizes. Typically, surface relaxivity is evaluated by matching a T_2 distribution's mode to that of a pore-aperture distribution obtained using mercury intrusion capillary pressure analysis. However, this approach is not entirely problem free, since pore-space geometry and the mode value are practically insensitive to temperature variations in the order of 50°C, while the value of T_2 relaxation distribution mode would noticeably change. It is well known that surface relaxivity is sensitive not only to pore geometry, but also to types of solid and fluid pair, wetting state, and temperature. The latter is also important in the context of calibration of field data with laboratory acquired data. Godefroy et al. [6] developed, and supported experimentally, a theoretical framework describing a surface relaxivity temperature dependency. They established that with increase of the temperature the surface relaxivity of quartz to water in a water-wet sandstone is increasing. Similar observations were reported by [7]. In this work we seek to test further the rarely reported temperature-dependent surface relaxivity behavior in sandstone rocks with the aid of an inverse problem formulation enabling to deduce surface relaxivity and accounting for several temperature effects unaccounted for in the past.

Micro-CT based random walk NMR relaxation simulation allows to easily incorporate effects of faster fluid diffusion at higher temperature, accounts for corresponding change of diffusion governed magnetization exchange, longer bulk water relaxation time etc., as long as corresponding inputs are known (measured). We here exclude from consideration the temperature effect on magnetic susceptibility primarily for the sake of simplicity, but also reasonably assuming

* Corresponding author: c.arns@unsw.edu.au

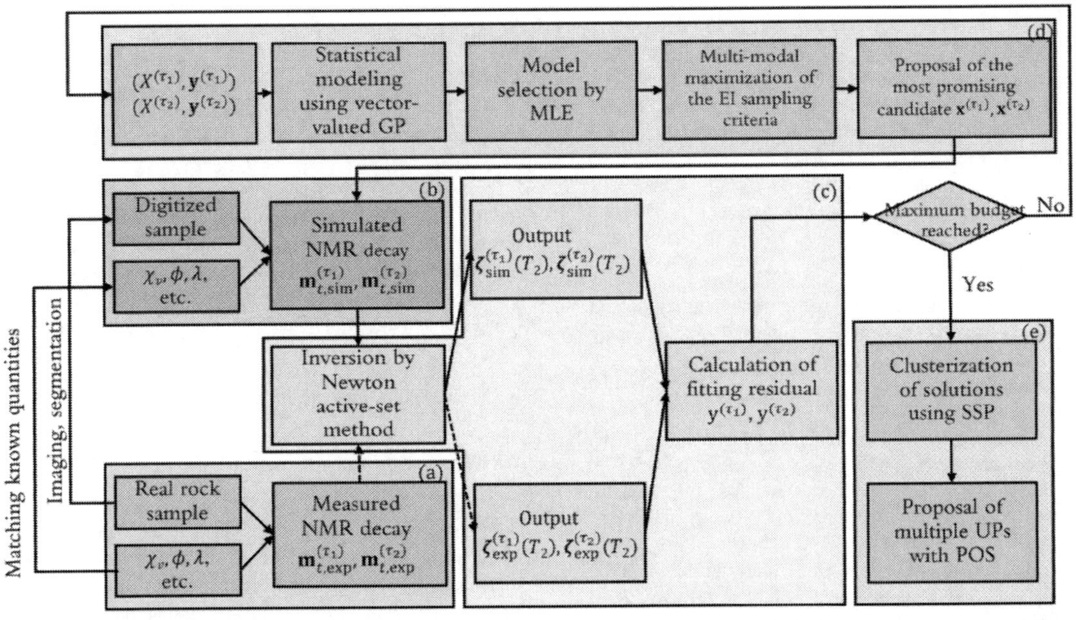

Fig. 1. Flowchart of CO-OPT for the estimation of intrinsic physical parameters from T_2 relaxation responses at two different temperatures.

insignificance of associated effects since the sample rock is diamagnetic, made of mostly of quartz. The magnetic susceptibility of the latter is practically constant in the temperature interval of interest [8]. We note that numerical simulations assign mineral-specific intrinsic properties like surface relaxivity at the pore scale [9-11] while matching distribution modes of experiments results in an effective [12] or apparent surface relaxivity [13; 14].

Given the significant advances in numerical techniques, the temperature-dependent intrinsic parameters can be regarded as unknown and later determined when good matches are achieved between simulations and measurements but is hindered by a practical issue: expensive function evaluations prohibit the enumeration of all configurations of unknown physical parameters in the multi-dimensional solution space. Luckily, the issue of computational burden has been successfully addressed by Bayesian optimization, for which only statistically optimum solutions are proposed for function evaluations, while balancing exploration and exploitation in the solution space. The advantage of Bayesian optimization has been leveraged to recover physical, geometrical, or structural parameters where expensive cost functions are involved [15-17], and in particular, complex multi-physics, multi-parameter, multi-modal problem arising in NMR relaxometry [11]. Furthermore, Bayesian optimization has been integrated with transfer learning for multi-objective optimization, and in particular, for by simultaneous optimization of T_1 and T_2 distributions [18]. In both studies, a multi-modal search strategy, comprising a multi-start L-BFGS-B optimizer searching for local optimum solutions, and a global optimizer social-learning particle swarm optimizer (SL-PSO) for global optimum solutions, are applied to recover all (major) local optimal solutions, i.e., potentially identifying multiple physically valid solution sets.

In this study, we adapt the dual-task inverse solution workflow (DT-ISW) developed for simultaneous fitting of T_1 and T_2 distributions in [18] to the context of simultaneous optimization of T_2 distributions associated with various values of temperature, referred to as concurrent optimization (CO-OPT). We demonstrate the performance of CO-OPT on a Bentheimer sandstone sample by the identification of five intrinsic parameters controlling the NMR relaxation response: one shared quantity, i.e., the tortuosity in clay regions, together with four unique quantities, i.e., transverse surface relaxivity of quartz and effective transverse relaxation time in clay regions, at two values of temperature τ_1 and τ_2. We also provide the SEP as the reference for which the physical quantities related to two values of temperature are separately identified ignoring the constraints that tortuosity is a constant value for a given structure. Finally, we provide the Pareto optimal solutions (POS), which allows operators to balance the preference of T_2 data fits at two values of temperature for the slightly conflicting objectives.

2 Methodology

2.1. Concurrent optimization

As adapted from DT-ISW introduced in [18], CO-OPT is composed of five components: (a) observables, (b) NMR forward solver, (c) cost-function, (d) Bayesian optimization, and (e) solution analysis. Fig. 1 demonstrates CO-OPT using two values of temperature as an example. We briefly summarize these components as itemized list below, while detailed descriptions follow in the next section:

a) Measure the NMR transverse relaxation decays $\mathbf{m}_{t,\exp}^{(\tau_1)}$ and $\mathbf{m}_{t,\exp}^{(\tau_2)}$ at temperatures τ_1 and τ_2.

b) Simulate the decays $\mathbf{m}_{t,\mathrm{sim}}^{(\tau_1)}$ and $\mathbf{m}_{t,\mathrm{sim}}^{(\tau_2)}$ for the chosen sets of unknown intrinsic parameters $\mathbf{x}^{(\tau_1)}$ and $\mathbf{x}^{(\tau_2)}$ on the segmented tomographic images (example: transverse surface relaxivities of the resolved phase: quartz, tortuosity of clay region, effective transverse relaxation times of clay regions), while honoring known quantities like porosity ϕ , volumetric magnetic susceptibility χ_v, bulk physical properties of water, and matching the physical kernel (see Eq. (3)) and regularization parameter λ.

c) Acquire the T_2 distribution using both simulated and measured decays, i.e., $\boldsymbol{\zeta}_{\mathrm{sim}}^{(\tau_1)}(T_2)$ and $\boldsymbol{\zeta}_{\mathrm{exp}}^{(\tau_2)}(T_2)$. Evaluate the cost-function and calculate the sum of the L_2 norm of the fitting residuals for the T_2 distributions.

d) Jointly model the T_2 objective functions at the two values of temperature using the vector-valued Gaussian Process (GP) with correlation captured, followed by inference of model hyperparameters using maximum likelihood estimation; then the EI acquisition function is maximized to identify the promising candidates $\mathbf{x}^{(\tau_1)}$ and $\mathbf{x}^{(\tau_2)}$. Finally, solutions within sparsely explored regions are evaluated using exact simulation.

e) Model the objective functions with the updated model hyperparameters upon depletion of the optimization budget and divide the solution space into various unique partitions (UPs) associated with different modes of the objective function using the solution space partitioning (SSP) introduced in [11]. For each UP the top three solutions and the POS are reported.

2.2 Materials, measurements, and NMR simulations

2.2.1 *Rock Sample*

Experimental reference NMR responses were obtained for a Bentheimer sandstone cylindrical core plug (25 mm diameter and 50 mm long) of 23.9% porosity saturated with 3 wt.% NaCl brine. This rock is composed primarily by quartz (96%) with remaining fraction represented equally by feldspar and kaolinite. The sample has volumetric magnetic susceptibility -7.45×10^{-6} (SI units).

2.2.2 T_2 *Reference Data*

We acquire transverse relaxation response of saturated sandstone rock using the standard Carr, Purcell, Meiboom and Gill (CPMG) pulse sequence [19; 20]. The NMR measurements were carried out using a Magritek Rock Core Analyzer operating at 2 MHz proton resonance frequency. Acquired echo-trains were 40,000 echoes long, with echo-time interval 250 us. Summation of 4 scans following phase cycling sequence resulted in a signal-to-noise ratio of around 50. The magnetization decays are treated as multi-exponential sum with unknown distribution of amplitudes $\zeta(T_2)$representing a Fredholm integral equation of the first kind:

$$m_{T_2}(t) = \int_0^\infty \zeta(T_2)\; e^{-t/T_2}\mathrm{d}T_2 + \eta_n(t), \qquad (3)$$

where $m_{T_2}(t)$ is the transverse magnetization decay as function of time t and η_n is additive white noise, and the objective is the estimation of the probability density function of T_2, i.e., $\zeta(T_2)$.

Practically, the T_2 distribution are obtained using a non-negative least squares (NNLS) algorithms, which minimizes the residual sum between the known left-hand size signal and right-hand side prediction, at the same time retaining a degree of control over distribution shape (smoothness) with a penalty/weight term (known as regularization parameter) [21-23]. All experimental and simulated decays were inverted using the algorithm [21] with resulting solutions identically binned across 256 logarithmically spaced T_2 intervals.

2.2.3 *Digitized Image*

A Bentheimer sandstone plug of 5mm diameter was imaged in double helix mode on the UNSW Tyree X-ray CT facility with a total acquisition time of 12h, resulting in a reconstructed tomogram with a resolution of $\epsilon = 2.16$ µm. The grey-scale tomogram was segmented into pore-space, quartz, clay region (a micro-porous effective phase), feldspar and an iron-rich dense mineral phase. For details of the segmentation process see [**Error! Reference source not found.**], which uses the same sample. Fig. 2 depicts a slice through the tomogram and resultant phase segmentation. Statistics of the Bentheimer sandstone segmentations for the calculation of the internal magnetic field at lattice resolution ϵ that follows are listed in Table 1 and Table 2. The micro-CT image was segmented taking into account XRD data reported in [24], though there is a slight difference in between possibly due to mineralogy variation at the mm scale. Basic mineral composition from XRD data and segmented CT image (in brackets) are quartz 96% (96.6%), feldspar 2.4% (1.7%), kaolinite 1.4% (1.6%), others 0.2% (0.15%).

Table 1. Statistics of the Bentheimer sandstone segmentations. Segmented image porosity used in simulation (ϕ_{sim}) is calculated as $f_p + 0.5 f_c$ [24], domain size is in voxel, ϵ denotes voxel size, and f_i are phase fractions (f_p: resolved porosity, f_c: clay region, f_q: quartz, f_f: feldspar, f_h: high-density).

phase fractions					
ϕ_{sim}	f_p	f_c	f_q	f_f	f_h
0.2373	0.2252	0.0241	0.7369	0.0126	0.0011

Assuming quartz, feldspar and kaolinite are all iron-free and having similar elemental composition are likely to exhibit similar magnetic susceptibility, susceptibility of bulk sandstone is decomposed into four separate values. Quartz and feldspar are indeed typically reported in a similar range, e.g., Hunt reported $-13\sim-17 \times 10^{-6}$ SI for both [25]. We measured χ_v of fluid and bulk rock only. In particular, χ_v for distilled water and 3% NaCl brine are -8.90 and -9.12×10^{-6} SI, respectively, while for convenience we typically use in simulations -9.0×10^{-6} SI for aqueous phase [26]. The balance value required to match the average bulk is attributed to a volumetrically small fraction of the high-density minerals assumed to be iron-rich.

Table 2. Measured average volumetric magnetic susceptibilities χ_v, known values for NaCl brine, clay region, quartz, feldspar, and inferred value for high-density using material balance.

volumetric magnetic susceptibilities (SI, multiply by 10^6)					
χ_v	$\chi_{v,w}$	$\chi_{v,c}$	$\chi_{v,q}$	$\chi_{v,f}$	$\chi_{v,h}$
-7.45	-9	$(\chi_{v,w} + \chi_{v,k})/2$	-12	-12	1900

2.2.4 T_2 *Simulation*

Given the naturally discretized segmented tomogram, the NMR responses are simulated via a lattice random walk algorithm following [9-11]. The simulation is carried out in multiple stages: first, the internal magnetic field is calculated in the dipole approximation via a convolution of the susceptibility field as defined by the individual phase susceptibilities given in Table 1 with a dipole field; second, the random walks are carried out on the segmented tomogram with a sub-lattice spacing of $\epsilon/10$ (low temperature) or $\epsilon/15$

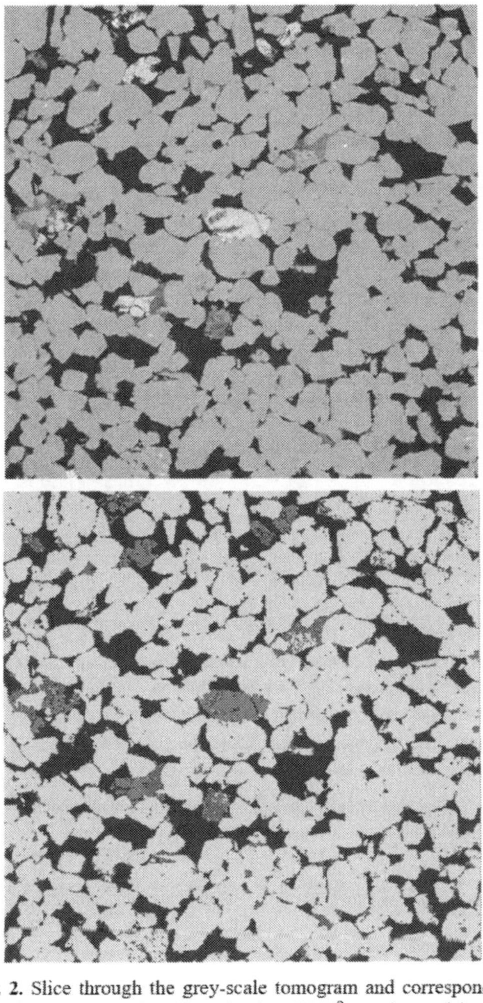

Fig. 2. Slice through the grey-scale tomogram and corresponding five-phase segmentation. Domain size: 1240^3 voxel, voxelsize ϵ = 2.16 μm. Blue: pore, indigo: clay, green: quartz, brown: feldspar, red: high-density phase.

(high temperature); random walks are started with a probability linearly related to the porosity of each voxel with each simulation consisting of about 80,000 random walks, dephasing according to the internal fields, with surface interactions lumped into the surface relaxivity for resolved phases, or into a local effective relaxation time for clay regions. Local internal magnetic field values at the required discretization level are derived via tri-linear interpolation from the coarse-scale internal field (resolution of ϵ). Third, the resultant magnetization decay is transformed to the T_2 distribution via an inverse Laplace transform (ILT) [22]. In the optimization framework the phase susceptibilities and the bulk properties of the saturation brine (diffusion coefficient and T_2 bulk relaxation time, see Table 3) are considered as

Table 3. Bulk diffusion coefficient and transverse relaxation time of water for the two temperatures considered.

τ / Celsius	$D_0/(\mathrm{cm}^2/\mathrm{s})$	$T_{2b,w}$ / s
20	2.03×10^{-5}	2.27
60	4.76×10^{-5}	5.26

constants. The other pore-scale physical parameters, namely the surface relaxivity of quartz $\rho_{2,q}$, the effective T_2 relaxation time of clay regions ($T_{2e,c}$), and the effective diffusion coefficient of clay regions ($D_{e,c}$) at different temperatures, in the following noted by superscripts (τ_1) and (τ_2), are the target of the proposed optimization effort. The surface relaxivity of minority phases is set to the surface relaxivity of quartz.

2.3 Minimization of T_2 objective under various temperatures

For notational simplicity we describe only two temperatures here, i.e., τ_1 and τ_2. The problem of minimization of the sum of the L2 norm of the fitting residuals for T_2 distributions acquired at τ_1 and τ_2 can be formulated as:

$$\text{minimize} f^{(\text{obj})}(\mathbf{x}) = f^{(\tau_1)}(\mathbf{x}^{(\tau_1)}) + f^{(\tau_2)}(\mathbf{x}^{(\tau_2)}), \quad (4)$$
$$\text{where} \quad \mathbf{x} = [\mathbf{x}^{(\tau_1)}, \mathbf{x}^{(\tau_2)}], \quad (5)$$
$$\mathbf{x}^{(\tau_1)} = \left[\rho_{2,q}^{(\tau_1)}, T_{2e,c}^{(\tau_1)}, D_{e,c}^{(\tau_1)}\right], \quad (6)$$
$$\mathbf{x}^{(\tau_2)} = \left[\rho_{2,q}^{(\tau_2)}, T_{2e,c}^{(\tau_2)}, D_{e,c}^{(\tau_2)}\right], \quad (7)$$
$$\text{subject to} \quad D_{e,c}^{(\tau_1)}/D_0^{(\tau_1)} = D_{e,c}^{(\tau_2)}/D_0^{(\tau_2)}, \quad (8)$$
$$\mathbf{x}_l \le \mathbf{x} \le \mathbf{x}_u. \quad (9)$$

where $\mathbf{x}_l$ and $\mathbf{x}_u$ are the lower and upper bounds of the pre-specified search domain and $f^{(\tau_1)}$ and $f^{(\tau_2)}$ are L2 norm of the fitting residuals for T_2 distributions acquired at temperatures τ_1 and τ_2, respectively, expressed as

$$f^{(\tau_1)}(\mathbf{x}^{(\tau_1)}) = ||\zeta_{\text{sim}}^{(\tau_1)}(T_2|\mathbf{x}^{(\tau_1)}) - \zeta_{\text{exp}}^{(\tau_1)}(T_2)||_2^2, \quad (10)$$
$$f^{(\tau_2)}(\mathbf{x}^{(\tau_2)}) = ||\zeta_{\text{sim}}^{(\tau_2)}(T_2|\mathbf{x}^{(\tau_2)}) - \zeta_{\text{exp}}^{(\tau_2)}(T_2)||_2^2, \quad (11)$$

with weights for all data points of the T_2 distribution being equally 1. Each lower-dimension component $\mathbf{x}^{(\tau_i)}, i \in \{1,2\}$ is associated with a specific temperature and therefore Eq. (5) specifies the complete solution. The constraints Eq. (8) apply due to Archie's Law, i.e.,

$$\frac{1}{\mathcal{T}} = \frac{D_e}{D_0}, \tag{12}$$

where $\mathcal{T}$ is the tortuosity. We assume that for a given structure the $\mathcal{T}$ is independent of temperature and Eq. (12) holds under both values of temperature.

2.3.1 *Scalar-Valued Gaussian Processes*

GP is a distribution over functions, and is characterized by its mean function $m(\mathbf{x})$ and covariance function $k(\mathbf{x}, \mathbf{x}')$ as [11; 27]

$$f(\mathbf{x}) \sim \mathcal{GP}\big(m(\mathbf{x}), k(\mathbf{x}, \mathbf{x}')\big), \tag{13}$$

where $x = (x_1, x_2, \ldots, x_D)^\top$ is a candidate in $\mathcal{X}$ of dimension $\mathrm{D} \times 1$, $f(\mathbf{x})$ is the process evaluated at location x. In GP, the prior distribution over f is expressed as

$$\mathbf{f} \,|X \sim \mathcal{N}\left(\mathbf{m}_0, K_f(X, X)\right), \tag{14}$$

where $\mathcal{N}$ denotes normal distribution, $\mathbf{m}_0 \in \mathcal{R}^{n\times 1}$ denotes the mean vector for observations, $(X, \mathbf{y}) = \{\mathbf{x}_i, \mathrm{y}_i\}_{i=1}^{n}$ and $(\mathbf{x}^*, y^*)$ denotes observed and predicted candidates respectively, and $K_f(X, X)$ denotes the $n \times n$ covariance matrix for $\mathbf{f}$, with the (p, q) entry expressed by $k\big(\mathbf{x}_p, \mathbf{x}_q\big)$. Accommodating the noise using Gaussian likelihood function yields $\mathbf{y}|\mathbf{f} = \mathcal{N}(\mathbf{f}, \sigma_n^2 I)$, and the posterior distribution for a single predicted candidate becomes

$$y^* |\, \mathbf{x}^*, X, \mathbf{y}, \boldsymbol{\theta}_s \sim \mathcal{N}\big(\mu(\mathbf{x}^*), \sigma^2(\mathbf{x}^*)\big), \tag{15}$$

where $\mathcal{N}$ is normal distribution, $\boldsymbol{\theta}_s$ is the model hyperparameters for the scalar-valued GP, and

$$\begin{aligned} \mu(\mathbf{x}^*) &= m_0^* + K_f(\mathbf{x}^*, X)\big[K_f(X, X) + \sigma_n^2 I\big]^{-1}(y \\ &\quad - m_0), \\ \sigma^2(\mathbf{x}^*) &= K_f(\mathbf{x}^*, \mathbf{x}^*) + \sigma_n^2 \\ &\quad - K_f(\mathbf{x}^*, X)\big[K_f(X, X) \\ &\quad + \sigma_n^2 I\big]^{-1} K_f(X, \mathbf{x}^*). \end{aligned} \tag{16}$$

2.3.2 *Vector-Valued Gaussian Processes*

GP with multiple outputs can be represented by the vector-valued GP indicates. $Q = 2$ applied throughout this study, indicating that there are two correlated tasks. The GP kernel in Eq. (13) has to be replaced by a vector-valued GP kernel taking into account the correlation between the Q outputs of $\mathbf{f}$, with $\mathbf{f}$ expressed as [28; 29]

$$\mathbf{f}(\mathbf{x}) \sim \mathcal{GP}\big(\mathbf{m}(\mathbf{x}), \mathbf{K}(\mathbf{x}, \mathbf{x}')\big), \tag{17}$$

where $\mathbf{m}(\mathbf{x}) \in \mathcal{R}^{Q\times 1}$ denotes the stacked mean function $\{\mathbf{m}_q(\mathbf{x})\}_{q=1}^{Q}$ for each output and K denotes a matrix-valued function whose entries $\big(\mathbf{K}(\mathbf{x}, \mathbf{x}')\big)_{q,q'}$ express the correlation between the input-pairs $\mathbf{x}$ and $\mathbf{x}'$, and between the outputs-pairs $f_q(\mathbf{x})$ and $f_{q'}(\mathbf{x})$. We account for the noisy observations using a Gaussian likelihood, i.e., $\mathbf{y}|\mathbf{f} = \mathcal{N}(\mathbf{f}, \Sigma_n \otimes I)$, where $\Sigma_n = \mathrm{diag}(\sigma_n^2)$ is a $Q \times Q$ diagonal matrix incorporating task-dependent noise levels, and $\sigma_n^2 = \big(\sigma_{n,1}^2, \ldots, \sigma_{n,Q}^2\big)^\top$ is a vector of noise variances. Similar with the case of scalar-valued GP, the predicted distribution for a single predicted candidate $\mathbf{x}^*$ is expressed as

$$\mathbf{y}^* |\, \mathbf{x}^*, X, \mathbf{y}, \boldsymbol{\theta}_v \sim \mathcal{N}(\boldsymbol{\mu}_{\mathbf{x}^*}, \Sigma_{\mathbf{x}^*}), \tag{18}$$

where $\mathbf{y}^* = \big[y_1^*, \ldots, y_Q^*\big]^\top$ concatenates Q outputs for $\mathbf{x}^*$, $\boldsymbol{\theta}_v$ is a vector of model hyperparameters (v stands for vector-valued GP), and

$$\begin{aligned} \boldsymbol{\mu}_{\mathbf{x}^*} &= \mathbf{m}_0^* + \mathbf{K}_f(\mathbf{x}^*, X)\big[\mathbf{K}_f(X, X) + \Sigma_n \otimes I\big]^{-1}(\mathbf{y} \\ &\quad - \mathbf{m}_0), \\ \Sigma_{\mathbf{x}^*} &= \mathbf{K}_f(\mathbf{x}^*, \mathbf{x}^*) + \Sigma_n \\ &\quad - \mathbf{K}_f(\mathbf{x}^*, X)\big[\mathbf{K}_f(X, X) + \Sigma_n \\ &\quad \otimes I\big]^{-1} \mathbf{K}_f(X, \mathbf{x}^*). \end{aligned} \tag{19}$$

2.3.3 *Candidate Proposal and Solution Analysis*

We use the standard EI acquisition function, which for the qth task is

$$a_{\mathrm{EI},\tau_q}(\mathbf{x}^*) = \int_{-\infty}^{y_q(\mathbf{x}_q^-)} \big(y_q\big(\mathbf{x}_q^-\big) - u\big)p(u)\, \mathrm{d}u, \tag{20}$$

where $\mathbf{x}_q^-$ is the current best observed candidate for the qth objective and $u = y_q^*|X, \mathbf{y}, \mathbf{x}^*, \boldsymbol{\theta}_v$ is the Gaussian distributed variable given by Eq. (18).

A promising candidate is proposed by joint maximization of a_{EI,τ_1} and a_{EI,τ_2} in the form of $\sum_{q=1}^{2} a_{\mathrm{EI},\tau_q}$ under equality constraints using sequential quadratic programming (SQP) [30]. The initial guesses are placed randomly in the feasible domain, ensuring that the multi-modal nature of the objective function is captured.

When the optimization budget depletes, the model hyperparameters are updated and the T_2 objectives associated with various values of temperature are jointly modelled using new hyperparameters settings. The solution space of Eq. (4) is then divided into various unique partitions (UPs) corresponding to various modes of the objective function. We referred to the solution with the lowest fitness value in each UP is as the local minimum (LM). For details see [11]. Meanwhile, the two objectives are expected to be slightly conflicting since measurements are corrupted by noise so that there is no solution that simultaneously achieves good matches for each objective. As a result, we provide the POS of the T_2 data fits for the slightly conflicting objectives, which is defined as the set of solutions that is not dominated by any other solutions. Mathematically, $\mathbf{x}_1$ dominates $\mathbf{x}_2$ if $\mathbf{x}_1$ beats or ties $\mathbf{x}_2$ for each of the two objectives, i.e., $f^{(i)}(\mathbf{x}_1) \le f^{(i)}(\mathbf{x}_2)$, $\forall i \in \{1,2\}$, and $f^{(i)}(\mathbf{x}_1) < f^{(i)}(\mathbf{x}_2)$, $\exists i \in \{1,2\}$.

3 Results

In this study, we determined the surface relaxivity of quartz and the effective transverse relaxation time and tortuosity for the clay region at two temperatures: $\tau_1 = 20^\circ\mathrm{C}$ and $\tau_2 = 60^\circ\mathrm{C}$, respectively. We compared the speed of convergence, non-unique solution sets as well as the optimum solutions provided by CO-OPT, where tortuosity is a shared parameter for both temperatures and SEP - tortuosity at each

https://doi.org/10.1051/e3sconf/202336701002

Table 4. The top three candidates in terms of $f^{(\mathrm{obj})}$ (or $f^{(\tau_1)}$ and $f^{(\tau_2)}$ if $f^{(\mathrm{obj})}$ does not exist) within each UP, identified using the two approaches. UPs are sorted in ascending order of the best identified $f^{(\mathrm{obj})}$ within the UP. T_2 distributions corresponding to the LMs are displayed Fig. 3. Both $log_{10}\mathcal{T}$ and $\mathcal{T}$ are reported for convenience. Numbers in the *R* column indicate the rankings of the solutions out of 200 evaluated candidates in terms of $f^{(\mathrm{obj})}$, whereas numbers in the *Step* column indicate at which step those solutions are proposed.

UP	R	$\rho_{2,\mathrm{q}}^{(\tau_1)}$ (μm/ s)	$\rho_{2,\mathrm{q}}^{(\tau_2)}$ (μm/ s)	$\log_{10}[T_{2\mathrm{e,c}}^{(\tau_1)}(s)]$	$\log_{10}[T_{2\mathrm{e,c}}^{(\tau_2)}(s)]$	$\log_{10}\mathcal{T}$	$\mathcal{T}$	$f^{(\tau_1)}$ $\times 10^{-4}$	$f^{(\tau_2)}$ $\times 10^{-4}$	$f^{(\mathrm{obj})}$ $\times 10^{-4}$	Step
						CO-OPT					
1	1	11.169	9.820	-1.889	-2.304	0.535	3.424	2.52	24.23	26.75	131
	2	11.509	10.132	-1.722	-2.143	0.697	4.980	5.61	28.49	34.10	97
	3	9.744	9.978	-1.811	-1.905	0.901	7.956	10.54	27.12	37.66	141
2	4	12.075	9.398	-1.934	-2.530	0.341	2.194	6.45	32.15	38.61	140
	5	12.058	12.103	-2.445	-2.602	0	1	13.13	27.03	40.16	60
	6	12.042	12.100	-2.449	-2.602	0	1	13.31	27.03	40.34	58
						SEP, $\tau_1 = 20°\mathrm{C}$					
1	1	10.892	-	-1.923	-	0.625	4.213	2.87	-	-	84
	2	11.008	-	-1.943	-	0.593	3.919	3.44	-	-	74
	3	10.895	-	-1.875	-	0.643	4.393	3.85	-	-	196
2	29	11.915	-	-2.406	-	0	1	9.91	-	-	72
	30	11.949	-	-2.397	-	0	1	9.91	-	-	67
	31	12.005	-	-2.389	-	0	1	9.95	-	-	69
						SEP, $\tau_2 = 60°\mathrm{C}$					
1	1	-	9.068	-	-2.085	0.783	6.067	-	22.64	-	174
	2	-	10.061	-	-1.916	0.895	7.848	-	23.11	-	142
	3	-	8.741	-	-2.101	0.786	6.112	-	24.75	-	195
2	6	-	12.089	-	-2.601	0	1	-	27.03	-	169
	7	-	12.130	-	-2.594	0	1	-	27.05	-	149
	8	-	12.003	-	-2.606	0	1	-	27.08	-	172

temperature is an independent parameter for optimization. Practically the base-10 logarithm of the effective relaxation time and of the tortuosity are employed since they usually vary across orders of magnitude. The solution space is subject to the following bounds:

$$0 \le \rho_{2,\mathrm{q}}^{(\tau_1)}/(\mu\mathrm{m/\,s}), \rho_{2,\mathrm{q}}^{(\tau_2)}/(\mu\mathrm{m/\,s}) \le 24,$$
$$-3 \le \log_{10}\left(T_{2\mathrm{e,c}}^{(\tau_1)}/\,s\right), \log_{10}\left(T_{2\mathrm{e,c}}^{(\tau_2)}/\,s\right) \le 0,$$
$$0 \le \log_{10}\mathcal{T} \le 1.6.$$

The solution space $\mathcal{X}$ composed by such constraints is so large that a grid search or random search is intractable in consideration for the expensive simulation. The optimization budget is 200 function evaluations.

For both approaches here, the multi-modal nature of the objective functions is well-catered, and the non-unique solution sets are identified using solution space partitioning (SSP), clustering all solutions to different UPs. Table 4 shows identified parameter values of the LMs for the two approaches. LMs with $f^{(\mathrm{obj})}$ above 1×10^{-2} are not listed. The problem of non-unique solution sets is encountered in both methods: UP 2 is identified as a mathematically sound solution set, though it is physically less likely since the inferred value for tortuosity reaches the lower bound of 1. Within the physically valid solution set, UP 1, we observe two trends for the variation of surface relaxivity with temperature: Both CO-OPT and SEP consistently show two different response patterns to increase of temperature. Group 1 shows a decrease of surface relaxivity $\rho_{2,\mathrm{q}}$ by about 15%, while group 2 shows a nearly constant surface relaxivity (increase of about 1%). Specifically, for CO-OPT there is a noticeable decrease in $\rho_{2,\mathrm{q}}$, as shown by UP 1, e.g., a decrease from 11.6 μm/s to 9.8 μm/s (averaged over three solutions) while there is a light to no increase in $\rho_{2,\mathrm{q}}$, as shown by the UP 2, e.g., an increase from 12.0 μm/s to 12.1 μm/s. SEP shows the same trend, with surface relaxivity decreasing from 10.9 μm/s to 9.3 μm/s, or increasing from 11.3 μm/s to 11.4 μm/s. We note that the solutions are of low fitness value for both objective functions and thus are considered mathematically correct. The trend is similar for SEP for which there is no constraint to keep tortuosity constant with temperature. As a result, the uncertainty in non-unique solution sets cannot be reduced by examining the quality of fit or the fitness value.

Fig. 3 shows the fit corresponding to the LMs in Table 4. From appearance, the fits for the T_2 distribution at $\tau_1 = 20°\mathrm{C}$ are better than $\tau_2 = 60°\,\mathrm{C}$ for both methods, leading to approximately 10 times larger fitness values at higher temperature. We assume that the larger misfit at higher temperature is mainly a function of increasing diffusion coupling emphasizing the effect of small micro-structural

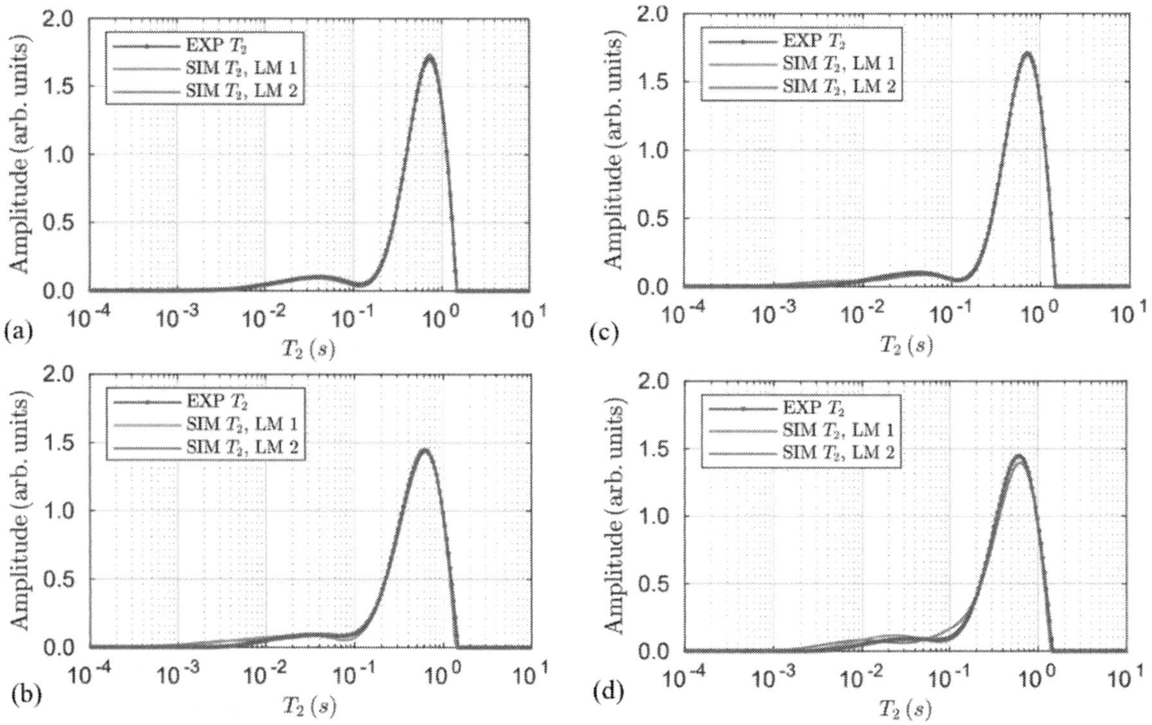

Fig. 3. The T_2 distributions corresponding to the LMs listed in Table 4: (a) CO-OPT, $\tau_1 = 20°C$, (b) CO-OPT, $\tau_2 = 60°C$, (c) SEP, $\tau_1 = 20°C$ and (d) SEP, $\tau_2 = 60°C$.

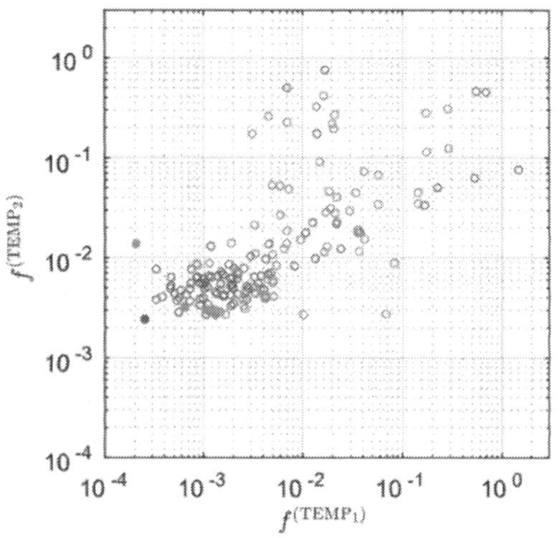

Fig. 4. Minimization of $f^{(\tau_1)}$ and $f^{(\tau_2)}$ using CO-OPT. Candidates in the bottom-left corner are preferred. Solid dots are Pareto optimum solutions; blue and orange dots correspond to UP 1 and UP 2 of CO-OPT in Table 4, and the circles are the dominated solutions.

features in the model.

Here we did not address the mathematical validity for the selection of an optimum solution since the two objectives are slightly conflicting: the $f^{(\tau_1)}$ value of LM 2 is inferior to the second-best solution though it achieves better value for $f^{(\tau_2)}$, as shown in Fig. 4, where candidates in the bottom-left corner are preferred and considered as feasible solutions. Such conflict is also reflected in the result of SEP from Table 4 that if $f^{(\tau_1)}$ and $f^{(\tau_2)}$ are individually optimized, $\mathbf{x}^{(\tau_1)}$ and $\mathbf{x}^{(\tau_2)}$ will neither recover the same value of $\log_{10} \mathcal{T}$ nor satisfy the constraints of Eq. (7). As a result, in situations where the uncertainty in measurement is high, or the conflict between the two objectives cannot be neglected, an optimum solution cannot be solely determined mathematically.

Since it is difficult to visualize the possible solution sets in five-dimensional space, we show in Fig. 5 the isosurface plot of the two objective functions $f^{(\tau_1)}$ and $f^{(\tau_2)}$ for the CO-OPT solution both in the three-dimensional space, as well as the correlation plot between $\rho_{2,q}$ and $\log_{10} T_{2e,c}$. From both isosurface plots and correlation plots we observe that there are multiple solution sets corresponding to both $f^{(\tau_1)}$ and $f^{(\tau_2)}$, which needs additional physical constraints to reduce such uncertainties.

4 Discussion and Conclusions

In this paper we analyzed the temperature dependent NMR relaxation responses and associated intrinsic physical properties of Bentheimer sandstone, namely the surface relaxivity of quartz and two effective clay parameters, the effective bulk relaxation time $T_{2e,c}$ and the effective diffusion coefficient $D_{e,c}$ of clay, the latter of which was reported as tortuosity to account for its geometric origin and accounting for the expectation that tortuosity remains a constant with respect to temperature. To achieve these aims we introduced

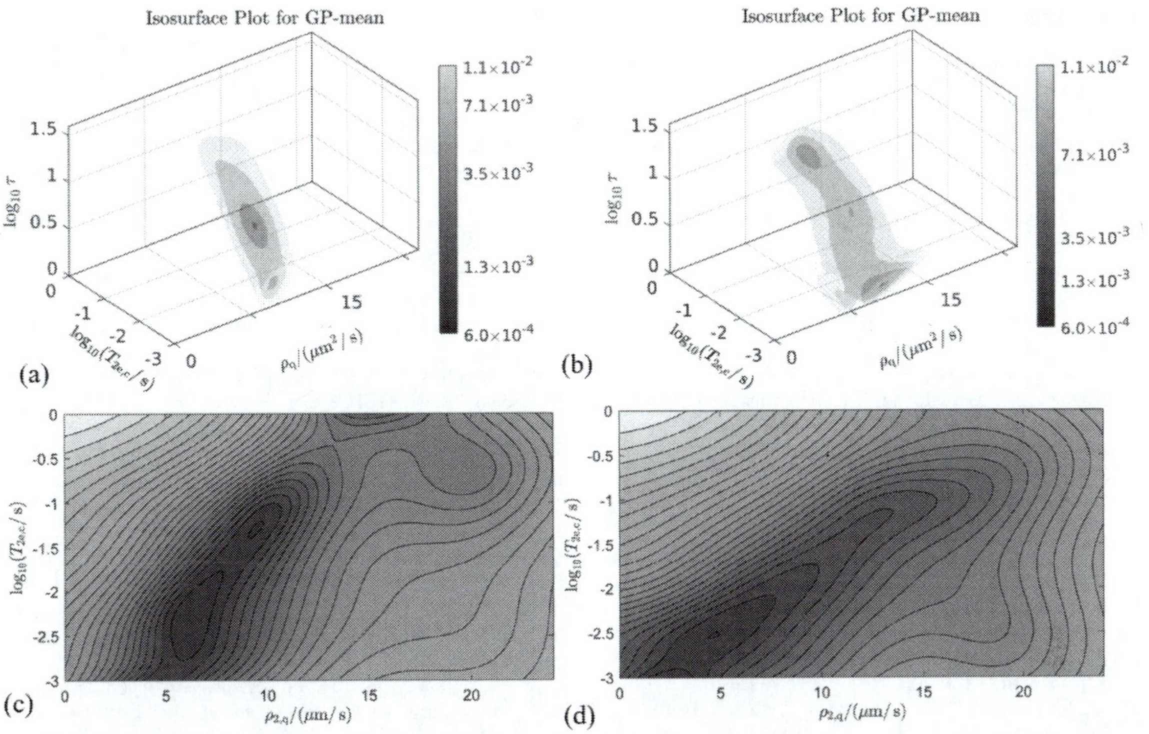

Fig. 5. The isosurface plot of (a) $f^{(\tau_1)}$ and (b) $f^{(\tau_2)}$ each in the three-dimensional space evaluated after the budget of 200 function evaluations is depleted, with contour plots (c) and (d) showing the correlation between $\rho_{2,q} - \log_{10} T_{2e,c}$ as projections of the iso-surface plot on to the $\rho_{2,q} - \log_{10} T_{2e,c}$ plane. We display five levels of isosurfaces, corresponding to the minimum 0.01%, 0.1%, 1%, 3%, 5% of the scalar field. Blue colors indicate regions with higher probability that mathematically good solutions can be found.

the CO-OPT workflow, which determines the temperature dependent transverse surface relaxivities and effective transverse relaxation time by joint minimization of the deviation between simulated and measured T_2 distributions at two values of temperature: five parameters were identified by computationally producing T_2 distributions almost identical with experiments for two values of temperature.

The experimentally observed enhanced relaxation in Bentheimer sandstone with temperature increase agrees with literature reports. However, contrary to these reports, the mechanisms behind the observed faster relaxation does not require the involvement of a higher surface relaxivity of quartz. We established two solutions, one is that of practically constant surface relaxivity of quartz and another is negatively correlating to temperature (15% lower if temperature is 40 °C higher). Both solutions enable matching the shift between experimental T_2 distributions, apparently either solely due to faster bulk water diffusion or additionally aided by magnetization coupling between micro- and macro-porosity.

The effective relaxation time of clay decreases by 30-40% with temperature increase. Though we are unable to conclude on specific values since multiple values within a fairly wide interval seem possible (approximately 3 to 15 ms), depending on the analysis method utilized.

The prediction of clay tortuosity is the least certain out of five properties, primarily due to an inherent difficulty of the effective medium approach when applied to dispersed clay in sandstone (in particular for smaller clay pockets/regions). The effect of incorrect magnetization exchange due to unrealistic discontinuity between effective and bulk diffusion at the clay-macropore interface increases with faster diffusion, i.e. is also "temperature-dependent", and may require a further refinement of numerical simulation.

The results of this study are contrary to what we expected upon starting on this analysis, the target of which was the derivation of the temperature dependence of intrinsic physical properties relevant to NMR relaxation-diffusion responses in rock samples.

We plan to further analyze this behavior in the future with resolved clay [29] and feldspar model structures and increase the resolution of micro-CT images, avoiding the introduction of an effective diffusion coefficient entirely. This should allow us to understand sensitivities and pinpoint the mechanism responsible for the observed behavior. We remark that resolving this issue may have direct implications to the interpretation of NMR responses in the context of wettability alteration, a subject of current intensive study [30].

CHA acknowledges the Australian Research Council for funding through ARC discovery project DP200103548. This work was supported by computational resources provided by the Australian Government through gadi under the National Computational Merit Allocation Scheme (grant m65).

References

1. M. G. Prammer,SPE Annual Technical Conference and Exhibition (1994).
2. W. F. J. Slijkerman, J. P. Hofman, W. J. Looyestijn, Y. Volokitin, Petrophysics - The SPWLA Journal of Formation Evaluation and Reservoir Description, **42**(04) (2001).
3. M. Fleury, F. Deflandre, Magn. Reson. Imaging, **21**(3), 385-387 (2003).
4. A. Timur,SPWLA 9th Annual Logging Symposium (1968).
5. H. C. Torrey, Phys. Rev., **104**(3), 563-565 (1956).
6. S. Godefroy, M. Fleury, F. Deflandre, J. P. Korb, J. Phys. Chem. B, **106**(43), 11183-11190 (2002).
7. K. Katika, H. Fordsmand, I. L. Fabricius, Microporous Mesoporous Mater., **269**, 122-124 (2018).
8. A. Ali, D. K. Potter, Geophysics, **77**(3), WA211-WA221 (2012).
9. C. H. Arns, T. AlGhamdi, J.-Y. Arns, New J. Phys., **13**(1), 015004 (2011).
10. Y. Cui, I. Shikhov, R. Li, S. Liu, C. H. Arns, J. Pet. Sci. Eng., **202**, 108521 (2021).
11. R. Li, I. Shikhov, C. H. Arns, Phys. Rev. Appl., **15**(5), 054003 (2021).
12. J. Chen, G. J. Hirasaki, M. Flaum, J. Pet. Sci. Eng., **52**(1), 161-171 (2006).
13. M. Müller-Petke, R. Dlugosch, J. Lehmann-Horn, M. Ronczka, Geophysics, **80**(3), D195-D206 (2015).
14. S. Costabel, C. Weidner, M. Müller-Petke, G. Houben, Hydrol. Earth Syst. Sci., **22**(3), 1713-1729 (2018).
15. J. Mockus, J. Glob. Optim., **4**(4), 347-365 (1994).
16. D. R. Jones, M. Schonlau, W. J. Welch, J. Glob. Optim., **13**(4), 455-492 (1998).
17. J. Snoek, H. Larochelle, R. P. Adams (2012), in *Proceedings of the 25th International Conference on Neural Information Processing Systems - Volume 2*, edited, pp. 2951–2959, Curran Associates Inc., Lake Tahoe, Nevada.
18. R. Li, I. Shikhov, C. H. Arns, SPE J. (2022).
19. H. Y. Carr, E. M. Purcell, Phys. Rev., **94**(3), 630-638 (1954).
20. S. Meiboom, D. Gill, Rev. Sci. Instrum., **29**(8), 688-691 (1958).
21. C. L. Lawson, R. J. Hanson, Prentice-Hall Inc.
22. P. C. Hansen, Inverse Probl., **8**(6), 849-872 (1992).
23. D. P. Bertsekas, SIAM J. Control Optim., **20**(2), 221-246 (1982).
24. I. Shikhov, D. S. Thomas, C. H. Arns, Energy & Fuels, **33**(9), 8184-8201 (2019).
25. C. P. Hunt, B. M. Moskowitz, S. K. Banerjee (1995), in *Rock Physics & Phase Relations*, edited, pp. 189-204.
26. I. Shikhov, Phd Thesis (2015).
27. C. E. Rasmussen, C. K. I. Williams (2005), The MIT Press.
28. E. V. Bonilla, K. M. A. Chai, C. K. I. Williams (2007), in *Proceedings of the 20th International Conference on Neural Information Processing Systems*, edited, pp. 153–160, Curran Associates Inc., Vancouver, British Columbia, Canada.
29. K. Swersky, J. Snoek, R. P. Adams (2013), in *Proceedings of the 26th International Conference on Neural Information Processing Systems - Volume 2*, edited, pp. 2004–2012, Curran Associates Inc., Lake Tahoe, Nevada.
30. P. T. Boggs, J. W. Tolle, Acta Numer., **4**, 1-51 (1995).

 https://doi.org/10.1051/e3sconf/202336701003

Numerical study of NMR relaxation responses in synthetic clayey sandstone by dual-scale modeling

Yingzhi Cui, *Igor* Shikhov, and *Christoph* Arns*

School of Minerals and Energy Resources Engineering, University of New South Wales, Sydney, Australia

Abstract. Nuclear Magnetic Resonance (NMR) relaxometry is a common technique for petrophysical characterization of sedimentary rocks. The standard interpretation of NMR relaxation response assumes that the fast diffusion limit is valid for the whole pore space, allowing to translate transverse relaxation components into pore apertures. However, porous media naturally exhibit multiple length scales. The diffusion between different sized pores may modify the transverse relaxation rate, weakening the relationship with corresponding pore size populations. Focusing on sandstones, we investigate the impact of diffusion coupling on transverse relaxation depending on kaolinite amount, spatial distribution and temperature. A series of synthetic clayey sandstone models with different clay amounts and morphological distributions (pore-lining, pore-filling and laminated) are generated based on a micro-CT image of an actual Bentheimer sandstone. A dual-scale random walk NMR relaxation simulation with resolved multi-porosity kaolinite models is utilized to avoid problems in near to interface exchange regions typical for effective medium representations. Simulations provide spatially resolved dynamics of magnetization exchange between different porosity populations. The results indicate that increased temperature and kaolinite clay amount with lower micro-porosity allows higher magnetization exchange between micro- and macro-porous regions. Pore-lining clay demonstrates stronger diffusional coupling effects, leading to an overestimation of micro-porosity. We further discuss the impact of diffusion coupling on NMR-estimated permeability via SDR and Coates models.

1 Introduction

Nuclear magnetic resonance (NMR) well logging is an important formation evaluation technique widely utilized for estimation of reservoir petrophysical properties, such as fluid types, saturations, and permeability. The interpretation of NMR data assumes that diffusing protons sample isolated pores (the so-called fast diffusion limit), in which case the observed relaxation rate is related to the surface-to-volume of individual pores [1]. However, molecular diffusion between different pore populations, referred to in literature as diffusional coupling, may lead to erroneous predictions if being unaccounted for. In fact, diffusional coupling appearing as magnetization exchange in NMR data is closely related to the connectivity between micro- and macro-porosity. It has been shown that depending on spatial distribution in rocks, microporosity may be a significant contributor to transport properties and may be responsible for a significant increase of the electrical conductivity [2], higher than anticipated hydraulic flux [3], and may control grain dissolution [4] or wettability alteration through asphaltenes deposition rate. While in this work we focus on the impact of diffusional coupling in the context of NMR interpretation, specifically on the relationship between magnetization exchange and transverse relaxation (T_2), the developed analysis is applicable to all mentioned transport problems.

The diffusional coupling phenomenon in respect to NMR data analysis has been extensively studied in synthetic porous media [5, 6], or sedimentary rocks, namely carbonate [7 - 10] and shaly sandstone [6, 11]. It has been reported in literature that two key petrophysical properties estimated from T_2 responses (irreducible water saturation and permeability) are affected if magnetization exchange between distinctive pore populations is significant. The standard practice in NMR core analysis is to relate T_2 values to pore size, i.e. the short relaxation time components of the T_2 distribution can be attributed to irreducible or capillary bound water (CBW), while the longer time components separated by the cutoff value ($T_{2,cutoff}$) correspond to mobile fluid in larger pores (known also as free fluid index FFI) [12]. However, it was demonstrated that calculated BVI values from NMR T_2 responses of sedimentary rocks exhibiting a well-coupled pore system are prone to large errors approaching 50% [15]. Interestingly, it was observed that at high temperature these effects were far less severe than at ambient temperature. The exact reason of this observation was not established since temperature dependence of governing physical properties (increase of self-diffusion coefficient of fluids and surface relaxivity [16]) suggests exactly the opposite. It is fair to acknowledge the limitation of the numerical simulation model used by [15], excluding internal field gradients effects from consideration. In addition, the temperature dependence of surface relaxivity is still poorly understood – both increases and decreases with temperature are reported in literature [6]. In the case of complete coupling, the micro- and macro-pores cannot even be identified from T_2 distribution resulting in the erroneous estimation of irreducible saturation [11].

NMR based permeability is estimated using one of two correlations: T_{2lm} - based SDR equation (after Schlumberger

* Corresponding author: c.arns@unsw.edu.au

Doll Research) [17] and $T_{2,cutoff}$ - based Coates model [18]. The magnetization exchange tends to shift one of the two coupled components – the one having a longer relaxation time – towards a shorter relaxation time [19], affecting NMR-related components of both models (T_{2lm}, CBW/FFI) and creating poor estimates. For instance, [8] demonstrated that permeability estimations based on SDR model for different rocks could exhibit relative errors up to 15%. [20] shows that T_2 shifts to shorter times because of intermediate diffusion coupling between micro- and macro-pores in chlorite-coated sample leading to inaccurate permeability estimation. In addition, a pore-connectivity parameter, introduced for the Coates model, offers more precise permeability and movable fluid volume estimates in complex coupled pore geometry [21]. The study [22] further extended to directional pore-connectivity factor to accurately characterize anisotropic directional permeability of sedimentary rock with NMR measurements.

Numerically diffusional coupling was analyzed in the context of relaxation in ideal pore networks [23]. Three geometrical models were presented to describe magnetization decay in carbonates with mixed intra- and intergranular porosity [24]. They reported that diffusional coupling may causes a shift of the longer relaxation peak and distort the amplitudes of the T_2 distributions. [7] further investigated coupling effects based on a 3D sphere model; constant internal magnetic gradient assumed. The results indicate that inter-pore diffusion has a strong impact on NMR measurements. More recently, simulations based on 3D periodic micro-porous grain consolidation models support the interpretation of NMR measurements in terms of magnetization exchange between dual-scale pore system [25 - 27]. The simplified models miss the morphological and geometrical information of the sedimentary rocks. To have an explicit representation of porous media, the X-ray micro-CT tomography was employed. Pore-pore coupling was analyzed by tracking random walkers within the partitioned pore space, which demonstrates that diffusion between pores occurs to a significant degree [28]. A numerical study on fractured carbonates [29] established that diffusional coupling may lead to underestimation of intergranular porosity by more than 10%. Although micro-CT images can offer more accurate internal structures of rock, micro-porosity often cannot be fully resolved due to either resolution limitations or simultaneous requirements of resolution and field of view. The non-resolved intermediate phase is then treated either as effective medium [28] or merged into the solid phase [30]. Yet, scanning electron microscopy (SEM) micrographs even of relatively simple rocks like Bentheimer sandstone (Fig. 1) indicate that the realistic micro-structure is complicated. Clay particles may fully or partially fill the pore space with heterogeneous local micro-porosity for each clay pocket. More generally, clay may also coat solid grains or appear as thin laminations [31]. These differences in the spatial distribution of clay as well as varying clay porosity and connectivity invite more complex coupling effects.

In this work we consider the impact of the spatial distribution and amount of clay on NMR relaxation responses. A simulation-based methodology is employed, which enables the quantification of the diffusional coupling phenomenon, in particular the evolution of magnetization across clay interface and impact on NMR relaxation derived

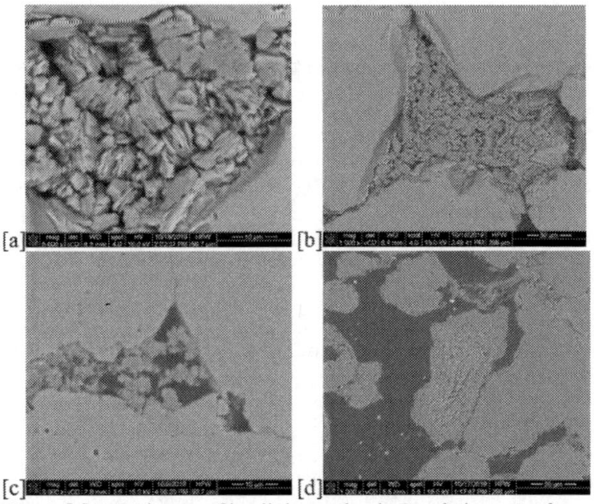

Fig. 1. [a] SEM image of kaolinite pocket in Bentheimer Sandstone; [b] Low porosity full-filled kaolinite pocket. [c] Medium porosity partial-filled kaolinite pocket. [d] Coated clay around quartz.

petrophysical properties such as irreducible water saturation (BVI) and permeability. A series of synthetic clayey sandstone models are generated with different clay amounts (6 vol% - 16 vol%) and synthetic morphological distributions (loosely termed coated, laminated, pore-lining and pore-filling) combined with resolved multi-porosity kaolinite models to track magnetization exchange between macro- and micro-pores. The evolution of magnetization exchange between different porosity populations is captured as function of observation time and penetration depth, together with the increase in the diffusion coefficient due to temperature; other intrinsic physical parameters like mineral-specific surface relaxivities are set constant in this fundamental study. Accounting for these effects from clay consequently improves NMR-estimated permeability via SDR and Coates models and enables more robust petrophysical interpretations.

2 Methodology

The objectives of this study require generation of clay distributions at different scales ranging from dispersed (pore-lining and pore-filling) to coated and laminated clay types. In the following we first introduce the methods for the coarse-scale distribution of clay regions, followed by fine-scale clay structure modeling, associated NMR simulations, and regional analysis fields embedded in the NMR solver enabling the local tracking of magnetization exchange. This is followed by a recapturing of NMR permeability correlations and the determination of permeability from tomographic images using the lattice Boltzmann technique.

2.1 Synthetic Clayey Sandstone

The synthetic clayey sandstones 3D models utilized in this study are based on the micro-CT image of Bentheimer sandstone shown in Fig. 2 and the SEM images (Fig. 1). The latter shows that pores of the original rock can be either fully or partially filled with kaolinite aggregates. Furthermore, those clay pockets exhibit different aggregate density and

 https://doi.org/10.1051/e3sconf/202336701003

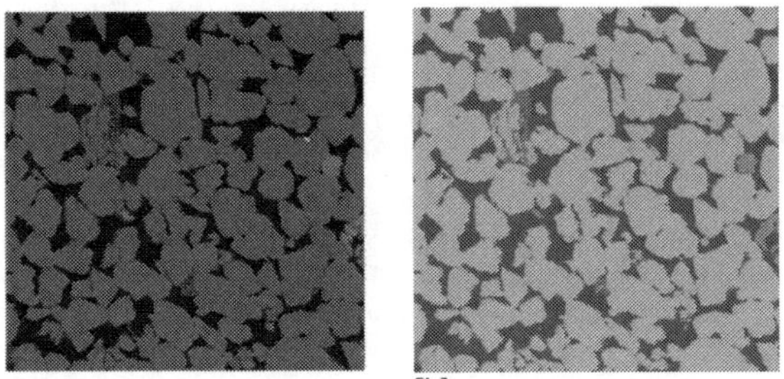

[a] [b]

Fig. 2. [a] Slice through an 800^3 subdomain of a tomogram of Bentheimer sandstone at a resolution of 2.2 μm voxel and [b] corresponding five phase segmentation. The colors are: dark blue - pore, green - quartz, cyan - clay, brown - feldspar, red - high-density minerals.

thus clay region porosity. This translates to clay regions with different intensity in the micro-CT images, where clay particles are not resolved (Fig. 2a). It is assumed that the intensity of kaolinite particles is similar to quartz. Therefore, the intensity in each clay region (I_c) is volume-weighted by the pore intensity (I_p) and quartz intensity (I_q). The intensity distribution of clay pockets (Fig. 3a) shows that the major values are around 1.2×10^4 to 1.3×10^4. The porosity of each clay pocket is estimated by $\phi_k = (I_q - I_c)/(I_q - I_p)$. The corresponding distribution shows that clay region porosity falls mainly into the range of 0.3 - 0.6 (Fig 3b). Consequently, we allowed for four discrete clay porosity labels corresponding to clay porosity of $\phi_c \in \{0.3, 0.4, 0.5, 0.6\}$.

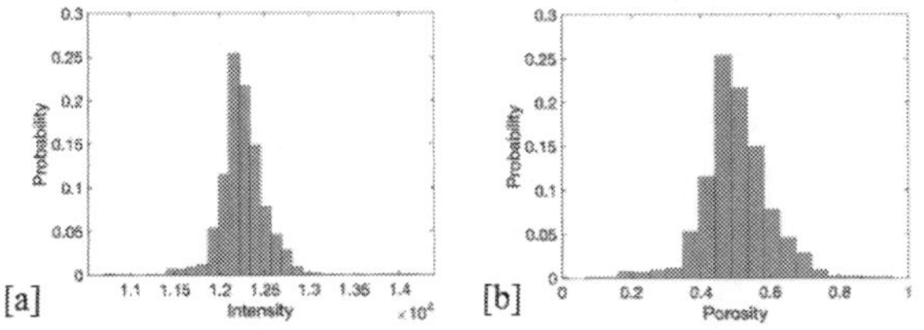

Fig. 3. [a] Intensity probability distribution of all clay pockets in the Bentheimer micro-CT images; [b] corresponding porosity distribution of clay pockets.

Four different clay distributions are constructed: pore-filling, pore-lining, coated and laminated. The models are generated with different clay volume fractions of 6 vol%, 11 vol% and 16 vol% to investigate the effects from clay amount. For pore-filling clay pockets we divided the resolved pore space into convex regions or pores. This is achieved by performing a watershed transformation [33] on the Euclidean distance map [34, 35] (Fig. 4a). We randomly assign the clay phase to the partitioned pores until the clay volume fraction meets the requirement. For pore-lining clay pockets, we determine the center for each pore (Fig. 4b) and preserve the pore fraction corresponding to the maximal inscribed sphere as pore space.

The remaining pore space is assigned as clay, resulting in a partial-filled clay layer of varying thickness. The coated clay phase is assigned only based on the distance threshold of EDT map. For laminated clay, we generated different layered system according to kriging interpolation method with random range, sill, and nuggets for mimicking the clay distribution.

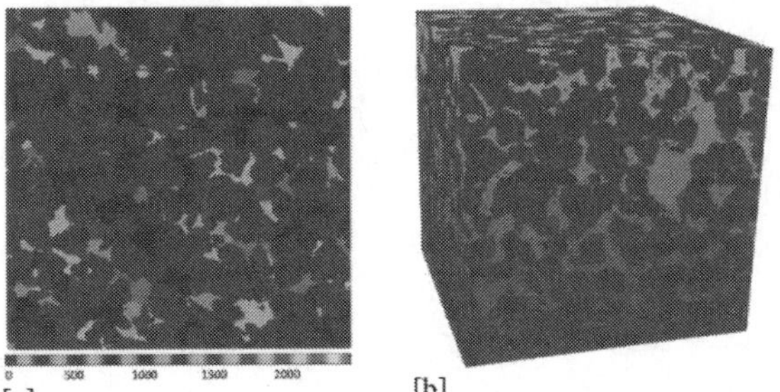

[a] [b]

Fig. 4. [a] Slice through the 800^3 watershed pore partitioning – region boundaries are the watersheds of the EDT, and [b] a 3D visualization of the centers in 800^3 watershed pore partitioning. Resolution: 2.2 μm/voxel.

To satisfy the clay volume fraction, the mean thickness values of each horizon are different. To reduce discretization effects from the image and force an accurate calculation of the internal magnetic fields [32], a tri-linear interpolation is applied with factor 4 refinement. Thus, the original 800^3 domain is refined to 3200^3 voxels with voxel size of 0.55 μm/voxel and smoother solid surface shown in Fig 5.

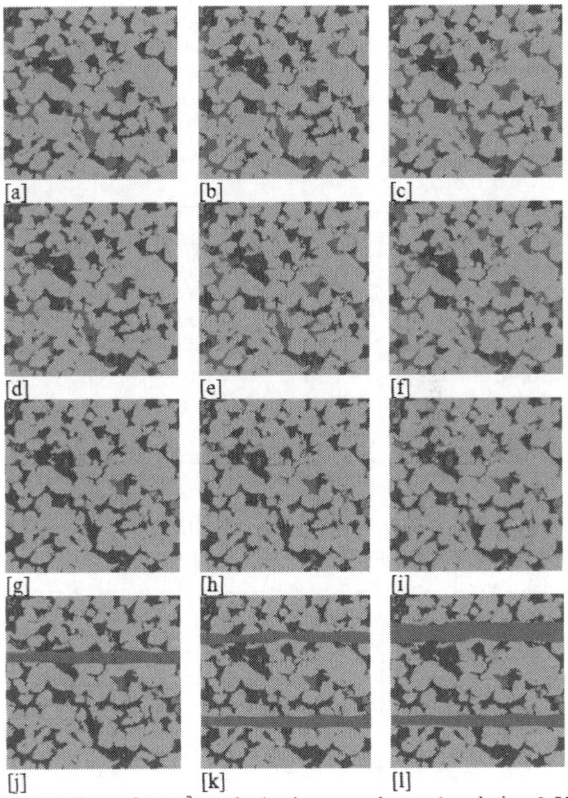

Fig. 5. Slices of 3200^3 synthetic clayey sandstone (resolution 0.55 µm/voxel). [a-c] pore-filling clay; [d-f]: pore-lining clay; [g-i]: coated clay; [j-l]: laminated clay. The left column with 6 vol% clay phase, the middle column with 11 vol% clay phase and the right column with 16 vol% clay phase. The colors are: dark blue - pore, green - quartz, cyan - clay, brown - feldspar, red - high-density minerals.

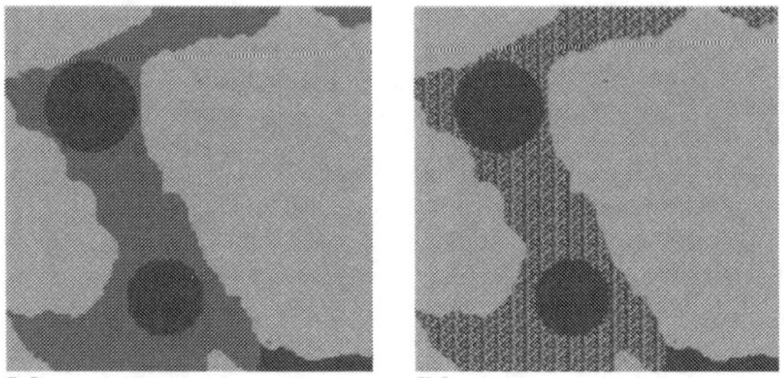

[a] [b]

Fig 6. [a] A 400^2 subsection of the 3200^3 synthetic clayey sandstone (16 vol%) with unresolved clay phase and voxel size of 0.55 μm. [b] Corresponding 10000^2 cross-section of [a] where each clay pocket is replaced by the corresponding kaolinite model. The resolution of the fully resolved model is 22 nm/voxel. The color schemes is: dark blue - pore, green - quartz, cyan - clay, brown - feldspar, red - high-density minerals.

2.2 Kaolinite Model

Kaolinite is one of the most common clay mineral constituents of sedimentary rocks, especially in sandstones. However, the microstructure cannot be fully resolved in the micro-CT image due to resolution limitations (Fig. 2). To resolve the microstructure within kaolinite pockets, we reconstruct 3D volumetric kaolinite models following the method detailed in [36]. In this study, we discretize the kaolinite microstructure as a periodic domain of 400^3 voxel with resolution of 22 nm, which is 1/100 of micro-CT image resolution. Four realizations with average porosity $\phi_c \in \{0.3, 0.4, 0.5, 0.6\}$ are constructed. Surface area S_v and T_{2eff} of the clay models are reported in Table 1 and are in agreement with published surface area measurements [6, 37] and T_2 values [38]. Here, the density for kaolinite we used is 2.6 g/cm^3. The surface relaxivity (ρ_2) is 1.2 μm/s. For NMR simulations the unresolved clay regions of the micro-CT images are replaced by the corresponding kaolinite model according to clay porosity. Fig. 6 depicts a subsection of the coarse scale model with unresolved clay regions, and the corresponding high-resolution model with fully resolved structure and different clay pocket porosity.

Table 1. Surface area (S_v) and effective relaxation time (T_{2eff}) of different porosity kaolinite model.

ϕ_c	S_v (m^2/g)	T_{2eff} (ms)
0.3	23.5	13.2
0.4	19.7	18.7
0.5	15.1	24.5
0.6	12.3	30.3

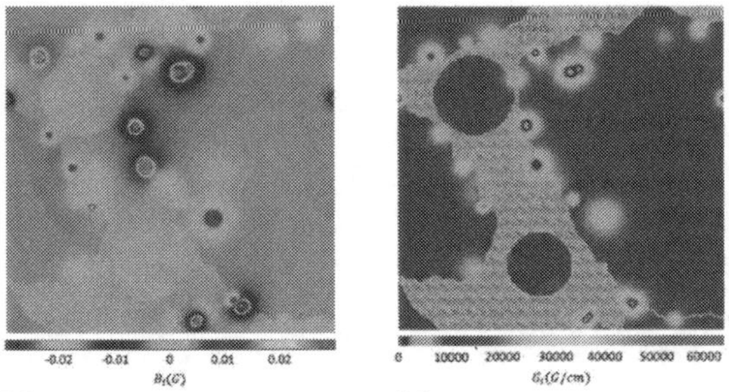

[a] [b]

Fig 7. Cross-section of Fig 6 corresponding 10000^2 subset of [a] internal magnetic field distribution $B_i(\mathbf{r})$ and [b] internal magnetic gradient distribution $G_{in}(\mathbf{r})$ at 22 nm/voxel.

2.3 Dual-scale NMR Relaxation Simulation

The dual-scale NMR relaxation response simulation has been described in detail previously [32]. In this work the simulation approach is extended to clay regions with different porosity. Furthermore, we implement the tracking of magnetization for dual-scale simulations, following the approach of [39]. In the previous section we detailed our approach of combining the resolved clay regions with the micro-CT based segmented tomogram. This resulted in a micro-structure where the interpolated coarse-scale segmentation (0.55μm resolution) is mapped to a voxel size of 22nm, the same as the clay region models. Thus, a random walk could now in principle be carried out completely in the resolved pore space. However, for the simulation of the NMR transverse relaxation response the internal magnetic field at every location is required. The calculation of high-resolution internal magnetic fields (B_i) is achieved by a separation of near- and far-field effects in a dipole-approximation of the internal field with dual-scale samples [32]. This allows calculating the coarse-scale internal magnetic fields via Fourier methods. The fine-scale internal magnetic fields of the periodic clay models are calculated separately. Then, the fine resolution internal magnetic field can be calculated as a combination of the coarse scale field interpolated to a voxel size of 22nm and a perturbation from the fine-scale field. The latter is calculated by deducting the coarse scale trend from the fine-scale fields (ensuring zero mean at tomogram scale). Like the fine-scale microstructure, B_i is never stored in memory at full resolution (at $80{,}000^3$ voxel it would be prohibitively large) but constructed locally when required for either a random trajectory during NMR simulations, to visualize parts of the high-resolution internal magnetic field, or to extract statistics. An example of high-resolution B_i distribution and corresponding spatial internal gradient distribution (G_i) is shown in Fig. 7.

Once the magnetic field distribution is generated, the random walk algorithm is carried out on a regular cubic lattice with fine-scale resolution (ϵ = 22 nm). The corresponding time step τ_i of a step i is given by $\tau_i = 6D_0/\epsilon^2$, where D_0 is the free diffusion coefficient of water, which we here vary with temperature. Bulk and surface relaxation are implemented as signal weighting factors for each step, with $S_i = S_b S_s$. Here $S_b = \exp(-\tau_i/T_{2,b})$ is the contribution by bulk relaxation, and $S_s = 1 - 6\rho_2\tau_i/(\epsilon A)$ the contribution by surface relaxation, where ρ_2 is the surface relaxivity of solid. For steps within the same fluid $S_s = 1$. A is a correction factor of order 1, which accounts for the details of the random walk implementation [40]. For imaged structures, this value is typically close to A = 3/2. Besides, the contribution of the internal field to dephasing is calculated explicitly written as $\Delta\phi$. The bulk relaxation, surface relaxation and phase variation caused by diffusive relaxation are then incorporated into the calculation of each walker's magnetization, which becomes

$$M_w(t_j) = M_w(0)\cos\,[\Delta\phi(t_j)]\prod S_i \qquad (1)$$

Averaging over random walks and recording at $t = nt_E$ (n= 1,2,...N) results in the magnetization decay M(t). The relaxation time distribution is obtained by fitting a multi-exponential decay to M(t).

In this study, the NMR transverse relaxation response is simulated for a given Larmor frequency of f = 2MHz and echo time t_E = 200 μs. For synthetic clayey sandstone, the surface relaxivity ρ_2 of the solid and the magnetic susceptibility χ of each phase are reported in Table 2.

Table 2. The surface relaxivity (ρ_2) and magnetic susceptibility (χ) of each phase assigned for NMR relaxation response simulation.

Phase	ρ_2 $(\mu m/s)$	χ (μSI)
Water	-	-8.9
Kaolinite	1.2	-11.0
Quartz	9.0	-12.0
Feldspar	9.0	-12.0
High-Density	60.0	5270

2.4 EDT Based Regional Magnetization Exchange

The magnetization exchange (or coupling) refers to a fraction of the spin population, which originates in one of the porosity types (e.g. macro-porosity) at the beginning of NMR signal acquisition, and found after a specific time interval (exchange time) in another porosity type (e.g. micro-porosity). The spin population in question should possess a measurable magnetization, thus defining a process sensitive to porosity types morphology and inter-connectivity.

Measuring the magnetization exchange between porosity partitions using random walk simulations is achieved by adding a phase label to the start and end positions of the walkers and employing the Euclidean distance transform (EDT). The former labeled phase can identify the diffusion of magnetization as macro-micro (start at macro-pore and end at micro-pore) and micro-macro (start at micro-pore and end at macro-pore). The latter EDT refers to a process that generates Euclidean distance fields for a 3D voxelized representation of samples. Valid distance values for each phase are calculated by assigning each voxel of the phase of interest the shortest distance to a voxel of another phase using distances between voxel centers. The implementation is a parallel version of algorithm 4 of [34]. An example of high-resolution EDT maps for macro- and micro-porous regions is depicted in Fig. 8. Given the distance information and phase labels, the quantification of the local magnetization evolution can now

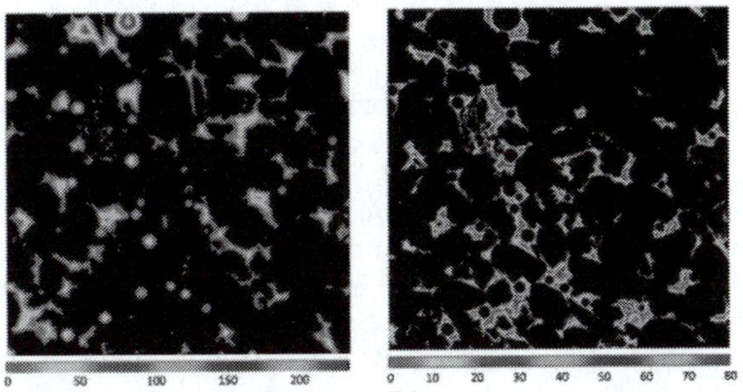

[a] [b]

Fig 8. A slice of Fig 5 [e] corresponding high-resolution EDT map for [a] macro-pores in voxel unit, 3200^2 in 0.55 μm/voxel; [b] micro-pores in voxel unit, 3200^2 in 0.55 μm/voxel.

be achieved by counting the walker strength in respect to a particular shell and phase. This information allows tracking the penetration depth (d) of magnetization diffusing across different pore populations. In addition, the magnetization exchange is counted at different observation time $t = nt_E$ (n= 1,2,...N). Therefore, The magnetization exchange can be characterized as a function of penetration depth and observation time $M_{ex}(d,t)$.

2.5 Permeability Estimation

In this section we briefly overview common NMR permeability correlations as well as the lattice Boltzmann technique to calculate permeability directly from tomographic images. Together these techniques are utilized to quantify the effect of clay content and spatial distributions with associated diffusion coupling on permeability predictions.

2.5.1 NMR Correlated Permeability Model

The NMR-estimated permeability by SDR [17] and the Coates models [18] are given by the following equations

$$k_{SDR} = a_s\phi^b T_{2lm}^c \qquad (2)$$

$$k_{Coates} = a_c\phi^b\left(\frac{FFI}{BIV}\right)^c \qquad (3)$$

where k is permeability in μm^2 (approx. mD), ϕ is porosity in %, with b and c exponents typically set to 4 and 2, respectively. For the SDR model (Eqn. 2), a_s is a parameter accounting for formation lithology. The SDR estimates mean hydraulic radius through the logarithmic mean relaxation time T_{2lm} and despite apparent limitations has proven applicable to fully water saturated water-wet formations. For the Coates Model (Eqn. 3), a_c is the Coates permeability coefficient, which is often used as denominator of porosity. The fixed $T_{2,cutoff}$ divides the T_2 distribution into two fractions corresponding to bound water (BVI, shorter relaxation times) and free fluid (FFI, longer relaxation times). Generally, cut-off times of 33 ms and 92 ms are appropriate for sandstones and carbonates, respectively [18]. However, our previous study shows that $T_{2,cutoff}$ values determined based on the irreducible water saturations S_{wirr} vary in different sandstones [36]. The primary assumption behind these correlations is that the fast diffusion limit condition applies to all components of the relaxation spectrum.

The SDR correlation is essentially based on a general relationship between permeability k, characteristic length scale of a pore system L, and porosity ϕ, $k \propto \phi^m L^2$, reported by many authors [41 - 43]. We note that the definitions of the characteristic length scale are different. L can be obtained from capillary pressure measurements in the study [41]. On the other hand, L is defined as dynamically connected pore size in the study from [42]. While the length scale offered from NMR T_2 distribution is related to pore volume-to-surface ratio V_p/S. $k \propto \phi^{1.8}(V_p/S)^2$ for the grain consolidation model is presented in [43]. More recently, the correlation between permeability and four different types of pore diameters is demonstrated as $k \propto \phi^4 \langle d \rangle^2$ based on micro-CT image [35]. Eqn. 2 can be easily expressed in the latter form noting in the fast diffusion limit $T_2 \sim L/(6\rho)$. Since the observable physical property is relaxation time and numerical simulation offers (and requires as input) surface relaxivity, we apply the squared proportionality factor 6ρ to the lithological parameter a_s, making it dimensionless: $a_s' = a_s/(6\rho)^2$. Typically, in laboratory core analysis the only unknown in the SDR equation is the lithological parameter, which has to be established for log calibration. Below we discuss the effect caused by the diffusional coupling in presence of different amount of clay on the modified lithological parameter a_s'.

2.5.2 Lattice Boltzmann Method

The Lattice Boltzmann Method (LBM) is a mesoscopic approach to fluid mechanics utilizing particle densities via a collision and streaming algorithm. The LBM has become an effective method for estimation of fluid flow in porous media because of its simplicity and flexibility to various flow geometries [44 - 46]. It is employed in this study to provide reference permeability values evaluate the enhancements in the NMR permeability estimations. The fluid is treated as continuous medium, and the solution of a discretized Boltzmann equation can match the Navier-Stokes equation. The D3Q19 (3-dimensional lattice with 19 possible velocity directions) lattice [47] is selected. The method starts with the initialization of distribution functions. The macroscopic fluid properties can be computed by

$$\rho(\boldsymbol{x}, t) = \sum_{i=1}^{19} f_i(\boldsymbol{x}, t) \quad (4)$$

$$\rho \boldsymbol{v}(\boldsymbol{x}, t) = \sum_{i=1}^{19} \boldsymbol{e}_i f_i(\boldsymbol{x}, t) \quad (5)$$

where ρ is the macroscopic fluid density, $\boldsymbol{v}$ is the velocity, $\boldsymbol{e}_i$ are the microscopic velocity vectors and f_i is the particle density distribution function. After applying boundary conditions, the algorithm for density evolution which includes the stream and collide operation is given by

$$f_i(\boldsymbol{x} + \boldsymbol{e}_i \delta t, t + \delta t) = f_i(\boldsymbol{x}, t) - \frac{f_i(\boldsymbol{x}, t) - f_i^{eq}(\boldsymbol{x}, t)}{\tau} \quad (6)$$

where f_i^{eq} is the equilibrium distribution function and τ is the relaxation time. In the simulation, the BGK (Bhatnagar-Gross-Crook) collision kernel is applied to simulate the single phase fluid behavior with bounce-back condition at the fluid-solid interface [48] The pressure gradient is simulated by a body force [46].

3 Magnetization Exchange Analysis

In this section we analyze the exchange of magnetization between different compartments of the clayey sandstone samples as function of clay distribution, clay region porosity, and temperature. Change in those parameters affects the main observables, namely the shape of the T_2 distribution, the position of individual modes of the distribution, as well as the size of the spatial regions active in magnetization exchange, which are discussed in the following. Consequences for permeability estimation are analyzed in the section thereafter.

3.1 Clay Distribution and Micro-porosity Effects

Consider first the relationships between diffusional coupling, clay distribution, clay content, and clay region porosity (ϕ_c) at ambient conditions. NMR simulations were conducted on both the pore-lining and pore-filling samples for a range of clay fractions, where kaolinite clay models with constant porosity ϕ_c (average over the 400^3 clay model) are mapped to the respective clay regions. The corresponding T_2 distributions are given in Fig. 9. Stronger diffusional coupling effects can be observed for the coated and pore-lining clay distributions with higher clay porosity (ϕ_c). The part of the T_2 distribution associated with coated type of clay morphology exhibits faster relaxation time than other three clay morphology despite equal volumetric fraction in sandstone. Especially, with increasing clay amount, the T_2 distribution of the coated clay becomes unimodal and no longer represents pore size distribution. Diffusional coupling enhanced the relaxation rate and the slow relaxation time peak shifts to short time.

The correlation between clay amount and the transverse relaxation rate ($1/T_{2lm}$) is demonstrated in Fig. 10. $1/T_{2lm}$ of four clay morphological types demonstrate a linear relationship with clay content in different slopes. The micro-porosity has more impact on the relaxation rate of grain-coating and pore-lining clay morphology when the amount surpasses 10 vol% (Fig 10 [b, c]). The higher clay porosity and stronger coupling effects lead to the fast relaxation peak shifts for the longer relaxation times, resulting in a decreased slope and vice versa. The distortion of relative amplitudes of the fast relaxation peak and slow relaxation peak due to diffusional coupling may break down the correlation between the peak area ratio and the porosity fraction of micro-pores and macro-pores [24]. The micro-porosity may be overestimated with increasing clay amount. We note that both grain-coating and pore-lining clay morphological types with higher clay porosity exhibit larger errors. This is because of two effects -- the higher clay porosity (ϕ_c) allows more protons to diffuse between micro- and macro-pores and the pore-lining clay distribution pattern offers more clay-region to macro-pore interface area and thus lead to a higher diffusi

 https://doi.org/10.1051/e3sconf/202336701003

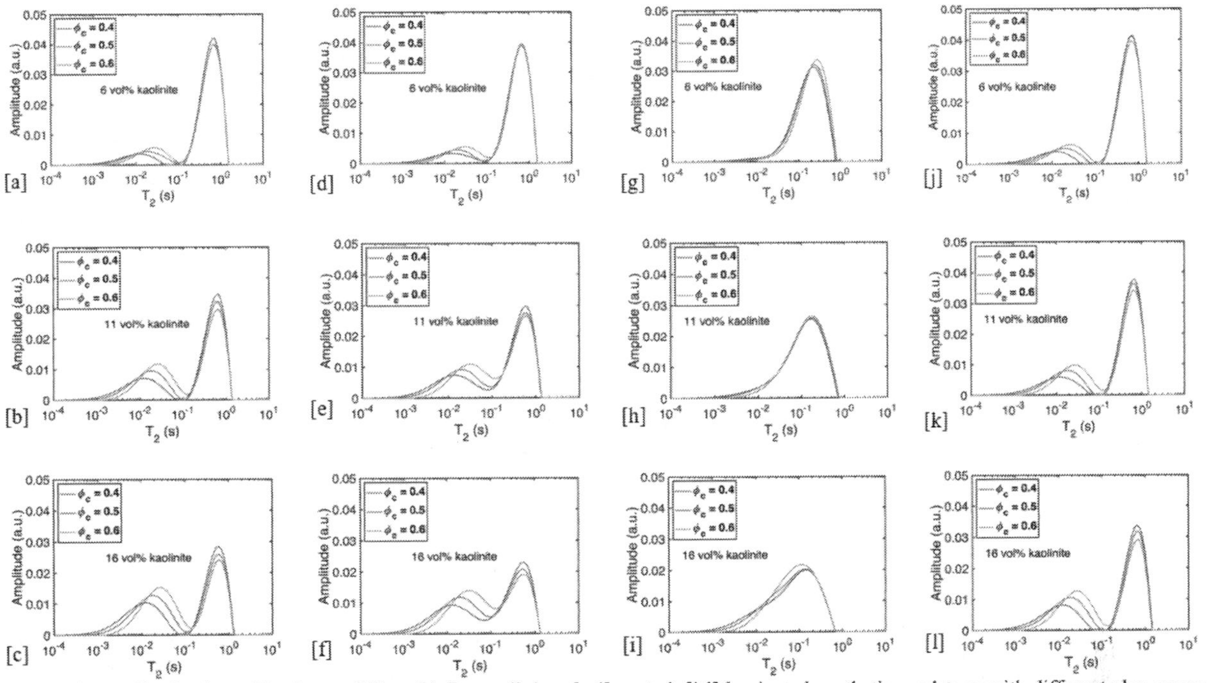

Fig. 9 The T_2 distribution of [a-c] pore-filling; [d-f] pore-lining; [g-i] coated; [j-l] laminated synthetic sandstone with different clay amount. *Top row:* 6 vol%, *middle:* 11 vol%, *bottom:* 16 vol%. $\phi_c \in \{0.4, 0.5, 0.6\}$ denotes the micro-porosity of kaolinite

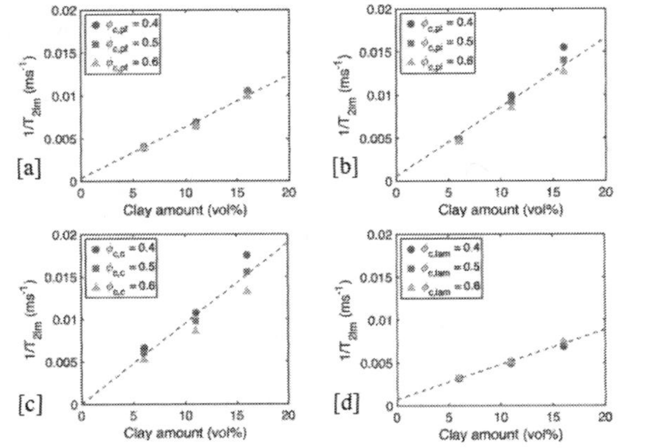

Fig. 10 Linear relationship between clay volume fraction and T_{2lm} for different clay distributions. [a] pore-filling, [b] pore-lining, [c] coated, [d] laminated. The purple dashed lines are the linear trend between T_{2lm} and clay amount.

3.2 Diffusional Exchange Between Different Compartments

We demonstrate the correlation between magnetization exchange ($M_{ex}(d,t)$) and time as well as Euclidean distance. Since the samples containing different clay amounts indicate the same trends, we chose a system with 11 vol% clay amounts as a representative example. $M_{ex}(d,t)$ as function of observation time is shown in Fig. 11, where $\sum d$ is the distance integration over each EDT shell. Similar magnetization exchange profiles are presented in both macro-micro and micro-macro directions. At the beginning, a growing number of walkers diffuse between micro- and macro-porous regions with time increasing. The diffusional exchange of magnetization reaches a maximum value at around 10-30 ms. After that, magnetization exchange decreases and there is almost no leaked magnetization between two pore populations when time exceeds 1s as defined by compartment association at start and end position of the exchanging spin packet. The magnetization exchange is significantly affected by micro-porosity for coated and pore-lining clay morphology when diffusion occurred in both two directions. These two clay distribution patterns with higher micro-porosity allow protons diffusing between micro- and macro-pores more frequently. For grain-coated morphological pattern, the magnetization exchange when $\phi_c = 0.6$ is two times stronger than $\phi_c = 0.4$. While, the $M_{ex}(\sum d, t)$ for pore-filling and laminated samples are much closer. Overall, the coated clay type exhibits five times higher magnetization exchange compared to other samples. The magnetization exchange of pore-lining clay type demonstrates two-fold compared to pore-filling and laminated one when clay porosity is 0.5 and 0.6. However, if porosity of the compared clay types reduced to 0.4, the difference in exchanged magnetization diminishes. This can be explained by stronger surface relaxation in the clay region due to an increase in surface-to-volume ratio.

Consider now the correlations between magnetization exchange and Euclidean distance presented in Fig 12. Here, we take examples of pore-filling and pore-lining samples. The overall trend demonstrates that magnetization exchange presents approximately exponential decay with Euclidean distance for both directions. The leaked magnetization from two porosity populations towards the other is reduced as the walkers moving far from the interface boundary.

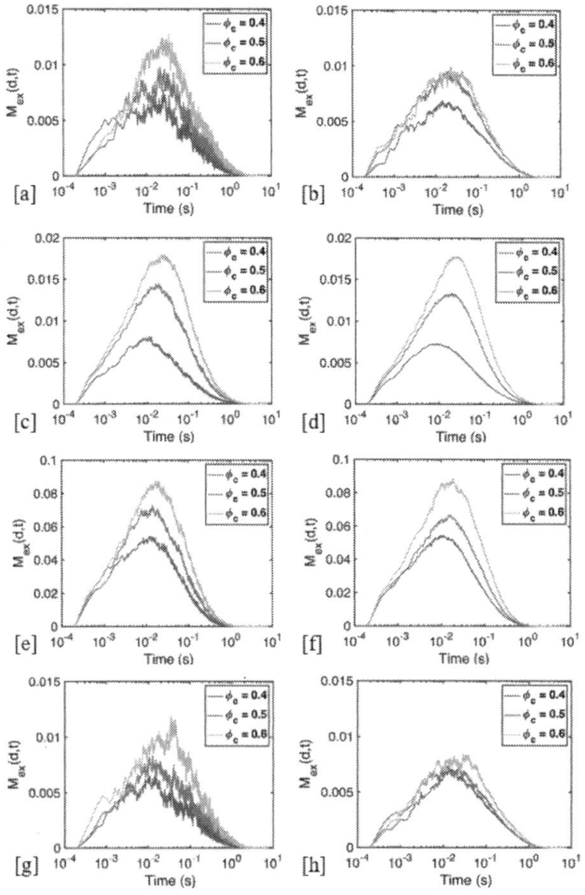

Fig. 11 The magnetization exchange $M_{ex}(d,t)$ as function of observation time for [a, b] pore-filling, [c, d] pore-lining, [e,f] coated, [g, h] laminated samples with clay amount 11 vol% in diffusion direction: left column: macro-micro; right column: micro-macro. Here, the distance is the integration of each EDT shell $d = \sum d$. ϕ_c is the micro-porosity for clay phase.

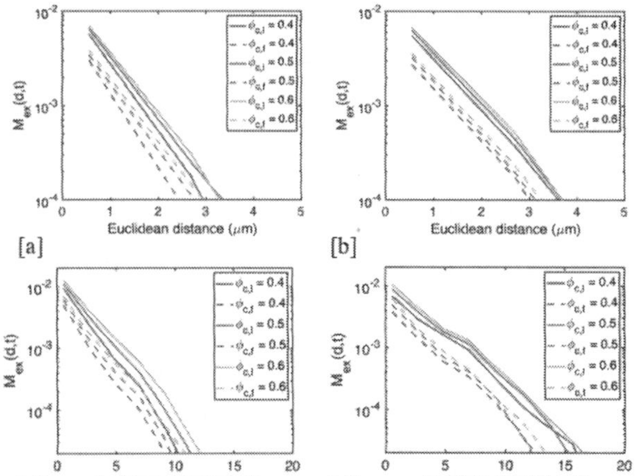

Fig. 12 The magnetization exchange $M_{ex}(d,t)$ of samples with 11 vol% clay from macro- to micro-pore (left column) and from micro- to macro-pore (right column) at various times [a, b] t = 5 t_E; [c, d] t = 50 t_E. The solid line and dash line represent for pore-lining and pore-filling clay distribution respectively.

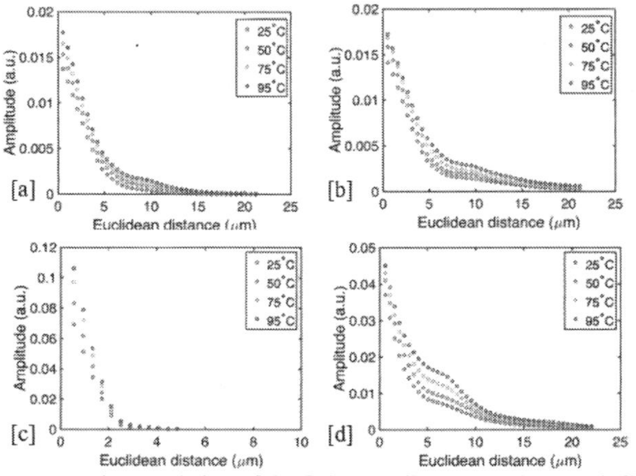

Fig. 13 The correlation of the fitting amplitude at each EDT shell and Euclidean distance in different directions macro-micro (left column) and micro-macro (right column) for [a, b] pore-lining samples, [c, d] coated samples at temperature 25 °C, 50 °C, 75 °C and 95 °C.

The corresponding maximum distance at short time (t = 5 t_E) is close to 3-4 μm. The distance increases to 10-15 μm at t = 50 t_E. The higher clay porosity allows more walkers travel to longer distance between micro- and macro-porous regions. The pore-lining sample with lower clay porosity presents less diffusion of magnetization than pore-filling with higher clay porosity. The reason for that may be, that strong surface relaxation is dominant in tight clay pockets. On the contrary, the pore-lining sample always presents higher $M_{ex}(d,t)$ values when walkers travel from micro-pore to macro-pore (Fig. 12b, d).

3.3 Temperature Effects

The diffusional coupling effects can be enhanced with the temperature increasing due to alteration of diffusion coefficient and surface relaxivity [6, 7]. In this section, we noted that we only focus on the contribution of diffusion to magnetization exchange. We simulated the NMR relaxation response at 25 °C, 50 °C, 75 °C and 95 °C with diffusion coefficient $2.15 \times 10^{-5} cm^2/s$, $3.97 \times 10^{-5} cm^2/s$, $6.10 \times 10^{-5} cm^2/s$ and $8.07 \times 10^{-5} cm^2/s$, respectively. The magnetization exchange between micro- and macro-porous regions with integrated time are recorded in respect to particular EDT shell $M_{ex}(d, \sum t)$, that provides the local magnetization evolution in approximate Gaussian distribution. The correlation of the fitting amplitude for $M_{ex}(d, \sum t)$ and Euclidean distance at different temperatures is presented in Fig. 13. Here we demonstrate pore-lining and grain-coated clay morphological types due to their stronger diffusional coupling effects.

As expected, the diffusion of magnetization is enhanced with increased temperature. For pore-lining sample, the diffusion of magnetization becomes weak when distance exceeds 20 μm, the maximum Euclidean distance employed here is 22.05 μm (around 40 voxels) from macro-pore to micro-pore. The amplitude of the magnetization exchange at 95 °C is around 1.3 - 1.4 times higher than that at 25 °C within the short distance. While, in the opposite direction, the walkers can penetrate slightly deeper into the macro-pore space. For the grain-coated type, since the distance values around the solid are small, the corresponding penetration distance from macro-pore to micro-pore is around 5 μm. On the contrary, the walkers can still invade macro-pores with distance around 22 μm.

4 Permeability Estimation

In this section, we analyze the coupling effects on permeability estimation for pore-filling and pore-lining samples which are the most common clay patterns in sandstones.

Table 3. The permeability simulated by LBM for pore-lining (K_l) and pore-filling (K_f) samples with different clay amount.

Clay (vol%)	K_l (mD)	K_f (mD)
6	639.35	746.41
11	226.59	355.07
16	63.91	109.79

We first calculated permeability by LBM for synthetic clayey sandstones, assuming the clay regions provide no contribution to transport (Table. 3). The permeability values decrease with increasing clay amount. The pore-filling samples exhibit higher permeability than corresponding pore-lining samples of same clay amount. The following analysis of coupling effects on permeability estimate using SDR and Coates models is performed in respect to corresponding prefactors.

4.1 SDR Model

The implications of temperature on the shape of the relaxation time distributions are shown in Fig. 14 [a-f]. We calculated the dimensionless lithological parameter a_s' following Eqn. 2 for both coupling-on and coupling-off scenarios by employing the absolute permeability in Table 3. We report the prefactor $a_{s,l}'$ for pore-lining sample and $a_{s,f}'$ for pore-filling sample with different temperatures in Table 4. The temperature effects are weak when the clay amount is small leading to limited contrast between the coupling-on and coupling-off scenarios. However, as clay amounts increase the temperature effects are enhanced, resulting in a prefactor change of around 1.6 times for the pore-lining sample with clay amount of 16 vol%. While, for pore-filling samples, $a_{s,f}'$ presents weak dependency of temperature due to less shifts and distortion of the T_2 distribution due to diffusion between micro- and macro-porous regions.

4.2 Coates Model

Consider now the permeability estimation via Coates model. According to the coupling-off T_2 distributions (Fig. 15 blue-dashed line), the separated fast relaxation peak corresponds to the bounded irreducible water (S_{wirr}). The normalized BVI is obtained by plotting the cumulative T_2 curves of S_{wirr} in Fig. 15. The $T_{2,cutoff}$ is determined by horizontal projection of S_{wirr} curve for $S_w = 100$ %. The cross-point is then project down to the x axis with relevant T_2 value, which is $T_{2,cutoff}$ [18]. The correlation between $T_{2,cutoff}$ and temperature for pore-lining and pore-filling samples as function of clay amount is given

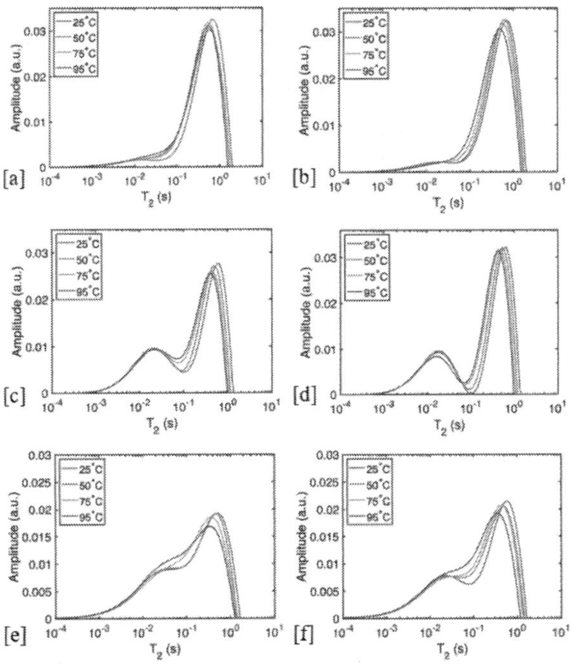

Fig. 14 The comparison of T_2 distributions of pore-lining (left column) and pore-filling (right column) for samples with clay amount [a, b] 6 vol%, [c, d] 11 vol%, [e, f] 16 vol% at the temperature 25 °C, 50 °C, 75 °C and 95 °C.

Table 4. The SDR model prefactor for pore-lining sample ($a_{s,l}'$) and pore-filling sample ($a_{s,f}'$) at temperature 25°C, 50°C, 75°C and 95°C with different clay amount.

Clay	T	$a_{s,l}'$		$a_{s,f}'$	
(vol)	(°C)	Coupling-on	Coupling-off	Coupling-on	Coupling-off
6	25	1.06	0.94	1.22	1.07
	50	1.34	0.94	1.54	1.09
	75	1.47	0.98	1.67	1.10
	95	1.66	0.98	1.91	1.10
11	25	3.66	1.89	4.36	2.96
	50	5.03	1.94	5.27	3.13
	75	5.39	2.03	5.50	3.16
	95	5.93	2.04	6.39	3.17
16	25	3.97	1.37	6.06	2.28
	50	4.44	1.40	6.47	2.28
	75	4.62	1.41	6.81	2.35
	95	6.51	1.41	7.10	2.37

in Fig 16. The overall trend shows that $T_{2,cutoff}$ values decrease as temperature increase. This may result from the decrease in amplitude of the slow relaxation peak, and the tendency of the bimodal T_2 distribution towards becoming more unimodal with increasing temperature, especially for higher clay amounts (Fig. 14). Since pore-lining clays exhibit stronger coupling effects, the related $T_{2,cutoff}$ values are lower than for pore-filling clays. As clay amounts increase, the contrast

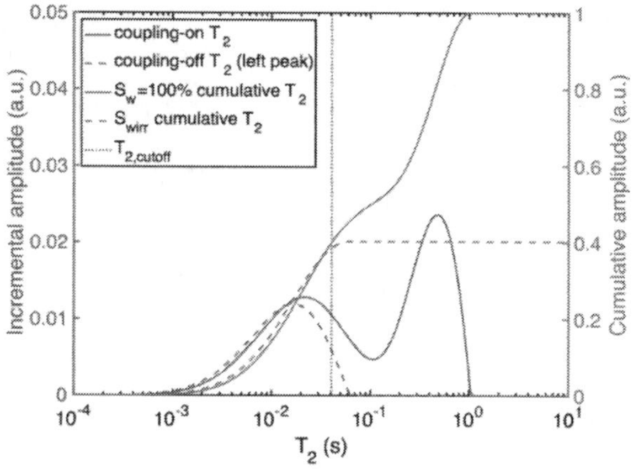

Fig. 15 The cumulative T_2 curves on fully saturated (S_w = 100\%) sample with 11 vol\% clay amount and on irreducible saturation S_{wirr}) are utilized to establish a $T_{2,cutoff}$ via BVI.

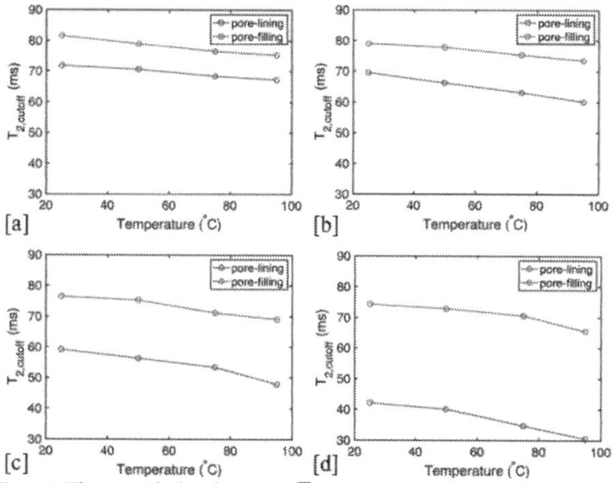

Fig. 16 The correlation between $T_{2,cutoff}$ and temperature for pore-lining and pore-filling samples in different clay amount [a] 2 vol% (original clay amount within Bentheimer sandstone), [b] 6 vol%, [c] 11 vol %, [d] 16 vol%.

Table 5. The Coates model prefactor for pore-lining sample ($a_{c,l}$) and pore-filling sample ($a_{c,f}$) at temperature 25 °C, 50 °C, 75 °C and 95 °C with different clay amount.

Clay (vol%)	T (°C)	$a_{c,l}$	$a_{c,f}$
6	25	10.03	10.09
	50	9.58	10.03
	75	8.85	9.71
	95	8.64	9.59
11	25	45.77	66.55
	50	43.42	63.01
	75	40.59	58.25
	95	36.13	50.73
16	25	92.37	114.82
	50	86.53	107.67
	75	77.39	99.19
	95	72.38	95.14

between the two clay distribution scenarios is growing. When clay amount reaches 16 vol%, the $T_{2,cutoff}$ for pore-lining sample decrease from 42 ms to 30 ms at temperature range from 25 °C to 95 °C. While the reduction for pore-filling samples is from 74 ms to 65 ms.

By applying various $T_{2,cutoff}$, the FFI/BVI ratio can be obtained via T_2 distribution of each sample. With known permeability, porosity and FFI/BVI ratio, we derived prefactor a_c in Coates model according to Eqn. 3. The a_c values are reported in Table 5. The results indicate that the prefactors are not affected much by temperature when little clay is present in the rocks. However, when the clay amount is higher than 10 vol%, both pore-lining and pore-filling samples show dependency on temperature. The smaller $T_{2,cutoff}$ values lead to higher FFI/BVI ratio, resulting in decreased a_c as temperature increasing. Therefore, the permeability may be overestimated if only fixed $T_{2,cutoff}$ is utilized without prefactor compensation.

5 Conclusion

This study evaluates the magnitude of diffusional coupling effects on petrophysical evaluation of clayey sandstone based on NMR transverse relaxation depending on clay volume fraction, distribution pattern, local porosity, and temperatures. The magnetization exchange between micro- and macro-porous regions with aid of EDT map is employed to characterize the diffusion as function of time and penetration depth. This in turn enabled an analysis of potential impacts on permeability predictions. We draw the following specific conclusions:

1. The rocks with coated and pore-lining clay distribution present stronger diffusional coupling effects than pore-filling and laminated one. The distortion of T_2 distribution can result in an overestimation of micro-porosity.
2. The magnetization exchange between micro- and macro-porous regions due to diffusion is demonstrated as exponential function of penetration depth (Euclidean distance) and Gaussian function of observation time.
3. The log-mean of transverse relaxation time demonstrates linear dependency on clay amount. The slope can be altered by different micro-porosity, especially for rocks with pore-lining clay distribution.
4. The diffusional coupling effects are magnified with increasing temperature. The prefactors of SDR model and Coates model exhibit temperature dependency in higher clay amount rocks. The NMR-related permeability can be overestimated if fixed $T_{2,cutoff}$ is utilized without consideration of prefactor compensation.

References

[1] K. Brownstein and C. Tarr, "Importance of classical diffusion in NMR studies of water in biological cells," *Phys. Rev. A* , vol. 19, p. 446–2453, 1979.

[2] H. Dai, I. Shikhov, R. Li, J. Arns and C. Arns, "Mechanisms of confining pressure dependence of resistivity index for tight sandstones by digital core analysis," *SPE J.,* vol. 26, p. 883–896, 2021.

[3] S. Apourvari and C. Arns, "Image-based relative permeability upscaling from pore scale," *Adv. Warer Res.,* vol. 95, pp. 161-175, 2016.

[4] J. Qajar, N. Francois and C. Arns, "Microtomographic characterization of dissolution-induced local porosity changes including fines migration in carbonate rock," *SPE J.,* vol. 18, p. 545–562, 2013.

[5] E. Grunewald and R. Knight, "A laboratory study of NMR relaxation times and pore coupling in heterogeneous media," *Geophysics* , vol. 74, p. 215–221, 2009.

[6] V. Anand and G. Hirasaki, "Paramagnetic relaxation in sandstones: distinguishing *T*1 and *T*2 dependence on surface relaxation, internal gradients and dependence on echo spacing," *J. Magn. Reson,* vol. 190, p. 68 – 85, 2008.

[7] E. Toumelin, C. Torres-Verdin, S. Chen and D. Fischer, "Analysis of NMR diffusion coupling effects in two-phase carbonate rocks: comparison of measurements with Monte Carlo simulations," in *SPWLA 43rd Annual Logging Symposium*, Oiso, Japan, 2002.

[8] E. Toumelin, C. Torres-Verdín, S. Chen and D. Fischer, "Reconciling NMR measurements and numerical simulations: assessment of temperature and diffusive coupling effects on two-phase carbonate samples," *Petrophysics,* vol. 44, 2003.

[9] N. Gomaa, A. Al-Alyak, D. Ouzzane, O. Saif, M. Okuyiga, A. Opco, D. Allen, D. Rose, R. Ramamoorthy and E. Bize, "Case study of permeability, Vug quantification, and rock typing in a complex carbonate," in *SPE Annual Technical Conference and Exhibition*, San Antonio,Texas, 2006.

[10] W. Trevizan, P. Netto, B. Coutinho, V. Machado, E. Rios, S. Chen, W. Shao and P. Romero, "Method for predicting permeability of complex carbonate reservoirs using NMR logging measurements," *Petrophysics,* vol. 55, p. 240–252, 2014.

[11] M. Fleury and J. Soualem, "Quantitative analysis of diffusional pore coupling from T_2-store-T_2 NMR experiments," *Journal of Colloid and Interface Science,* vol. 336, p. 250–259, 2009.

[12] C. Morriss, J. MacInnis, R. Freedman, J. Smaardyk, C. Straley, W. Kenyon, H. Vinegar and P. Tutunjian, "Field test of an experimental pulsed nuclear magnetism tool," in *SPWLA 34th Annual Logging Symposium*, Calgary, Alberta, 1993.

[13] C. Straley, D. Rossini, P. Tutunjian, H. Vinegar and C. Morriss, "Core analysis by low-field NMR," in *The Log Analyst 38*, 1997.

[14] G. Coates, J. Galford, D. Mardon and D. Marschall, "A new characterization of bulk-volume irreducible using magnetic resonance," in *The Log Analyst 39*, 1998.

[15] E. Toumelin, C. Torres-Verdín and S. Chen, "Modeling of multiple echo-time NMR measurements for complex pore geometries and multiphase saturations," *SPE Reser. Eval. Eng.,* vol. 6, p. 234–243, 2003.

[16] S. Godefroy, M. Fleury, F. Deflandre and J. Korb, "Temperature effect on NMR surface relaxation in rocks for well logging applications," *J. Phys. Chem. B,* vol. 106, p. 11183–11190, 2002.

[17] E. Kenyon, I. Day, C. Straley and F. Willemsen, " A three-part study of NMR longitudinal relaxation properties of water-saturated sandstones," *SPE Formation Evaluation,* vol. 3, p. 622–636, 1988.

[18] G. Coates, L. Xiao and M. Prammer, NMR logging: principles and applications, Houston: Haliburton Energy Services, 1999.

[19] G. Carneiro, A. Souza, A. Boyd, L. Schwartz, Y. Song, R. Azeredo, W. Trevizan, B. Santos, E. Rios and V. Machado, "Evaluating Pore Space Connectivity by NMR Diffusive Coupling," in *SPWLA 55th Annual Logging Symposium*, Abu Dhabi, United Arab Emirates, 2014.

[20] M. Jácomo, R. Trindade, E. de Oliveira, C. Leite, E. Montrazi, M. Andreeta and T. Bonagamba, "Nuclear magnetic resonance and pore coupling in clay-coated sandstones with anomalous porosity preservation, ÿgua Grande Formation, Rec & ocirc;ncavo Basin, Brazil," *Petrophysics* , vol. 59, p. 136–152, 2018.

[21] S. Chen, D. Jacobi, H. Kwak, M. Altunbay and J. Kloos, "Pore-connectivity based permeability model for complex carbonate formations," in *SPWLA 49th Annual Logging Symposium*, ustin, Texas, 2008.

[22] L. Chi and Z. Heidari, "Directional-permeability assessment in formations with complex pore geometry with a new nuclear-magnetic-resonance-based permeability model," *SPE J.,* vol. 21, p. 1436–1449, 2016.

[23] M. Cohen and K. Mendelson, "Nuclear magnetic relaxation and the internal geometry of sedimentary rocks," *J. Appl. Phys.,* vol. 53, p. 1127–1135, 1982.

[24] T. Ramakrishnan, E. Fordham, W. Kenyon, L. Schwartz and D. Wilkinson, "Forward models for nuclear magnetic resonance in carbonate rocks," in *The Log Analyst 40*, 1999.

[25] Z. Zhang, D. Johnson and L. Schwartz, "Simulating the time-dependent diffusion coefficient in mixed-pore-size materials," *Phys. Rev. E,* vol. 84, p. 031129, 2011.

[26] L. Schwartz, D. Johnson, J. Mitchell, T. Chandrasekera and E. Fordham, "Modeling two-dimensional magnetic resonance measurements in

coupled pore systems," *Phys. Rev. E,* vol. 88, p. 032813, 2013.

[27] J. Mitchell, A. Souza, E. Fordham and A. Boyd, "A finite element approach to forward modeling of nuclear magnetic resonance measurements in coupled pore systems," *J. Chem. Phys.,* vol. 150, p. 154708:1–12, 2019.

[28] C. Arns, A. Sheppard, R. Sok and M. Knackstedt, "NMR petrophysical predictions on digitized core images," in *46th Annual Logging Symposium, Society of Petrophysicists & Well Log Analyst*, New Orleans, Texas, 2005.

[29] L. Chi and Z. Heidari, "Diffusional coupling between microfractures and pore structure and its impact on nuclear magnetic resonance measurements in multiple-porosity system," *Geophysic,* vol. 80, pp. 31-42, 2014.

[30] C. Arns, Y. Melean, A. Sheppard and M. Knackstedt, "A comparison of pore structure analysis by NMR and Xray-CT techniques," in *9th Annual Logging Symposium, Society of Petrophysicists & Well Log Analyst*, Edinburgh, 2008.

[31] M. Tankiewicz, "Experimental Investigation of Strength Anisotropy of Varved Clay," *Procedia Earth Planet. Sci.,* vol. 15, p. 732–737, 2015.

[32] Y. Cui, I. Shikhov and C. Arns, "NMR relaxation modeling in porous media with dual-scale resolved internal magnetic fields," *Transport in Porous Media,* 2022.

[33] A. Sheppard, R. Sok and H. Averdunk, "Techniques for image enhancement and segmentation of tomographic images of porous materials," *Physica A* , vol. 339, p. 145–151, 2004.

[34] T. Saito and J. Toriwaki, "New algorithms for Euclidean distance transformation of an n-dimensional digitized picture with applications," *Pattern Recognition* , vol. 27, p. 1551–1565, 1994.

[35] C. Arns, A. Sakellariou, T. Senden, A. Sheppard, R. Sok, W. Pinczewski and M. Knackstedt, "Digital core laboratory: Reservoir core analysis from 3D images," *Petrophysics,* vol. 46, no. 260–277, 2005.

[36] Y. Cui, I. Shikhov, R. Li, S. Liu and C. Arns, "A numerical study of field strength and clay morphology impact on NMR transverse relaxation in sandstones," *J. Pet. Sci. Eng,* vol. 202, p. 10852, 2021.

[37] J. Mbey, F. Thomas, A. Razafitianamaharavo, C. Caillet and F. Villiéras., "A comparative study of some kaolinites surface properties," *Applied Clay Science,* vol. 172, p. 135–145, 2019.

[38] A. Matteson, J. Tomanic, M. Herron, D. Allen and W. Kenyon, "NMR relaxation of clay-brine mixtures," *PE Reser. Eval. Eng.* , vol. 3, p. 408–413, 2000.

[39] Y. Zheng, I. Shikhov, L. Gbayan, J. Arns and C. Arns, "About the connectivity of dual-scale media based on micro-structure based regional analysis of NMR flow propagators," *J. Contam. Hydrol.,* vol. 212, p. 143–151, 2018.

[40] D. Bergman, K. Dunn, L. Schwartz and P. Mitra, "Self-diffusion in a periodic Porous medium: a comparison of different approaches," *Phys. Rev. E,* vol. 51, p. 3393–3400, 1995.

[41] A. Katz and A. Thompson, "Quantitative prediction of permeability in porous rock," *Phys. Rev. B* , vol. 34, p. 8179–8181, 1986.

[42] L. Johnson, J. Koplik and L. Schwartz, "New pore size parameter characterizing transport in porous media," *Phys. Rev. Lett.* , vol. 57 , p. 2564–2567, 1986.

[43] J. Banavar and L. Schwartz, "Magnetic resonance as a probe of permeability in porous media," *Phys. Rev. Lett.,* vol. 58, p. 1411–1414, 1987.

[44] U. Frisch, B. Hasslacher and Y. Pomeau, "Lattice-gas automata for the Navier-Stokes equation," *Phys. Rev. Lett.* , vol. 56, p. 1505–1508, 1986.

[45] D. Rothman, "Cellular-automaton fluids: A model for flow in porous media," *Geophysics* , vol. 53, p. 509–518, 1988.

[46] B. Ferreol and D. Rothman, "Lattice-Boltzmann simulations of flow through Fontainebleau sandstone," *Trans. Porous. Med.,* vol. 20, p. 3–20, 1995.

[47] N. Martys and J. Hagedorn, "Multiscale modeling of fluid transport in heterogeneous materials using discrete boltzmann methods," *Materials and structures,* vol. 35, p. 650–658, 2002.

[48] N. Martys and H. Chen, "Simulation of multicomponent fluids in complex three-dimensional geometries by the lattice Boltzmann method," *Phys. Rev. E,* vol. 53, p. 743–750, 1996.

[49] M. Elsayed, G. Glatz, A. El-Husseiny, A. Alqubalee, A. Adebayo, K. Al-Garadi and M. Mahmoud, "The effect of clay content on the spin–spin NMR relaxation time measured in porous media," *ACS Omega,* vol. 5, p. 6545–6555, 2020.

[50] D. Johnson and L. Schwartz, "Analytic theory of two-dimensional NMR in systems with coupled macro- and micropores," *Phys. Rev. E* , vol. 90, p. 032407, 2014.

Prediction of centrifuge capillary pressure using machine learning techniques

Brandon Jeremy Bursey[1], *Erfan* Mohagheghian[1], *Edison* Sripal[1] and *Lesley Anne* James[1,*],

[1]Department of Process Engineering, Memorial University of Newfoundland, St. John's, NL, Canada

Abstract. In current literature in the petroleum industry, machine learning has been used to predict capillary pressure only on the centrifugal data points and not the complete capillary pressure curves generated from existing correlations after analysis. This paper will present novel information that will benefit the petroleum industry as it shows machine learning techniques can be used to obtain the complete capillary pressure curve which is the end goal in undertaking an SCAL centrifuge experiment.

This research involves testing core samples using a centrifuge set up to produce capillary pressure data points. Then, using a commercial SCAL interpretation software, the collected data is utilized to generate complete capillary pressure curves based on developed literature correlations. RCAL data for the core samples is also obtained to be used with the machine learning techniques. The machine learning models are then applied to the collected data to predict the capillary pressure curves. Optimization of the different machine learning techniques is done to improve the predictions.

The results show the machine learning techniques perform very well on the validation set after being trained on the training set. The machine learning models also provide reasonable prediction of the complete capillary pressure curves on the testing data set. Changing of the machine learning technique parameters also shows the effect on the overall precision and the improvements that can be made. Further research can be done to see the effectiveness of using machine learning techniques to predict other SCAL properties such as relative permeability. This can then greatly reduce the time needed to obtain these extremely important properties for reservoir characterization.

1 Introduction

Researchers continue to find new ways to reduce the time and improve the accuracy of reservoir characterization [1]. Specialized core analysis (SCAL) examines important reservoir characteristics such as capillary pressure and relative permeability to estimate incremental oil recovery during secondary water or gas flooding and tertiary enhanced oil recovery. Arguably, SCAL is more time consuming and has more uncertainty compared to routine core analysis and fluid characterisation. Capillary pressure (P_c) is the pressure between two immiscible phases in a capillary tube or, in this case, within the rock's pores [2]. Capillary pressure is measured in a lab through spontaneous imbibition, gravity drainage, and forced imbibition tests. P_c is used directly in reservoir simulators to account for unrecoverable oil and it is further used to estimate relative permeability or the relative conductivity of the rock to more than one fluid phase simultaneously. The three main test methods used to determine capillary pressure curves are centrifuge, porous plate, and mercury injection [2]. The advantages and disadvantages of these methods are summarized in Table 1. It is worth pointing out that mercury injection capillary pressure (MICP) does not respond to wettability components in the pores since Hg is an ideal non-wetting fluid. Hence MICP reflects only the pore geometry and therefore differs slightly from other tests that use oil/brine/gas.

The oil and gas industry has always been highly competitive and somewhat unpredictable. In recent years, companies have begun using machine learning techniques to combat several challenges and issues in data processing and handling in various oil and gas activities such as reducing risk factors and cost of maintenance [3]. In current literature, machine learning has been used to predict the experimental capillary pressure data points obtained from centrifugal and mercury injection tests with varying results [4-6]. For the case that did use centrifugal data, correlations were not applied to the experimental results to obtain a complete capillary pressure curve. This research aims to obtain complete capillary pressure curves generated from existing correlations after analysis of centrifugal data and then test the novelty in using machine learning techniques to predict the experimental capillary pressure curve and water saturation. Three machine learning techniques, including the artificial neural network, support vector regression and random forest regression were used to predict the capillary pressure from centrifuge test samples with routine core analysis data (porosity, permeability, bulk density, and irreducible water saturation) being used as inputs. ML-based approaches generate a multi-point capillary pressure curve as a function of fluid saturation using the provided input parameters. The ML prediction is not just a single value just as the experimentally measured and calculated capillary pressure values are recorded at discrete fluid saturation points. Although experimental techniques are broadly used to generate the capillary pressure curve of core samples, they are expensive and require core samples to be in reservoir conditions that is very difficult to achieve. As a result, researchers ponder whether data-driven

* Corresponding author: ljames@mun.ca

methods can estimate SCAL data. Can one develop accurate machine learning methods to predict SCAL data such as capillary pressure of core samples from routine core analysis data without a need for time consuming and finicky laboratory experiments?

Table 1. Pc Methods – Advantages and Disadvantages [2]

Method	Advantages	Disadvantages
Centrifuge	- Applicable to most rock types - Simple and quick - Non-destructive test - Uses synthetic reservoir fluids	- Most expensive method - Raw production data needs to be corrected by the test lab
Porous Plate	- Best method to achieve a uniform saturation profile - Uses synthetic reservoir fluids - Non-destructive test	- Time consuming to achieve complete Pc curve - Time allowed to achieve equilibrium can completely govern resultant curve
Mercury Injection	- Low-cost technique - Rapid tests	- Not representative of reservoir fluids - Test is destructive (mercury remains in sample)

1.1 Machine Learning Technique Types

Machine learning is the broad study of programming algorithms that can learn through experience from the use of a variety of data [7]. Machine learning techniques are categorized into either supervised or unsupervised learning with supervised learning being further broken into classification and regression as shown in Fig. 1.

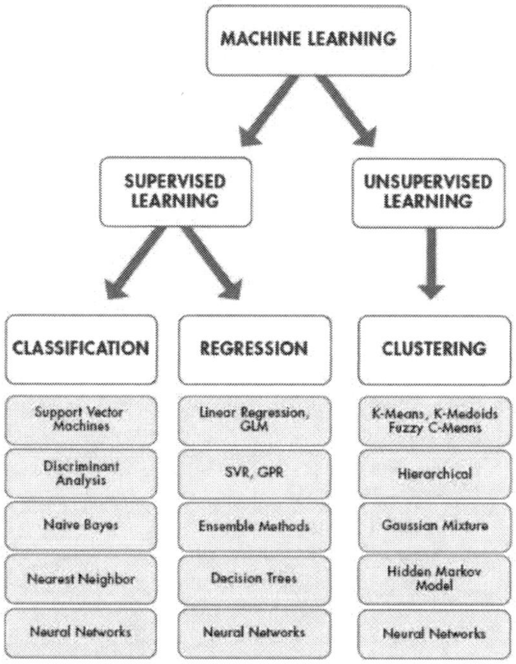

Fig. 1. Machine Learning Algorithm Breakdown [8]

For the purposes of this research, supervised learning regression techniques that were used are briefly explained as follows.

1.1.1 Support Vector Regression

Support vector regression (SVR) is a popular technique where support vector machines are adapted for regression with a quantitative response [9]. The technique provides flexibility in defining the acceptable error for the model and determines an appropriate line or hyperplane fit to the data [10]. For SVR, the objective function is to minimize the L2-norm of the coefficient vector. One strength of SVR is a variety of kernel functions being available to select or modify depending on the requirements for your own predictions [11]. The kernel function transforms data that are not linearly separable to linearly separable in a higher dimensional space. As well, given its dependence on support vectors, not all training data is needed for predictions. The support vectors determine the decision boundaries between the data from different trends that are used to do regression. SVR is a supervised technique in which data are fed into the model in (input, output) pairs.

1.1.2 Artificial Neural Network

Artificial neural network (ANN) is another regression technique that learns from processing (input, output) data pairs supplied during the training-testing split of input data. The more examples and variety of inputs the model is given, the more accurate the model typically will predict outputs [9]. Fig. 2 below shows the architecture of an ANN model, where the X nodes are the input layer, the Z nodes are the hidden layer, and the Y nodes are the output layer. For this study, X consists of porosity, permeability, bulk density, and irreducible water saturation and Y consists of capillary pressure and water saturation. The number of hidden layers and epochs (number of cycles through the training data) can be varied to see their effect on the predictions. The input to every hidden layer is the linear combination of the outputs of the previous layer passed through a non-linear function, also called the activation function. Activation function can be one of hyperbolic tangent (tanh), sigmoid, Rectified Linear Unit (ReLU), linear, etc.

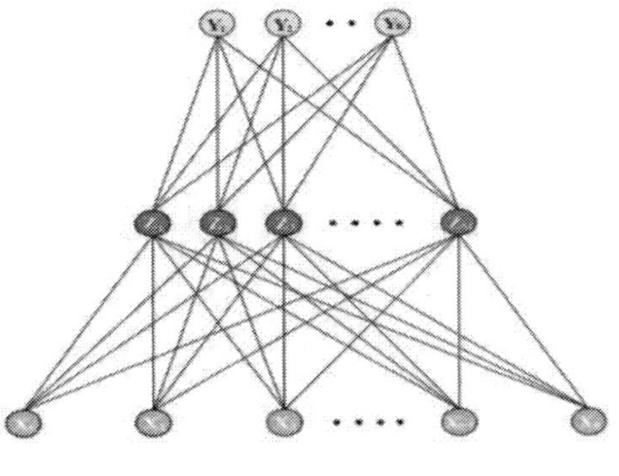

Fig. 2. Single Hidden Layer ANN [8]

1.1.3 Random Forest Regression

Random forest regression (RFR) is a supervised learning technique that utilizes ensemble learning for regression analysis. The method uses predictions from multiple learning algorithms to obtain a more accurate prediction [12]. RFR is also one of the most powerful regression techniques and is useful for non-linear relationships. A typical setup of a RFR model is where multiple decision trees are constructed on training data and the mean of the trees is outputted as the prediction. Each tree processes the input parameters, including the porosity, permeability, bulk density, and irreducible water saturation to predict capillary pressure and water saturation. Therefore, capillary pressure and water saturation are the output of each tree. The final output will be the average of all trees outputs in the forest.

However, it should be pointed out that capillary pressure curves generated by machine learning methods and experimental methods are often not accurate and seldom match the distribution of wetting and non-wetting phases sensed by imaging techniques. An ML-based method can predict water saturation and capillary pressure curve much faster than laboratory methods and potentially with no more uncertainty. Also, ML-based methods can be generalized to different fields if enough samples are used to train the models. In this manner, the need for core samples from new fields in reservoir conditions is mitigated.

2 Literature Review

2.1 Centrifuge Capillary Pressure Correlations

The centrifuge capillary pressure problem involves solving the challenge of determining local saturations along the length of the core based on the averaged saturations measured during the centrifuge experiment [13]. To solve the centrifuge capillary pressure problem numerous correlations have been developed that take on the general form of Eq. 1.

$$S(P_{c1}) = a_1\hat{S}(P_{c1}) + a_2 P_{c1}\frac{d\hat{S}(P_{c1})}{dP_{c1}} + E \quad (1)$$

where a_1, a_2, and E will vary depending on the correlation used [14]. The main assumptions are hydrostatic equilibrium being reached at each phase and the boundary condition of the capillary pressure being zero at the outflow face [13]. $S(P_{c1})$ is the local saturation and $\hat{S}(P_{c1})$ is the average saturation measured during the centrifuge experiment. P_{c1} is determined from Eq. 2.

$$P_{c1} = \frac{1}{2}\Delta\rho\omega^2(r_2^2 - r_1^2) \quad (2)$$

Where $\Delta\rho$ is the difference in the fluid densities and r_1 and r_2 are the radii at the inner and outer faces of the sample in the centrifuge core holder and ω is the rotation speed.

2.1.1 Hassler Brunner Method

The simplest solution for the centrifuge capillary problem was developed by Hassler and Brunner and it is also the poorest solution as it neglects both radial and gravity effects [15]. These assumptions can be satisfied for very short and narrow samples spun far from the rotation axis. In the case of the general centrifuge capillary problem, the Hassler-Brunner equation reduces to the following:

$$S(P_{c1}) = \hat{S}(P_{c1}) + P_{c1}\frac{d\hat{S}(P_{c1})}{dP_{c1}} \quad (3)$$

where $a_1 = 1, a_2 = 1$ and $E = 0$.

Forbes first solution demonstrated that this solution is always significantly lower, in terms of saturation, than the true $S(P_c)$ solution [14].

2.1.2 Forbes First Solution

Forbes built upon previous solutions to the centrifuge capillary problem and came up with his own solution that still neglects both radial and gravity effects but incorporates the difference in the centrifuge core radii [13]. Forbes solution took the form shown in Eq. 4.s

$$S(P_{c1}) = \hat{S}(P_{c1}) + \frac{1}{1+\alpha}P_{c1}\frac{d\hat{S}(P_{c1})}{dP_{c1}} \quad (4)$$

where $a_1 = 1, a_2 = \frac{1}{1+\alpha}$ and $E = 0$.

$$\alpha = \frac{1-\sqrt{1-B}}{1-2\sqrt{1-B}} \quad (5)$$

$$B = 1 - \left(\frac{r_1}{r_2}\right)^2 \quad (6)$$

2.1.3 Spline Functions

Nordtvedt and Kolltvelt were first to approximate the wetting phase saturation as a function of the capillary pressure whereby the capillary pressure curve is approximated piecewise by a polynomial [16]. This parameter estimation technique is usually related to functions which are too simplistic to describe the shape of capillary pressure curves [15]. A linear system of equations is then obtained and the piecewise approximation polynomial takes on the form shown in Eq. 7.

$$S(P_{c1}) = a_{i,o} + a_{i,1}P_{c1} + a_{i,2}P_c^2 i = 1,2, \ldots n_I \quad (7)$$

Where i indicates the interval number and n_I refers to the total number of intervals.

2.2 Current Machine Learning Applications

Busaleh et al. used several types of machine learning techniques to predict capillary pressure in carbonate oil reservoirs from data comprised of mercury injection drainage capillary pressure data [4]. Their data consisted of 202 carbonate core samples that contained at minimum 70 capillary pressure versus saturation points in addition to the corresponding porosity, permeability, and grain density for the core. For their machine learning techniques, the inputs used were porosity, permeability, grain density, and water saturation with their output being capillary pressure. They used a training-testing split of 70-30 of the core sample data. With the variety of machine learning techniques came with mixed results. ANN-trainbr, ANN-trainlm, ANN-trainscg, ANN-trainoss refer to neural networks using Bayesian regularization backpropagation, Liebenberg-Marquardt optimization, scaled conjugate gradient, and one-step secant methods, respectively.

The decision tree algorithm performed the best on the training data however this is likely due to overfitting as it performed considerably worse on the testing set. The ANN-trainbr method performed the best overall; however, there is still significant room for improvement [4].

Jamshidian et al. obtained centrifuge capillary pressure data for 15 core samples each of which having 27 capillary pressure data points totalling 405 data points. They completed seven cases to see the effects of additional input parameters on the final predictions [5]. An ANN model with a Cuckoo optimization algorithm was used to predict capillary pressure as the only output which was then compared against the experimental centrifuge values. The training set consisted of 8 cores with the remaining 7 used for validation.

Table 2. ANN Models with Input Parameters [5]

ANN Modelling Case	Contributing Input Parameter(s)
Case 1	S_w
Case 2	S_w, k_{air}
Case 3	S_w, k_{air}, φ
Case 4	S_w, k_{air}, φ, $S_{w,irr}$
Case 5	S_w, k_{air}, φ, $S_{w,irr}$, k_w
Case 6	S_w, k_{air}, φ, $S_{w,irr}$, k_w, k_o @ $S_{w,irr}$
Case 7	S_w, k_{air}, φ, $S_{w,irr}$, k_w, k_o @ $S_{w,irr}$, ρ_{grain}

Table 2 above shows the different cases modelled with the addition of input parameters. Water saturation, air permeability, porosity, irreducible water saturation, water permeability, oil permeability at irreducible water saturation, and grain density were the contributing input parameters.

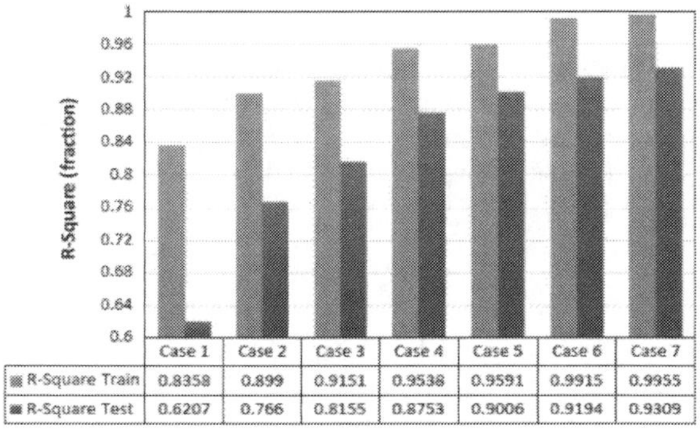

Fig. 3. Correlation Coefficients for the Different Cases [5]

From Fig. 3 the increase of input parameters correlates with an increase in the correlation coefficient for both the training and testing data sets. Case 1 through 7 correlates to the ANN modelling case and contributing input parameters shown in Table 2. Also, the greatest prediction comes with all the input parameters being used. Fig. 4 below highlights the accuracy of the predictions for core number 9 under the case 7 scenario.

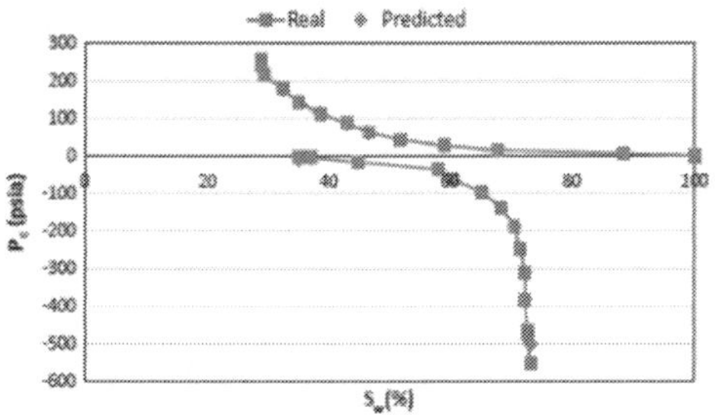

Fig. 4. Core Number 9 Experimental versus Predicted Values [5]

Kasha et al. developed a new method for capillary pressure estimation based on the Leverett J-function as shown in Eq. 8. This is a dimensionless function of water saturation describing capillary pressure also considering pore size and interfacial tension [6].

$$J(S_w) = \frac{P_c(S_w)\sqrt{\frac{k}{\phi}}}{\gamma \cos\theta} \quad (8)$$

$$r = 900.33\sqrt{\frac{k}{\phi}} \quad (9)$$

Eq. 9 above is the Carman-Kozeny equation which was used to determine the average pore size. The data consisted of 214 samples of mercury injection data with permeability and porosity data and the J versus water saturation for each curve was calculated. The workflow consisted of the formulation of grouping features from porosity, permeability, and pore throat-radius. The formed grouping features can be seen in Eq. 10 and 11.

$$F_1 = 0.01(\ln k \ln r^{0.5} - 3.23) \quad (10)$$

$$F_2 = 0.39\left(\frac{\phi^{0.5}\ln k}{\ln r^{0.5}} - 0.11\right) \quad (11)$$

This approach was compared against the flow zone indicator (FZI) method, which is a rock typing method grouping capillary pressure data into different rock types [6].

$$F_{1(FZI)} = 0.37\left(\log\frac{\phi}{1-\phi} + 2.71\right) \quad (12)$$

$$F_{2(FZI)} = 0.28\left(\log\sqrt{\frac{k}{\phi}} + 1.71\right) \quad (13)$$

Next an unsupervised machine learning technique (clustering) was used to group each sample based on the grouping features. A technique called the within-cluster sum of squares (WCSS) was used to determine the optimum number of clusters.

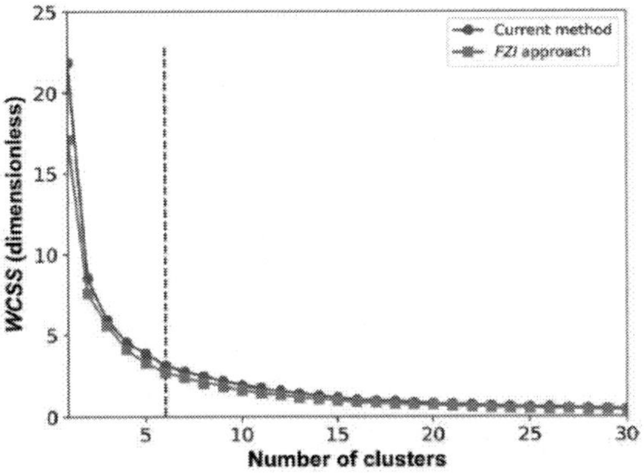

Fig. 5. WCSS Optimum Clusters [6]

The WCSS is the sum of squared deviations between each cluster's data points and the cluster's centroid. In this case, as shown in Fig. 5, four scenarios were considered based on the maximum curvature of 6 and three higher cluster scenarios of 10, 15 and 20 clusters.

The results of the cluster scenarios based on the grouping features can be seen in Fig. 6.

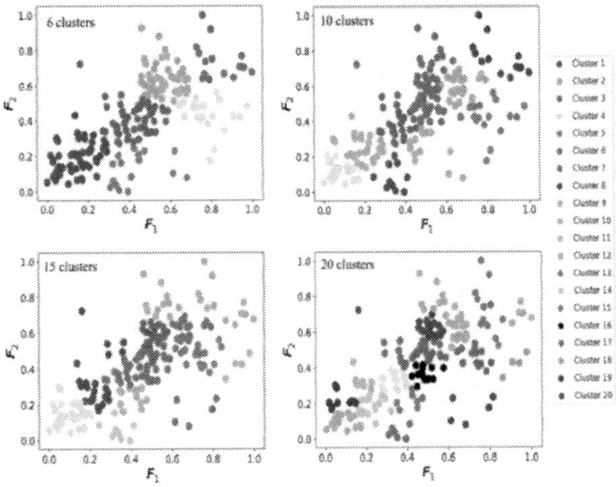

Fig. 6. Clustering Scenarios Based on Grouping Features [6]

From these generated clusters in each of the scenarios, an average J curve was calculated based on the J curves for each sample that were identified in the specific cluster as seen in Fig. 7.

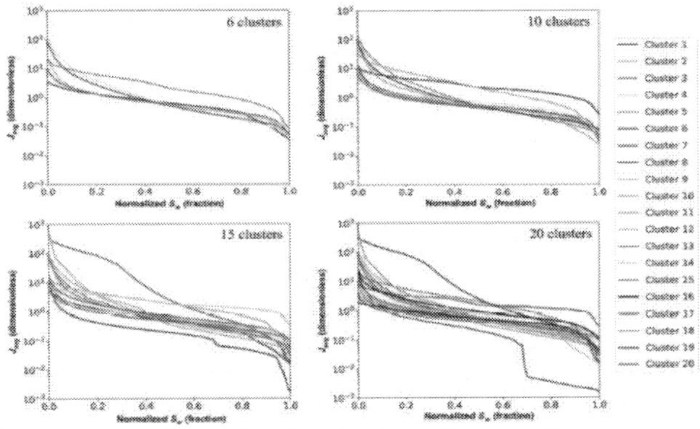

Fig. 7. Average J Curves Based on Clustering Scenarios [6]

An example of the average J curves calculated against the J curves for each sample identified in a cluster is shown in Fig. 8. This is showing the curves for cluster 1 and 2 for the 6 clusters scenario.

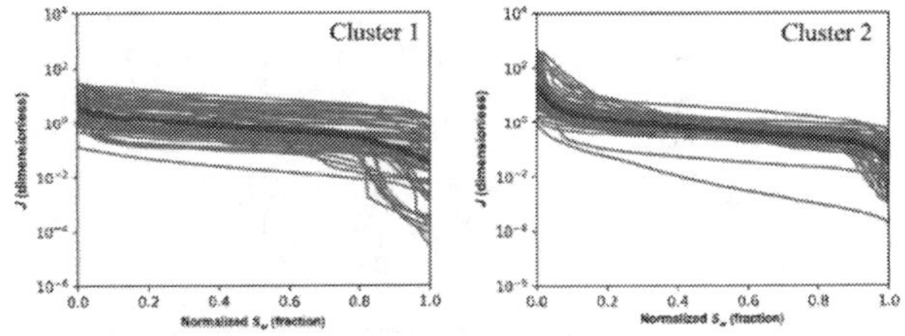

Fig. 8. Cluster 1&2 J Curves for 6 Cluster Scenario [6]

Next, classification algorithms (supervised machine learning) were used to develop maps for the prediction of corresponding clusters for new samples. The algorithms used were K-nearest neighbours (KNN), kernel support vector machine (KSVM), decision tree (DT), and random forest (RF) which were validated by using six carbonate rocks and comparing with the respective experimental data. The best predictions were for the 6 and 10 cluster scenarios with the accuracy for each algorithm found below in Table 3.

Table 3. Classification Algorithm Results [6]

Classification Algorithm	Current Method	
	6 Clusters (%)	10 Clusters (%)
KSVM	97.09	91.24
KNN	95.95	90.62
RF	95.29	87.75
DT	94.12	81.90

As shown, KSVM provided the highest accuracy in the predictions. This algorithm was then subsequently used to classify the validation set (6 carbonate rocks). The rock properties and the grouping features for each can be seen in Table 4 along with the identified cluster for each rock.

Table 4. Summary of Rock Properties and Validation Results [6]

Rock Information				Grouping Feature		Identified Cluster	
Rock Type	k (md)	Φ (%)	r (μm)	F1	F2	6 Clusters (KSVM)	10 Clusters (KSVM)
1	124.350	1.21	91271	0.91	0.10	4	5
2	26.710	11.87	13506	0.69	0.44	4	5
3	5.580	10.05	6709	0.57	0.39	3	6
4	0.586	8.64	2345	0.43	0.35	3	3
5	0.041	6.12	737	0.28	0.27	1	9
6	0.003	4.51	232	0.16	0.20	1	9

Busaleh et al. presented work completed on mercury injection data from carbonate rocks. Although they used a variety of machine learning techniques, results were poor, and more emphasis could have been placed on obtaining better results from a few techniques.

The work from Jamshidian et al. used centrifuge experimental data and an ANN with a Cuckoo optimization algorithm being used to predict capillary pressure with varying inputs. It is important to note here the predictions were completed on capillary pressure experimental centrifuge points with water saturation being used as an input. 7 different cases were conducted with each case involving an additional input parameter which in turn increased the accuracy of the predictions. The limitation here however was the use of the raw centrifuge experimental points. The use of an interpretation software was not used to apply literature correlations to obtain a complete capillary pressure curve.

Kasha et al. took a different approach in using an unsupervised algorithm, clustering, first followed by classification algorithms. The limitation here being the need for an abundance of mercury injection data for many samples to be clustered together and an accurate average J curve for those types of rocks to then be developed. Other approaches could have been taken as well with regards to the pore size distribution such as the Winland r_{35} values.

This research aims to show the novelty of using machine learning techniques to predict complete capillary pressure curves developed from conducting centrifuge tests followed by using SCAL interpretation software to apply literature correlations. This is turn can reduce the time to obtain a complete capillary pressure curve and help improve reservoir characterization which is the end goal in undertaking a SCAL centrifuge experiment. Although several researchers have used ML methods to predict SCAL data, they have focused on predicting a single value of the quantity of interest (relative permeability, capillary pressure, water saturation, etc.). This study assesses the capability of three machine learning algorithms in generating the capillary pressure curves that consists of both capillary pressure and water saturation. Therefore, the novelty of this paper is to simultaneously predict capillary pressure and water saturation values to generate the capillary pressure curve.

3 Methodology

3.1 Centrifuge Capillary Pressure Tests

For this research, multiple centrifuge capillary pressure tests were conducted. The setup consisted of a VINCI RC 4500 – capillary pressure refrigerated centrifuge along with a computer with the appropriate recording and interpretation software. The interpretation software used was CYDAR which offers a powerful solution for the interpretation of conventional and specialized core analysis experiments. CYDAR is also used to record the water saturation and speed versus time during the experiment. A separate VINCI software is used to set the rotational speeds and block times for the centrifuge as well as setting up the camera interfaces for the holding cups.

Core samples were first cleaned followed by drying out in an oven for approximately 1-2 days. Following drying the core samples, their length and diameter were recorded using a Vernier calliper. Next, the dry weight of the cores were recorded. The petrophysical properties of the cores used in the machine learning can be seen in Table 5 below.

Table 5. Core Petrophysical Properties

Core #	Porosity	Permeability (mD)	Bulk Density (g/mL)	Swirr
Core 1	0.222	300	2.011	0.045
Core 2	0.217	180	2.089	0.098
Core 3	0.317	2300	1.761	0.033
Core 4	0.316	2300	1.736	0.064
Core 5	0.308	2300	2.034	0.068
Core 6	0.308	2173	1.811	0.2
Core 7	0.285	2216	1.627	0.23
Core 8	0.265	2557	1.844	0.162
Core 9	0.261	1631	1.864	0.066
Core 10	0.296	2868	2.032	0.306
Core 11	0.2828	3591	1.792	0.058
Core 12	0.267	2900	2.033	0.185
Core 13	0.27	2200	1.686	0.23

Next, a synthetic brine was prepared for the simulation of formation water, the composition is shown in Table 6 below.

Table 6. Synthetic Brine Formulation

Synthetic Brine Formulation		
NaCl	50.397 g/kg	96.96 g/L
$CaCl_2*2H_2O$	13.202 g/kg	14.16 g/L
$MgCl_2*6H_2O$	2.760 g/kg	2.96 g/L
KCl	0.439 g/kg	0.46 g/L
Na_2SO_4	0.345 g/kg	0.37 g/L
Subtotal	107.132 g/kg	114.91 g/L
Add Water	892.868 g/kg	957.69 g/L
Total	1000 g/kg	1072.60 g/L

The cores were then loaded into a vessel where the approximately 1 L of brine was pumped in to saturate the cores for approximately one day. Due to the high permeability of the cores one day proved sufficient in saturating the cores. After saturation had occurred, the wet weight of the cores was measured. This data was then used to calculate the porosity and bulk density of each sample.

The cores were then loaded into centrifuge core holders where the top, bottom pipe, and sleeve were determined from the core length as shown in Fig. 9a.

 https://doi.org/10.1051/e3sconf/202336701004

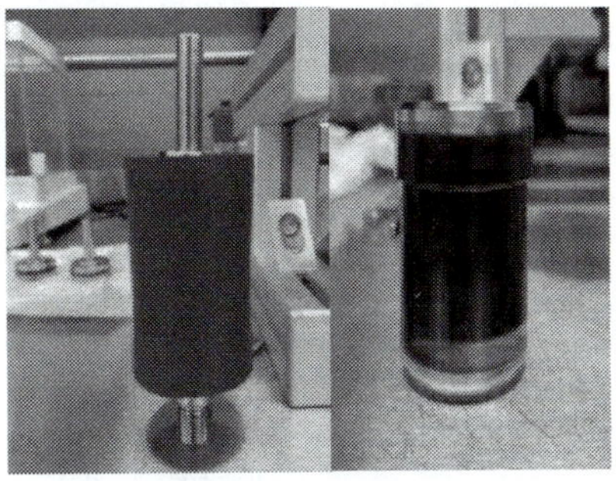

Fig. 9a) Assembled Core Sample (left) **9b)** Holding Cup (right)

The core holders were pressurized to 2000 psi using a hand pump and silicone oil. Small amounts of deionized water and oil were put into the holding cups to create an interface as seen in Fig. 17b. The core holders were also weighed to ensure symmetrical weight before being loaded in the centrifuge.

Once loaded in the centrifuge, the recorded data was inputted in CYDAR, and the recording software was loaded. Multi-step speeds (between 500 to either 3500 or 4500 rpm in increasing 500 increments) as well as step durations were specified, and the camera was set to record the oil-water interface for the respective cups as shown in Fig. 10. This figure shows an oil-water system with the right arrow indicating the water side and the left arrow indicating the oil side. The arrow pointing between the two indicates the oil-water interface. Throughout the drainage experiment as the oil is forced into the core and the water out, the interface will gradually move towards the left. The camera was also checked periodically to ensure it was still properly recording the interface.

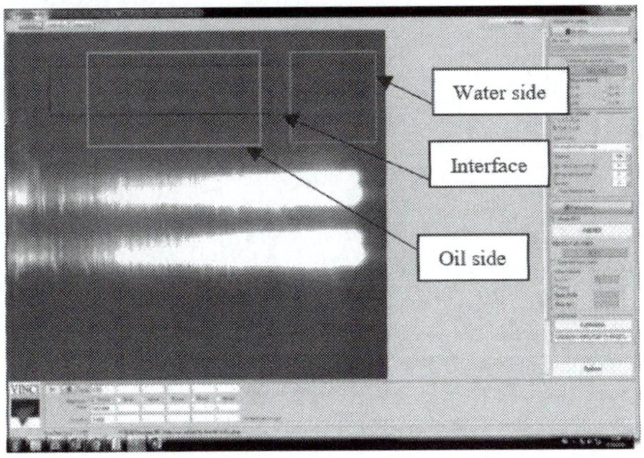

Fig. 10. Camera Interface

Upon completion of the centrifuge test, the next step was to process the recorded data. This required the use of CYDAR interpretation software. The water saturation versus time data was cleaned to exclude any outlier data that the camera may have recorded, and the block times and durations were recorded and entered for CYDAR to calculate the capillary pressure based on the step speeds. For each core, an analytical fit was calculated for the experimental data and Forbes first solution correlation [13] was applied to the experimental data. An example of a complete capillary pressure curve for a sample after processing can be seen in Fig. 11. A bi-exponential fit was the analytical method used in this case. As shown, both analytical and Forbes methods lack rigor to fit the experimental data.

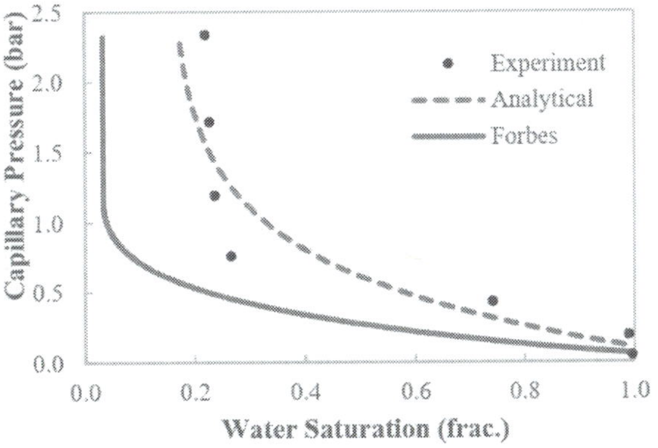

Fig. 11. Core 3 Processed Data (drainage capillary pressure)

For each of the 13 cores the Forbes first solution was applied to the experimental data. Using CYDAR, data from the Forbes curves was discretized to be used for the machine learning techniques in order to predict the Forbes curves.

3.2 Machine Learning Techniques

A spreadsheet was developed to collect all the test data for the machine learning techniques before pre-processing of the data from 13 core samples. The machine learning workflow followed in this case can be seen in Fig. 12.

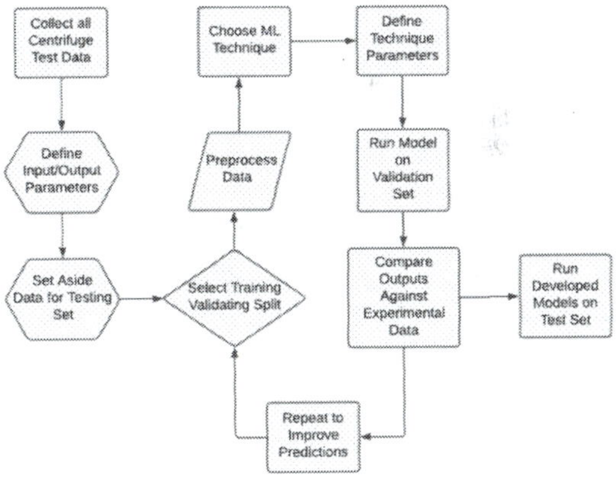

Fig. 12. Machine Learning Workflow

The inputs for the machine learning techniques as mentioned previously were the routine core analysis data of porosity, permeability, bulk density, and irreducible water saturation. The outputs were water saturation and capillary pressure; subsequently, multi-output regression models were developed. Data for 2 core samples were put into a separate file to be used as a test set after model development. Note that for these 2 samples, 1995 (input, output) data points were recorded. For the remaining 11 cores, a training-testing split of 70-30 was used to train the models. Using these 11 cores,

10,955 (input, output) data points were recorded and used to train and validate the ML models. Depending on the technique used, the parameters were defined for each; for ANN the epochs, activation function, hidden layers, and neurons, for RFR the number of trees, and for support vector regression the tolerance and max iterations. Standardization and normalization feature scaling were also implemented to see how they would affect overall performance. Eqs. 14 and 15 show the methods for standardization and normalization, respectively.

$$X' = \frac{X - \mu}{\sigma}, \tag{14}$$

where μ is the mean of the feature values and σ is the standard deviation of the feature values.

$$X' = \frac{X - X_{min}}{X_{max} - X_{min}}, \tag{15}$$

where X_{min} and X_{max} are the minimum and maximum of the feature, respectively.

Once the models were run on the training and testing data, their performance was measured using 2 measurements, including the mean squared error (MSE) and correlation coefficient (R^2) where n corresponds to the total number of data points.

$$MSE = \frac{1}{n} . \Sigma(actual - predicted)^2 \tag{16}$$

Finally, once all models were developed on the validation data set, they were then tested and evaluated on the unseen data set. All machine learning methods were implemented using Python 3.7 programming language along with Keras library with Tensorflow 2.3 backend.

4 Results and Discussion

The results from all ML algorithms on the validation data is shown in Table 7.

Table 7. ML Models Results – Test Data

ML Algorithm	MSE	R^2 (Validation Data)
Random Forest	2.09×10^{-5}	0.9997
Neural Network	0.00222	0.9937
Support Vector	0.00436	0.9543

All models performed well with random forest being the best with a correlation coefficient of 0.9997. Figs. 13 and 14 show the prediction versus experimental data for ANN for capillary pressure and water saturation, respectively.

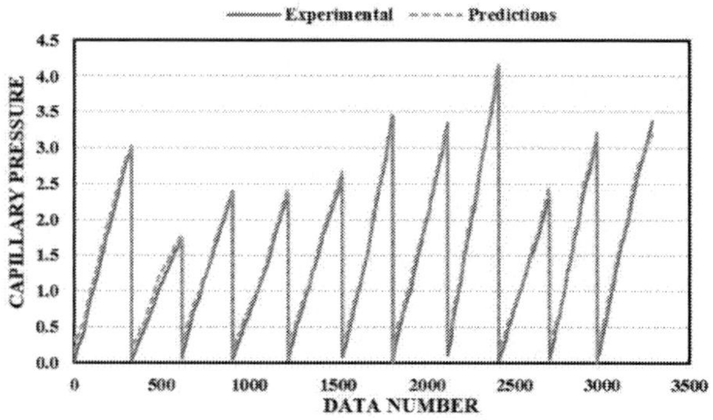

Fig. 13. ANN Validation Data Results – Capillary Pressure

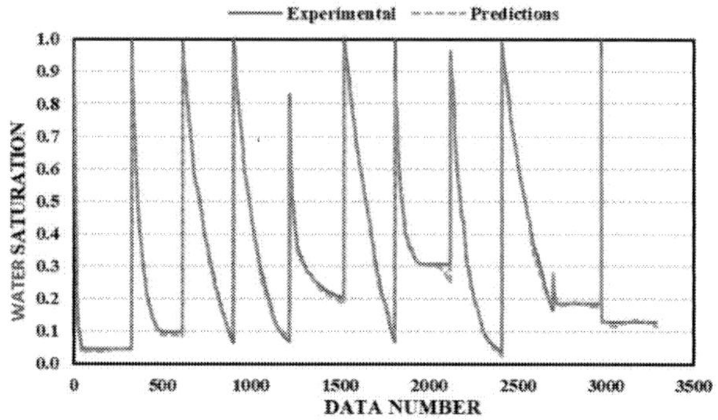

Fig. 14. ANN Validation Data Results – Water Saturation

Similarly, Figs. 15 and 16 highlight the results for the SVR model and Fig. 17 and 18 are the results for the RFR model on the testing data.

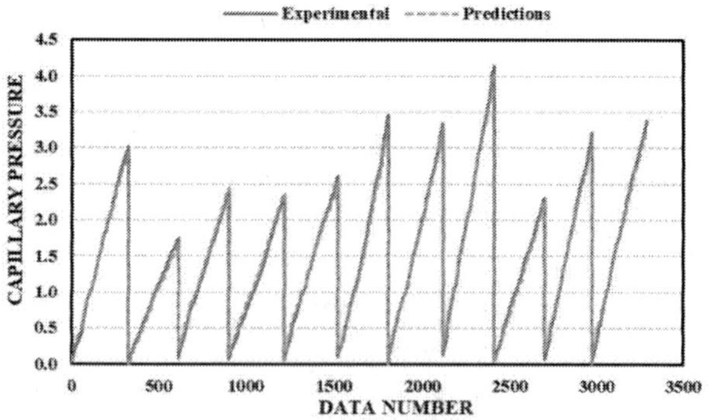

Fig. 15. SVR Validation Data Results – Capillary Pressure

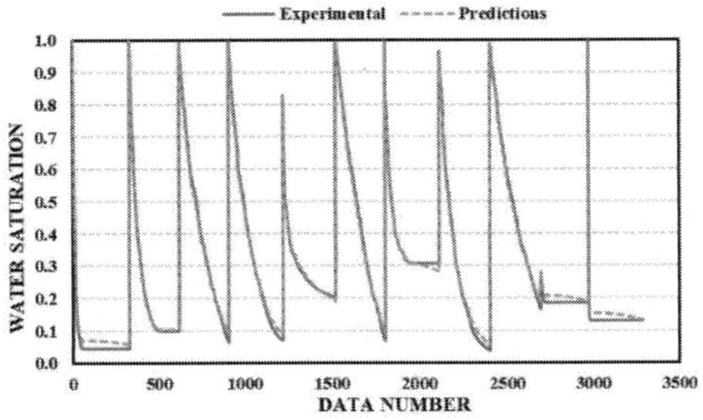

Fig. 16. SVR Validation Data Results – Water Saturation

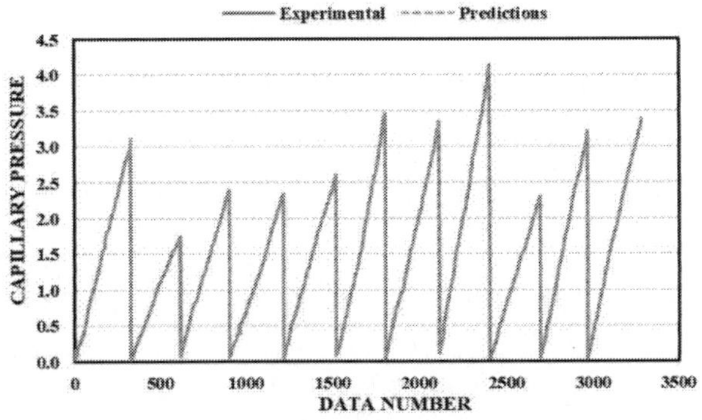

Fig. 17. RFR Validation Data Results – Capillary Pressure

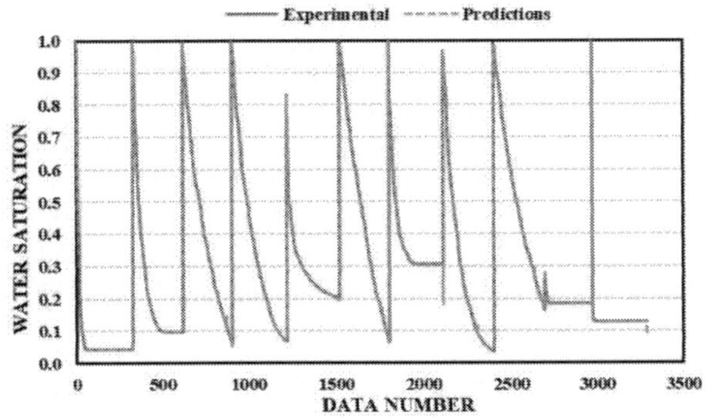

Fig. 18. RFR Validation Data Results – Water Saturation

A sensitivity analysis was also completed on the ANN parameters of hidden layers, neurons, and activation functions. The base case for the following graphs is a model with 32 neurons, 2 hidden layers and a rectified linear unit (ReLU) activation function (i.e. when seeing the effects of changing neurons, 2 hidden layers and ReLU activation function is used).

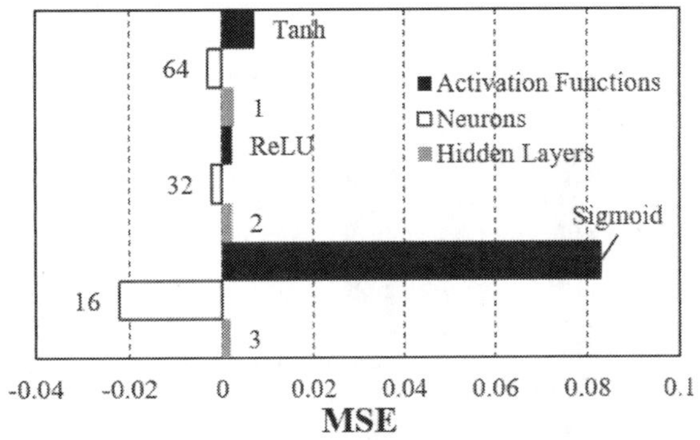

Fig. 19. Varying ANN MSE Results

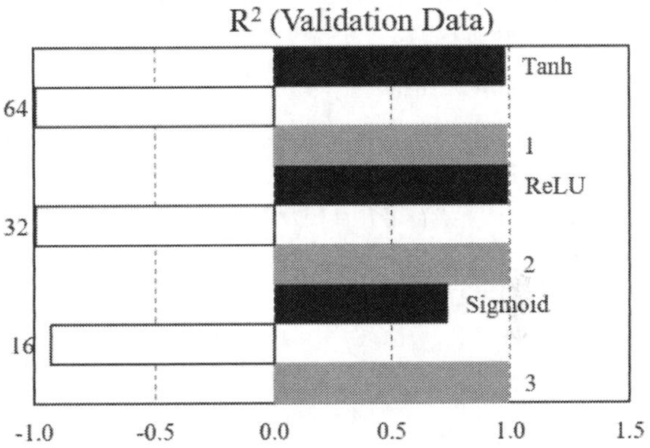

Fig. 20. Varying ANN R^2 Results

As can be seen from Figs. 19&20 the biggest change in prediction accuracy comes with the change in using Tanh as the activation function. Minimal changes occurred in the prediction accuracy with varying the hidden layers with 2 hidden layers performing the best. Finally, for the varying of neurons, the greatest error occurred when using 16, while the greatest prediction happened using 32.

With the ML models developed, the test set of the 2 cores that was set aside was then given to the models for predictions. The following graphs highlight the accuracy of each model showing the test data results with the results on the test set when using normalized and standardized feature scaling.

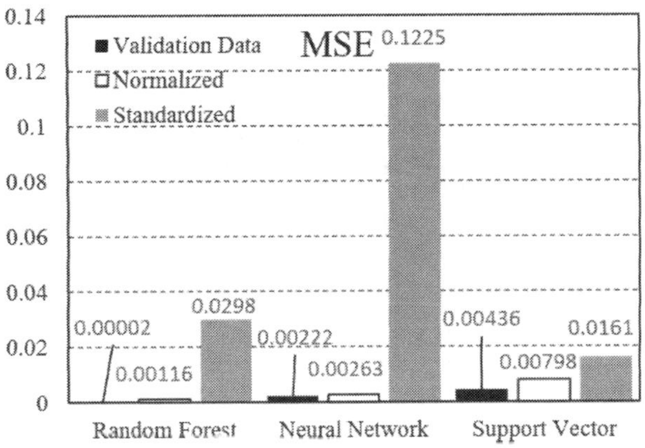

Fig. 21. MSE Results of Models for Test and Validation Data

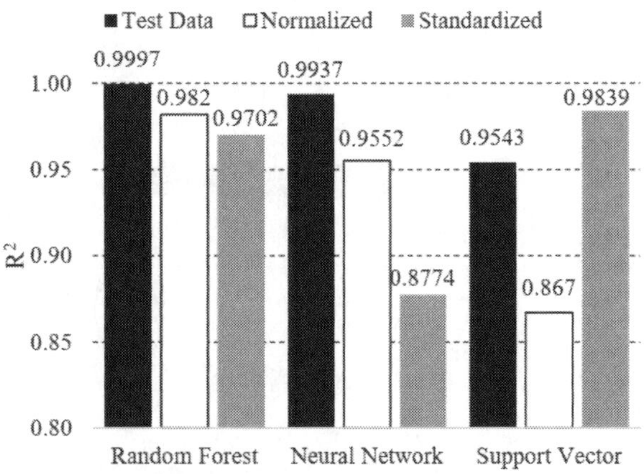

Fig. 22. R^2 Results of Models for Validation Data

Following the evaluation of the validation data, unseen data consisting of 2 cores was used to test the accuracy of the models. The results for each of the models can be seen in Table 8.

Table 8. ML Models Results – Unseen Data

ML Algorithm	MSE	R^2 (Unseen Data)
Random Forest	0.0248	0.9734
Neural Network	0.1034	0.8886
Support Vector	0.0438	0.9527

As shown in Fig. 23, although RFR performed the best overall on the validation data and unseen data with regards to the performance metrics, it is very noisy when predicting unseen data. SVR performed the greatest overall with a smoother curve, however the entry capillary pressure was predicted higher.

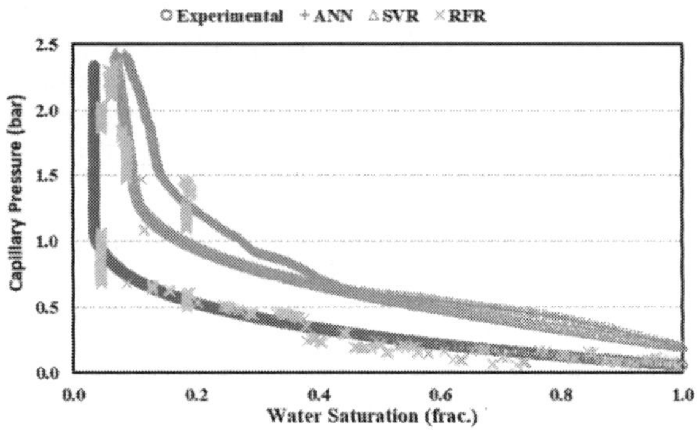

Fig. 23. Unseen Data Model Results – First Core

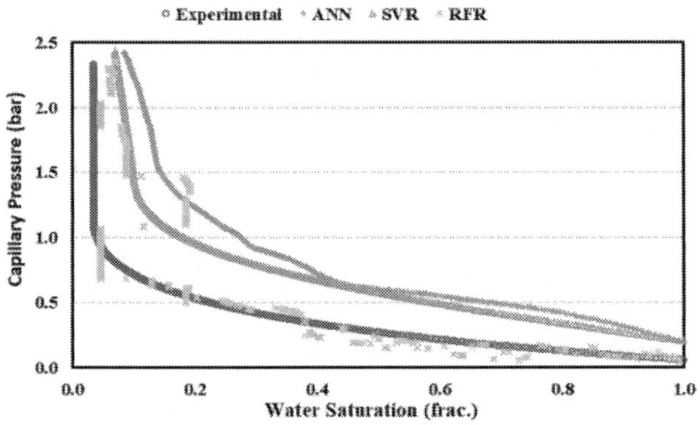

Fig. 24. Unseen Data Model Results – Second Core

Similar to the first core, RFR was noisy again in predictions for the second core. ANN and SVR curves were smoother but showed differences in the entry capillary pressure.

5 Conclusions

This work investigated various machine learning techniques in predicting complete drainage capillary pressure curves. Centrifuge capillary pressure tests were conducted with SCAL interpretation software being used to apply the Forbes First solution correlation. Data for 13 cores was obtained along with the respective RCAL data of porosity, permeability, bulk density, and irreducible water saturation for each core. Once all the data was collected, three machine learning models were developed using training and validation data sets with some of the collected data set aside to be used as a test data set. Some of the key findings from this work were that the three techniques used all performed very well on the validation data set as evident from the performance metrics used. Another takeaway is the effectiveness of conducting a sensitivity analysis as evident in the results from the ANN model when changing the neurons, hidden layers, and activation function used.

Some recommendations for future work are to conduct centrifuge experiments on more cores of different rock types to see how the models will perform when using a variety of rock types. As well as, incorporating additional input data such as pore size distribution to see the effects. Additional cores could also be added to the unseen data for the models to increase the accuracy of the predictions. With the success of applying machine learning techniques to capillary pressure, future work can also be to conduct other SCAL tests such as relative permeability to see the effectiveness of machine learning techniques in predicting other reservoir characteristics.

Acknowledgements

The authors would like to thank the Hibernia Management and Development Company (HMDC), Chevron Canada Ltd, Energy Research and Innovation Newfoundland and Labrador (ERINL), the Natural Sciences and Engineering Research Council of Canada (NSERC), the Province of Newfoundland and Labrador, and Mitacs for financial support.

References

1. S.B. Suslick, D. Schiozer, M.R. Rodriguez. Uncertainty and Risk Analysis in Petroleum Exploration and Production. Terrae **6**(1): 30-21 (2009). https://www.ige.unicamp.br/terrae/V6/PDF-N6/T-a3i.pdf
2. C. McPhee, J. Reed, I. Zubizarreta. Core Analysis – A Best Practice Guide. https://app.knovel.com/hotlink/toc/id:kpCAABPG03/core-analysis-best-practice/core-analysis-best-practice
3. A. Sircar, K. Yadav, K. Rayavarapu, N. Bist, H. Oza. Application of machine learning and artificial intelligence in oil and gas industry. Petroleum Research **6**, 379-391, (2021).
4. Y.R. Busaleh, A, Abdulraheem, T. Okasha. Prediction of Capillary Pressure for Oil Carbonate Reservoirs by Artificial Intelligence Technique. SPE Asia Pacific Oil and Gas Conference Exhibition. (2016). https://doi.org/10.2118/182173-MS
5. M. Jamshidian, M. Mansouri Zadehm M. Hadian, R. Moghadasi, O. Mohammadzadeh. A Novel Estimation Method for Capillary Pressure Curves Based on Routine Core Analysis Data Using Artificial Neural Networks Optimized by Cuckoo Algorithm – A Case Study. (2018). Fuel, **220**, 363–378. https://doi.org/10.1016/j.fuel.2018.01.099
6. A.A. Kasha, A. Sakhaee-Pour, I.A. Hussein. 'Machine Learning for Capillary Pressure Estimation.' SPE Res Eval & Eng **25**. (2022): 1-20. doi: https://doi-org.qe2a-proxy.mun.ca/10.2118/208579-PA
7. T. Mitchell. Machine Learning. New York: McGraw Hill. (1997). OCLC 36417892.
8. Mathworks Help Center. Machine Learning in MATLAB https://www.mathworks.com/help/stats/machine-learning-in-matlab.html
9. T. Hastie, R. Tibshirani, J. Friedman. The Eléments of Statistical Learning. Springer New York. https://doi.org/10.1007/978-0-387-84858-7_12
10. A. Shmilovici. Support Vector Machines. In : O. Maimon, L. Rokach. (eds) Data Mining and Knowledge Discovery Handbook. (2005). Springer, Boston, MA. https://doi-org.qe2a-proxy.mun.ca/10.1007/0-387-25465-X_12
11. Rodríguez-Pérez, R., Bajorath, J. Evolution of Support Vector Machine and Regression Modeling in Chemoinformatics and Drug Discovery. J Comput Aided Mol Des (2022). https://doi-org.qe2a-proxy.mun.ca/10.1007/s10822-022-00442-9
12. K. Fawagreh, M.M. Gaber, E. Elyan. Random forests: from early developments to recent advancements. System Science and Control Engineering. Volume 2. (2014). https://doi-org.qe2a-proxy.mun.ca/10.1080/21642583.2014.956265
13. P. Forbes. Simple and Accurate Methods for Converting Centrifuge Data into Drainage and Imbibition Capillary Pressure Curves. SPWLA-1994-V35n4a3, 35(04), 13. (1994).
14. Z.A. Chen, D.W. Ruth. Measurement and Interpretation of Centrifuge Capillary Pressure Curves-The SCA Survey Data. The Log Analyst, 36(05), 13. (1995).
15. P. Forbes. Centrifuge Data Analysis Techniques : An SCA Survey on the Calculation of Drainage Capillary Pressure Curves from Centrifuge Measurements. 20. (1997).
16. J.E. Nordtvedt, K. Kolltvelt. Capillary Pressure Curves From Centrifuge Data by Use of Spline Functions. SPE-19019-PA, 6(04), 497–501. https://doi.org/10.2118/19019-PA (1991).

Digital rock workflow to calculate wettability distribution in a reservoir rock

Ashraful Islam[1,*], *Rafael* Tio Salazar[1], and *Bernd* Crouse[1]

[1]Dassault Systèmes Simulia, USA

Abstract. Wettability has a strong influence on multi-phase flow behavior through reservoir rock. Reservoir rocks tend to have spatially varying wettability. Prior to contact with oil, rocks are almost always naturally water-wet. As oil invades the pore-space over geologic time, the initial water-wet state may be altered in certain locations due to adhesion of substances within the oil phase to the grains. Mechanisms of wettability alteration depend on various properties such as pressure, temperature, mineral chemistry, surface roughness and fluid composition. In this study wettability alteration in a reservoir rock is studied through direct simulation using multiphase Lattice Boltzmann method where the computational grid is constructed from segmented micro-CT images of the rock sample. The pore-grain interface is defined by a triangulated surface mesh for accurate fluxes near boundary and local curvature calculation. A capillary pressure drainage simulation is conducted in a water-wet Berea sandstone sample initially filled with water. When oil invades the pore space as the capillary pressure is increased, a fraction of the pore-grain surface is altered towards an oil-wet condition, as determined by a novel wettability alteration process. This process calculates local curvature at every surface element of the rock, obtains local capillary pressure from the simulation and assumes a disjoining pressure to determine water-film breakage at every location of the pore-grain surface. As a result, a spatially varying rock wettability is created. Using this new wettability distribution, the simulation is continued to allow the fluid phases to redistribute accordingly. The process is iteratively carried out until both fluid saturation and wettability distribution converged at a given applied capillary pressure. Afterwards, the pressure is ramped up to the next stage and the process is repeated again. It has been found that the wettability alteration is a slow dynamic process where the non-wetting phase can gradually invade finer pore space as the surrounding grain wettability is altered. In this study, it has also been found that wettability alteration of the reservoir rock produces lower connate water saturation during primary drainage compared to the simulation results without alteration. The resulting spatially varying wettability distribution from primary drainage is used for a subsequent water flooding simulation to calculate water-oil relative permeability curves. The methodology presented in this work can be leveraged to better understand and predict an improved mixed wetting conditions found in the reservoir rocks which is needed for more accurate displacement tests such as relative permeability simulations.

1 Introduction

Two-phase relative permeability describes how easily one fluid can be moved through a porous medium in the presence of another fluid. It is an important characteristic of hydrocarbon reservoir rocks and a crucial input to oil and gas reservoir simulation modeling. Relative permeability, and multiphase flow through porous media in general, is dependent on various characteristics of the fluid-fluid-rock system, including rock surface properties, physical properties of each fluid, and flow conditions. They can be expressed as dimensionless numbers such as viscosity and density ratio, capillary number, wettability and geometric state function [1]. Wettability is a measure of a fluid's tendency to adhere to a solid surface in the presence of other immiscible fluids. Macroscopically, it can be measured by the contact angle at the three-phase contact line and classified as water-wet (minimum 0°, maximum 60 to 75°), neutral-wet (minimum 60 to 75°, maximum 105 to 120°) and oil-wet (minimum 105 to 120°, maximum 180°) in the petroleum engineering context. Microscopically, it can be related to the molecular energy difference between the energy of the combined system and the combined bulk energies of the individual components.

The physics of multiphase flow at pore scale is very different for water-wet and oil-wet rocks and its impact can be found in capillary pressure or relative permeability curves [2-6], both critical inputs for reservoir simulation forecasting. Thus, an accurate wettability condition is an important input parameter for multiphase fluid displacement simulations and impacts the overall productivity of the hydrocarbon reservoir [7].

Reservoir rocks, unlike man-made materials, tend to have spatially varying wettability, i.e., the contact angle varies from location to location on the surface of the grains in contact with the fluids in the pore space of a rock [8-10]. The contact angle distribution is a result of the mineral composition and the geological history of a hydrocarbon-bearing rock as well as surface texture, chemical composition of fluids (e.g., water, oil) in contact with the different crystalline planes of the grain minerals present in the rock, etc. Prior to contact with crude oil, rocks are almost always naturally water-wet. As oil invades the pore-space over geological time, the initial water-wet character may be altered

* Corresponding author: ashraful.islam@3ds.com

in certain locations due to adhesion of surface-active materials from the oil phase. Mechanisms of wettability alteration depend on various local system properties such as pressure, temperature, mineral type, and fluids composition.

When rock samples are extracted from oil/gas reservoirs during the drilling process, they are often contaminated with mud and other fluids and chemicals. When those rock samples are subject to laboratory experiments to obtain two-phase relative permeability curves or capillary pressure curves, a common lab preparation procedure is to first clean the rock of fluids using solvents and then attempt to restore the rock's natural wettability. This process of attempting to reestablish fluid phase distribution and wettability distribution (e.g. contact angles) representative of the subsurface reservoir conditions is referred to as "aging". A typical aging procedure would involve cleaning the rock sample of original fluids and contaminants and treating the rock sample with chemicals that induce a water-wet condition. Next, the sample is saturated with brine, and in a "drainage" procedure oil is pushed in (hence brine is pushed out) using estimated reservoir conditions of temperature and pressure. The rock sample is then "aged" for a period of time, e.g. 4 weeks, allowing wettability alteration to occur, presumably in a fashion similar enough to what had occurred in the real subsurface rock formation so that a realistic wetting condition is restored.

With the advancement of imaging and computational fluid dynamics, it is now possible to perform a relative permeability or capillary pressure analysis numerically for rock samples. But the accuracy of these simulations also depends on the spatial distribution of the wetting condition of the grains among many other inputs [5]. Using x-ray micro-tomography (micro-CT), it is possible to measure in-situ contact angles [11-16] in an aged rock sample from laboratory procedure and use as input for numerical relative permeability analysis. Observations of the relative fluid coverage of rock mineral surfaces can also be used to create a 3D map of the wetting state of the rock [8-9]. However, establishing an initial fluid distribution and wetting conditions using numerical simulation remain challenging. It includes understanding the role of thin films in porous media. The type and thickness of films coating pore walls determines a reservoir rock wettability and whether or not a reservoir rock can be altered from its initial state of wettability [5, 17]. Thin film physics also depends on the grain surface mineralogy and the fluid pair. It is interrelated with the pore shape, especially pore curvature and surface roughness. As a result, the solution scope to this problem spans from molecular to microscopic scales, where molecular interactions determine the Derjaguin-Landau-Verwey-Overbeek (DLVO) forces [18, 19], and interfacial tension, while microscopic pore topology determines candidate surfaces for wettability alteration.

In this work, we describe a numerical Digital Rock workflow starting from pore scale 3D image acquisition to relative permeability fluid flow simulation, including a novel wettability alteration process. Wettability alteration is carried out in a digitally reconstructed 3D pore-grain model, paying attention to the micro scale interactions. The resulting spatially varying wettability distribution is used for relative permeability simulation. The next section describes the methodology, including image processing, an overview of the numerical modeling approach, and detailed description of the wettability alteration process. Afterwards, wettability alteration and relative permeability results for a Berea sandstone can be found in the results section, which is followed by conclusions to shed some light on future research in this regard.

2 Methodology

2.1 Sample imaging

This study uses a Digital Rock workflow [15] that starts with the acquisition of a 3D image of a Berea Sandstone sample using an x-ray micro-tomography (micro-CT). Fig 1(a) shows a 2D slice of the 3D scans captured perpendicular to the fluid displacement direction. Emphasis was given to acquiring images that are free of x-ray imaging artefacts such as beam-hardening, shadowing, ring artefacts etc. Next, a segmentation workflow is applied to that image to identify the pores and grains as shown in black and white respectively in Fig 1(b). This segmentation uses differences in grey values and intensity gradients to identify voxels along the solid-pore boundary and produce the resolved pore structure to be used in the flow simulation. Further reference for image acquisition and processing can be found in [20].

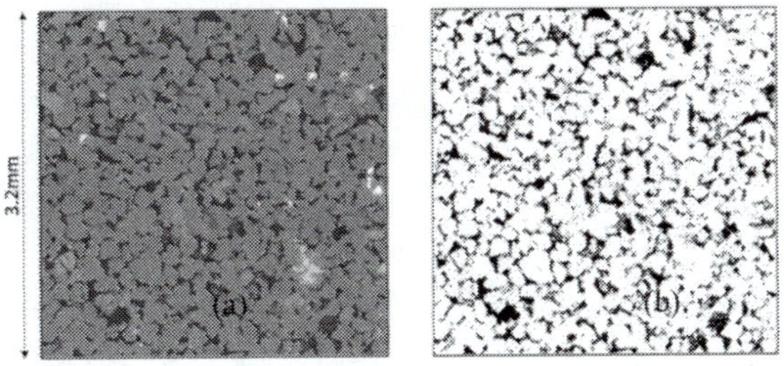

Fig. 1. Two-dimensional section of 3D scans for the Berea Sandstone used in this work: (a) original grey scale, (b) segmented image.

The stack of micro-CT images of a Berea sandstone was captured at $2.0\mu m$ per voxel resolution to ensure good mesh quality for subsequent fluid flow simulation. At this resolution, the resulting 3D image shows sufficient pore connectivity in all three directions. The pore bottleneck radii, which is defined by the radius of the largest rigid sphere that can be percolated across the 3d rock pore space without getting stuck, were calculated to be 3.0, 3.6 and 3.2 voxels in the x, y and z directions respectively. Note that the bottleneck radius needs to be large enough to ensure sufficient accuracy of the fluid flow simulation and that requirement is solver specific. In a previously published work [20], it has been demonstrated that the Lattice Boltzmann method (LBM) solver used for this study can provide accurate results at this resolution.

Attention was also paid to ensure that the field of view of the 3D image is large enough to capture a representative elementary volume. A size growing analysis was performed, where the entire analysis domain is divided into incrementally bigger, overlapping, and centered subdomains of equal aspect ratio. For each of the sub-domains, porosity and single-phase permeability values are calculated as shown in Fig. 2(a). Additionally, a coefficient of variation of permeability as a function of sub-domain size is calculated in Fig. 2(b) [21]. This chart represents likelihood that a certain subdomain permeability value will be one standard deviation of the mean value. Based on these results, we chose a subdomain of 800 voxels (corresponding to 1.6 mm) cubic size for subsequent multiphase simulation.

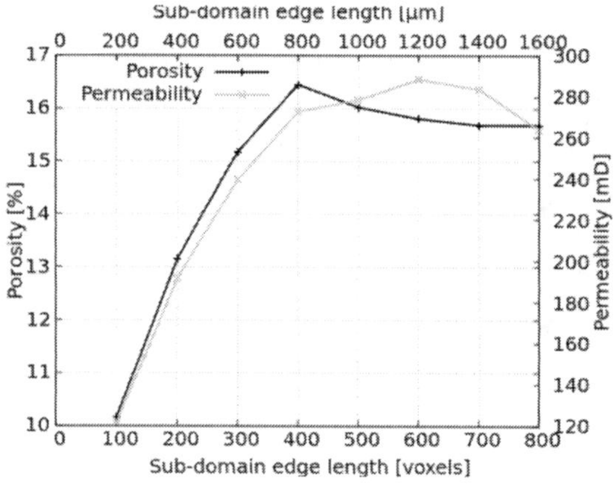

(a)

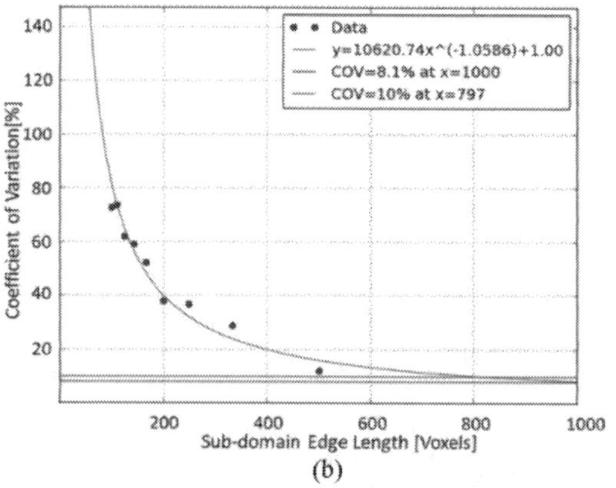

(b)

Fig. 2. Representative elementary volume calculation: a) size growing analysis for porosity and permeability b) Permeability coefficient of variation, defined as the ratio of standard deviation to mean value.

2.2 Numerical modelling approach

The multiphase fluid flow numerical simulation is conducted in the segmented 3D pore-space voxels of the selected rock domain using a multiphase LBM solver. The LBM is based on mesoscopic kinetic theory and solves a discrete form of the Boltzmann transport equations instead of Navier-Stokes equations [22-26]. In this method, most of the operations can be performed locally on a cubic lattice, resulting in highly parallel computational performance [27]. This simulation performance quality makes LBM an ideal candidate for porous media applications where typically a large domain with complex pore network needs to be simulated. The LBM has been successfully used for various porous media applications such as in hydrocarbon bearing rocks [20, 28], gas diffusion layers in fuel cells [29, 30], porous electrodes in lithium-ion batteries [31, 32] etc.

The LBM implementation used in the current study is based on the multiphase framework presented by Shan and Chen [33] that later went through various improvements for simulating porous media applications with sufficiently high accuracy at low resolution and small viscosity [27, 34]. This is a diffuse interface method where different particle species are used to generate separated fluid phases, and the interaction between the different particle species determines the inter-component interfacial tension. Thus, the interfaces between different components are automatically determined once the species' interactions are defined. As a result, the complexity of explicit interface tracking such as in traditional volume-of-fluid or level-set approaches can be avoided.

In order to capture correctly the physics of multi-component fluid flow, even at low resolution, the current LBM implementation uses, in addition to the voxel elements, triangulated surface elements also known as surfels, to accurately capture the pore-grain boundary as demonstrated in [35-37]. The use of surfels to define the exact wall location results in more accurate fluxes near boundaries, and increased accuracy at lower numerical resolution compared with operating directly on the voxel grid of the micro-CT scan. Usage of surfels also improves the accuracy of local curvature and curvature-based pressure calculation as described in section 2.3.3.

2.3 Numerical model setup

2.3.1 Capillary pressure displacement

A capillary pressure driven primary drainage (oil displacing water) simulation is performed in the selected rock domain to establish a fluid phase (oil/water) distribution for subsequent

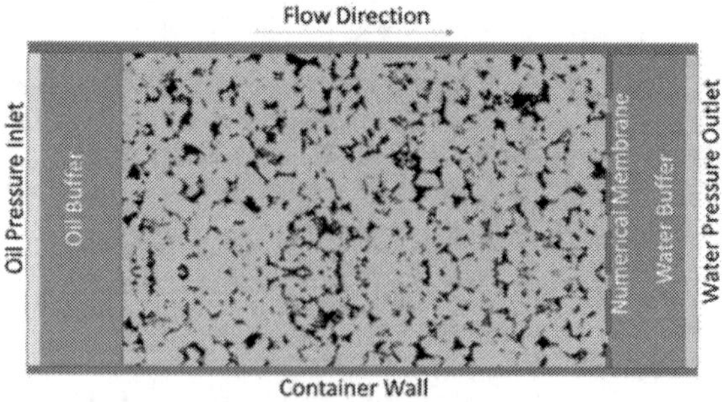

Fig. 3. Schematic of a capillary pressure drainage simulation setup mimicking a porous-plate method.

water flooding simulation. Initially, the simulated domain pore space is fully saturated with water and a uniformly distributed strong water-wet wetting condition is applied to all pore-grain boundary surfaces. In all previous Digital Rock multiphase flow studies, the initial wettability condition is maintained constant along the simulation. In the current study, the wetting condition is altered during this drainage simulation using a process described in section 2.3.5 and compared with a traditional no-alteration scenario.

As shown in Fig. 3, the rock sample is placed inside a container box. The container has solid walls in the x and y direction, while in z direction (flow direction), oil and water buffers are placed in the opposite sides of the rock domain to control the capillary pressure. In addition, a numerical membrane acting as a porous plate is placed in between the rock domain and water buffer. This numerical membrane behaves as a volume-free and massless barrier that prevents oil from entering into the water buffer. Water on the other hand, can flow through the membrane without any resistance. Thus, the usage of this numerical membrane serves as a capillary pressure barrier for the non-wetting phase, similar to the porous plate in capillary pressure lab analysis [38]. It is worth mentioning that this numerical membrane is a unique feature of the current LBM implementation we use, and that can practically replace the modeling of a finely resolved porous plate which can be computationally expensive and can potentially raise numerical issues.

During primary drainage, the pressure at the oil buffer is gradually ramped up so that oil starts to invade the pore space filled with water. At each pressure level the simulation continues until the saturation change is within a preset convergence value and a static equilibrium is achieved.

2.3.2 Relative permeability displacement

Using the initial fluid distribution, obtained from the previous capillary pressure drainage simulation, a multiphase unsteady-state k_r relative permeability fluid flow simulation is carried out. In this unsteady-state k_r method, a periodic boundary conditions and a driving force are applied in the flow direction, while no-flow boundary conditions are used in the directions perpendicular to flow. Further details of this procedure can be found in [20].

2.3.3 Curvature calculation

The pore-grain boundary, which consists of triangular surface elements, is used for both mean curvature calculation and to define wall boundary conditions for fluid flow. As described in [39], from a theoretical point of view, triangular meshes do not have any curvature at all, since all faces are flat and the curvature is not properly defined along edges and at vertices because the surface is not C^2 differentiable. One possible alternative is to think of a triangular mesh as a piecewise linear approximation of an unknown smooth surface and calculate mean curvature at the vertex. In the current study, we use an open-source, freely available tool "Visualization Toolkit" (VTK) [40] to calculate the mean curvature at every vertex:

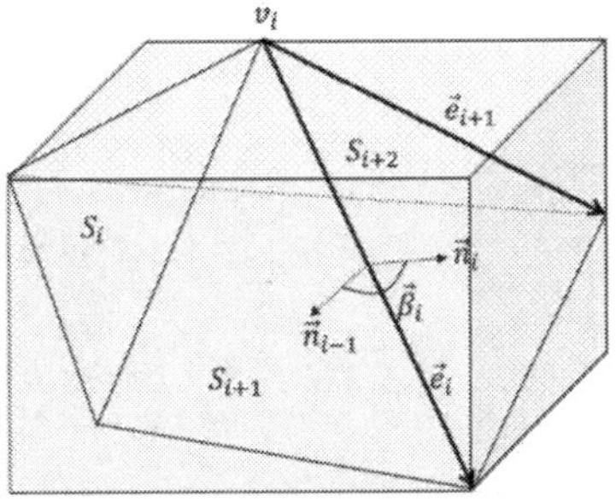

Fig. 4. Curvature calculation: cubic lattice voxel with more than one triangular surface elements inside.

$$|\bar{H}| = \int_S |H| = \frac{1}{4}\sum_{i=1}^{n} \|\vec{e}_i\| \, |\beta_i| \qquad (1)$$

where, $|H|$, $\vec{e}_i$ and β_i represent absolute mean curvature, edge shared by two triangular surface elements and dihedral angle between the normals of the adjacent triangles at an edge $\vec{e}_i$, respectively, as shown in Fig. 4 . This integral value is then normalized by the surrounding area of a vertex to calculate the curvature at the vertex.

In the current workflow, since we need to calculate the curvature for every voxel in the computational domain, the curvature at the triangular surface element is calculated by:

$$|H|_{S_i} = \frac{1}{3}[|H|_{v_i} + |H|_{v_{i+1}} + |H|_{v_{i-1}}] \qquad (2)$$

where, S_i and v_i represent the triangular surface element and one of the vertex in that element respectively. Since the surface elements do not cross over from one cubic lattice voxel to another, curvature for every voxel can be calculated by area weighted averaging:

$$|H|_{voxel} = \frac{1}{\sum_{i=1}^{n} A_{S_i}}\left[|H|_{S_i}A_{S_i} + |H|_{S_{i+1}}A_{S_{i+1}} + \cdots + |H|_{S_n}A_{S_n}\right] \qquad (3)$$

where, A_{S_i} *and* $|H|_{S_i}$ represent area and curvature of a triangular surface element while n represents number of elements in a given voxel.

2.3.4 Force balance

The basic mechanism that drives wettability alteration is related to achieving a local force balance in the fluid-fluid-solid interfaces. It involves three forces:

1) Capillary pressure(P_c), i.e. the pressure difference between non-wetting (oil) and wetting (water) fluids. Since the capillary pressure measurement and subsequent wettability alteration takes place when both of the fluid phases achieve static equilibrium, the difference between inlet oil pressure and outlet water pressure can also be considered as the capillary pressure of the system.

 https://doi.org/10.1051/e3sconf/202336701005

$$P_c = P_{inlet} - P_{outlet} \quad (4)$$

2) Curvature based pressure (P_v) is the pressure caused by solid geometry forcing the shape of the oil/water interface into the shape of the pore geometry:

$$P_v = \sigma |H|_{voxel} \quad (5)$$

where, σ is the interfacial tension coefficient.

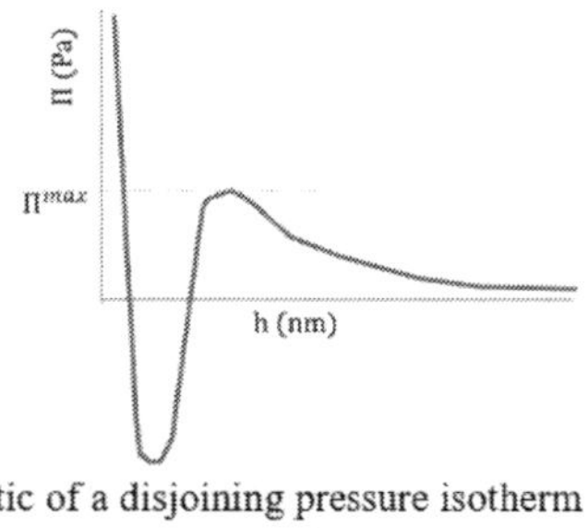

Fig. 5 Schematic of a disjoining pressure isotherm

3) Disjoining pressure (Π) is the pressure that prevents a thin water film from rupturing. As explained in [17], when oil invades a water-wet pore space, a thin water film is developed which coats and adheres to the solid surface. Disjoining pressure is a function of film thickness, h as shown in the schematic in Fig. 5. Positive disjoining pressure can be seen as the repulsive stabilizing force that keeps the oil away from the water-wet interface. Depending on the film thickness, either one of the following two repulsive forces can contribute to positive disjoining pressure: The first one is the electrostatic force that originates from the electrical double layers present at the solid/water and water/oil interfaces. The second one is a strong hydration force that plays dominant role when film thickness approaches molecular dimensions. On the other hand, dispersive van der Walls forces, are usually attractive and destabilizes thin water films contributing to the negative disjoining pressure.

The shape of the disjoining pressure isotherm, Fig. 5, suggests the existence of a local maximum disjoining pressure. When the applied capillary pressure exceeds the sum of the local curvature-based pressure and maximum disjoining pressure, the local thin water film can be considered ruptured making the adjacent grain surface a candidate for wettability alteration:

$$P_c > P_v + \Pi^{\max} \quad (6)$$

It should be emphasized that Eq. (6) i.e. the augmented Young-Laplace equation, considers both disjoining pressure and local pore shape for wettability alteration [5]. To illustrate the local pore shape dependency, consider a non-wetting fluid flowing through two separate micro channels, one with circular cross-section and other with star shaped cross-section. If they have the same bottleneck radius, as per Eq. (6), the thin film in a star shaped pore is more likely to be ruptured, even though the disjoining pressure and applied capillary pressure in star shaped pore is the same as the circular pore. Also, disjoining pressure is a necessary ingedient to explain capillary equilibrium for saturation high enough where only a smaller fraction of interfaces are really "free" interfaces but the majority of the oil-water interface follows the shape of the solid. In some instances the interfaces can even be flat, i.e. zero curvature. In order to sustain a force balance (Eq. 6), the phase pressure difference has to be taken up by the disjoining pressure.

2.3.5 Wettability alteration

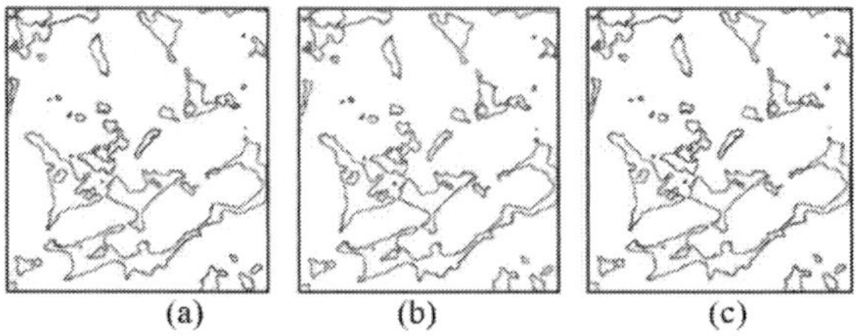

Fig. 6. Surface selection (red) for wettability alteration: (a) selection is adjacent to a non-wetting phase, (b) selection satisfies $P_c > P_v + \Pi^{max}$ condition and (c) selection is subset of (a) and (b).

The wettability alteration workflow starts by creating a digital representation of the pore-grain surface and setting up the scenario for capillary pressure drainage simulation as described earlier. Afterwards the following steps are implemented to change the local wettability within a multiphase fluid flow simulation:

1) The simulation is continued until static equilibrium is achieved for a certain capillary pressure(P_c).
2) The surface elements in contact with the non-wetting phase are identified as shown by the red lines in Fig. 6(a)
3) The curvature-based pressure, (P_v) is also calculated using Eq. (5) at every voxel containing the pore-grain surface boundary as described in section 2.3.3 and 2.3.4.
4) Then Eq. (6) is used to determine whether a local water-film breakage may occur as represented by the red lines in Fig 6(b).
5) Afterwards, a subset of surface elements from step 2) and 4) is identified as the candidate location for wettability alteration as shown by the red lines in Fig 6(c).
6) The wettability alteration can be a slow, dynamic process that occurs over many small discrete time intervals. In order to facilitate further change in fluid saturation, the simulation is continued with the updated wetting condition maintaining same capillary pressure until static equilibrium is achieved.
7) Afterwards the pressure is ramped up to the next value and the above steps are repeated again.

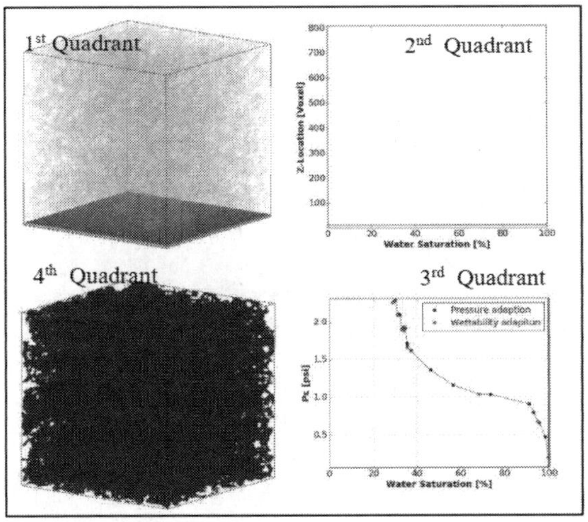

(a)

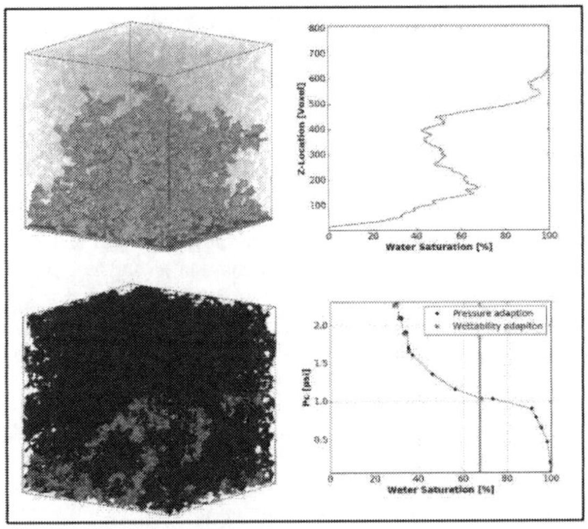

(b)

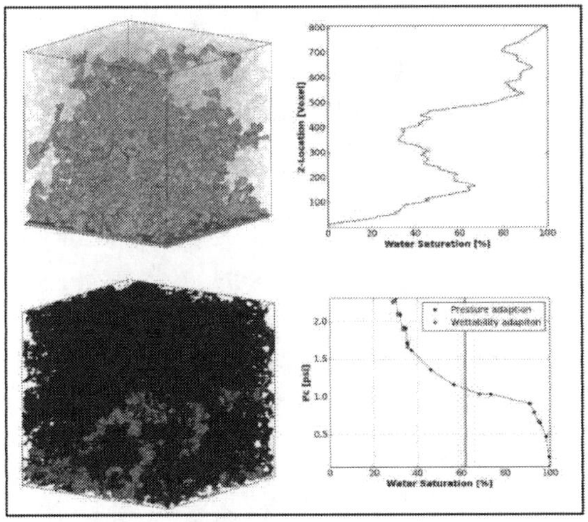

(c)

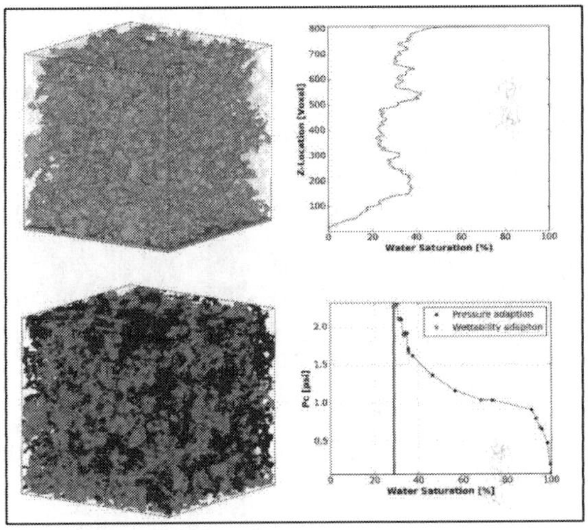

(d)

Fig. 7. Various stages of wettability alteration simulation. The 1st quadrant in each figure shows an iso surfaces of oil (beige) and a volume rendering of water (blue), 2nd quadrant shows the water saturation profile along the flow direction, 3rd quadrant shows capillary pressure versus water saturation chart and 4th quadrant shows oil wet (red) and water wet (blue) surface within the pore space in 3D.

Thus, it is ensured that the fluid saturation is converged not only for certain applied pressure but also due to the corresponding change in wetting condition. It should be noted that the above procedure takes disjoining pressure for different minerals and fluid pair as an input. In the current study, we assumed a positive, repulsive disjoining pressure of 5000 Pa. As a next step, the workflow can be extended to calculate disjoining pressure using Molecular Dynamics simulation or from the experiments.

3 Results

3.1 Wettability alteration in a Berea sandstone

Fig. 7 shows capillary pressure drainage simulation results for an oil/water system in a strongly water-wet Berea sandstone initially filled with water. The black and red measurement points in Fig. 7, 3rd quadrants represent equilibrium states after pressure and wettability adaption respectively. At the beginning of the simulation, the water saturation changes are small in response to the increase in oil pressure. But as the pressure is ramped up enough to overcome the capillary pressure imposed by the local pore throat radii, oil rapidly invades the pore space as can be seen in between 90% and 70% of the water saturation, which is a phenomena called *"Haines Jump"* [41]. As reported in [42], the Reynolds number calculated during Haines Jump are estimated to be in the order of one, indicating laminar flow, but the inertial

https://doi.org/10.1051/e3sconf/202336701005

forces are still relevant. The current simulation uses a Lattice Boltzmann formulation that represents the full transient Navier-Stokes equation which is essential to capture this phenomenon [18]. However, not all the pore space is filled during Haines jump, especially during a slow convergence to static equilibrium. Even though Haines Jump process is far from quasi-static, during the subsequent slow convergence stage the phase pressure difference is not significantly large enough for the non-wetting phase to rapidly invade and the flow of oil can be described as quasi static and capillary dominant. In such a scenario, oil droplets can be snapped off when passing through a narrow pore-throat, similar to *"Roof snap off"* phenomena as described in [43,44] which can also be seen in current simulation as illustrated in Fig 8.

As can be seen in Fig. 7(c), oil breakthrough occurs for this rock sample at around 60% of water saturation, when the numerical porous plate stops further advancement of oil into the water buffer. Thus, it allows the oil pressure to build up, so that the oil non-wetting phase can intrude smaller pores too. As a result, having porous plate also helps achieving higher pressure-saturation points as shown in Fig. 7(d).

(a) (b) (c)

Fig. 8. Oil blob "Roof" snap off during drainage simulation.

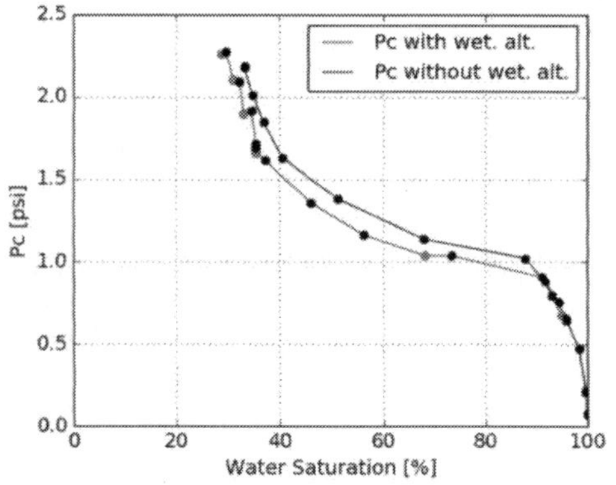

Fig. 9. Capillary pressure curves with (green) and without (blue) wettability alteration.

For this simulation we carry out the wettability alteration after every four pressure steps by following the methodology described in section 2.3.5. In Fig 7, the 4th quadrants show the water wet (blue) and oil wet (red) surfaces at different stages of the simulation. Only a small fraction of the surface area in contact with the oil non-wetting phase is subjected to alteration at low capillary pressures, because the applied capillary pressure is not large enough to overcome the repulsive disjoining pressure for thin film rupture. In contrast, alteration at higher capillary pressures creates larger fraction of oil-wet surface. By the end of simulation, almost 70% of

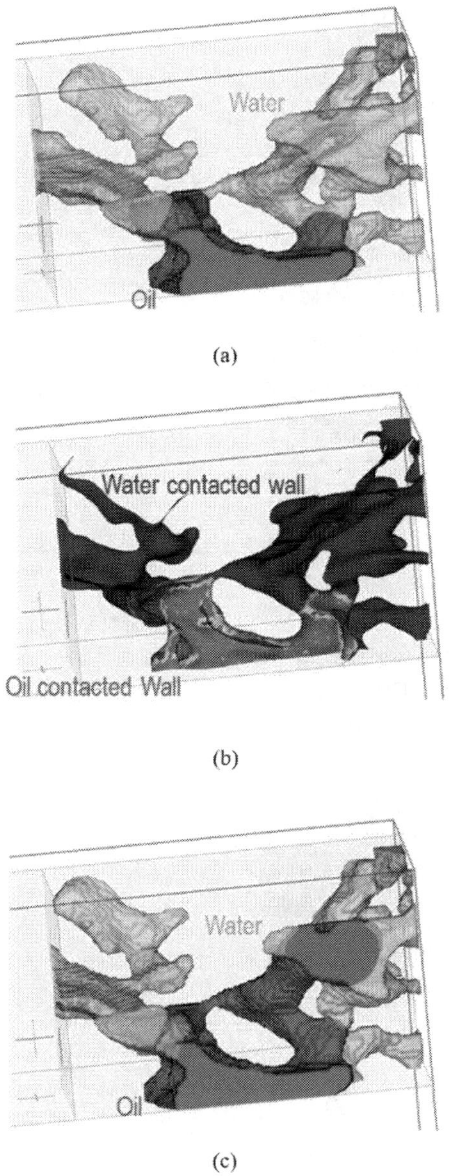

Fig. 10. A pore scale event showing the slow dynamics of wettability alteration: (a) flow of oil is blocked by a narrow pore-throat, (b) local surface wettability alteration, and (c) updated fluid distribution in response to the wettability alteration.

the pore-grain surface is altered from strongly water-wet to oil-wet condition.

Fig. 9 shows a comparison between two capillary pressure curves with and without wettability alteration. Drainage with wettability alteration exhibits smaller capillary entry pressure. Fig. 9 also confirms that the inclusion of the wettability alteration mechanism in the multiphase flow simulation leads to an overall lower water saturation in the drainage displacement. It can be hypothesized that for a given topology, the degree of observed difference between with and without alteration depends on the rock mineralogy. Different mineral and fluid pair can results in much higher or lower or even negative disjoining pressure which can potentially make the wettability alteration effect more pronounced.

This phenomenon can be explained by the slow dynamic impact of wettability alteration that allows non-wetting fluid to intrude into very small pores, as the wettability can be changed in such a way that capillary pressure does not need to be overcome. In other words, the iterative process between wettability alteration, the consequential change of capillary pressure and establishing a new capillary equilibrium from the updated capillary pressure allows the contact line to move into smaller pores than without wettability alteration. Fig. 10 illustrates an example of a local pore scale displacement where the flow of oil is blocked by a narrow water-wet pore-throat at certain applied capillary pressure. Next, the mechanism of wettability alteration of the pore-grain surface is carried out and certain portion of the surface in contact with oil is altered to more oil-wet condition, as shown in Fig 10(b). As the simulation is resumed with the updated wetting condition, but still at the same applied capillary pressure, the new oil wet surface helps oil to push through the small pore-throat and displace more water as shown in Fig 10(c). Wettability alteration shown in 10(b) is an outcome of procedure described in 2.3.5 that implements a steady-state solution of the augmented Young-Laplace equation. Once the wettability of the candidate surface is altered, the transient fluid flow simulation is resumed and the slow dynamic impact of alteration such as in Fig 10(c) is simulated in time domain.

From this study, it is clear that wettability alteration has an impact on the overall displacment, but the alteration has been modeled as a discrete event happening every after certain pressure interval. We plan to investigate further the impact of discrete pressure interval of alteration as a next step.

3.2 Relative permeability analysis

The initial fluid and wettability distributions obtained from the capillary pressure drainage simulation are used as inputs for a subsequent relative permeability simulation where water displaces oil (imbibition). Fig. 11 shows the relative permeability simulation results in the Berea Sandstone model, for both altered and un-altered wettability conditions. Wettability alteration creates a mixed wet condition that allows oil to penetrate deeper into the pore space. As a result, the initial water saturation is lower than the un-altered wettability case (water-wet). For the mixed wet case, a thin oil film develops onto the grains where wettability alteration took place and the flow of water during water-flooding kr simulation may not be in direct contact with the grain wall. As a result, water relative permeability is higher for the altered wettability than the unaltered case where there is no oil film. The film also creates persistent oil connectivity even at higher water saturation which leads to more oil recovery for the alteration induced mixed wet case.

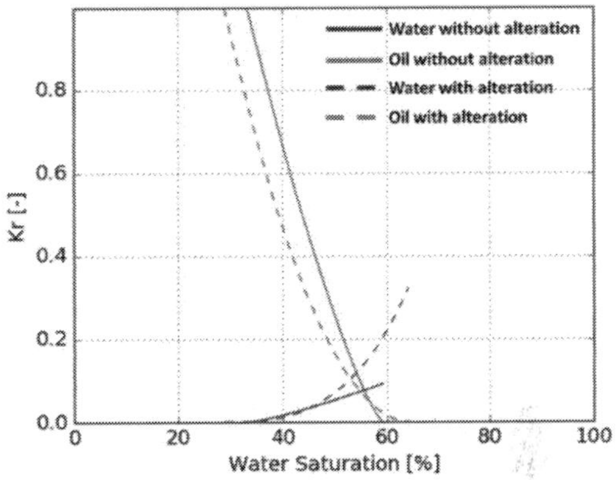

Fig. 11. Oil/water relative permeability curves with and without wettability alteration.

4 Conclusion

A numerical wettability alteration methodology based on a provided disjoining pressure isotherm has been established and implemented using a multiphase Lattice Boltzmann Method. In particular –

1. An iterative process for wettability alteration has been developed based on force balance between curvature-based pressure, applied pressures and disjoining pressure.
2. This work also offers a secondary oil displacement mechanism based on slow dynamic wettability alteration process. This theory can be helpful to understand how oil can get inside into very fine pore space even though capillary pressure is not sufficient.
3. The effect of wettability alteration on overall oil production has also been illustrated by performing a subsequent relative permeability simulation.

In the future, this work can be extended to encompass molecular models of thin film physics, mineralogy and fluid pairs. Using Molecular Dynamics simulation, it is feasible to calculate interfacial tension and wettability on first principles basis as demonstrated in [45-47]. Current Lattice Boltzmann workflow to calculate spatial wettability distribution can be integrated with a Molecular Dynamics workflow to obtain interfacial tension, contact angle and disjoining pressure isotherm inputs.

References

1. J. E. McClure, R.T. Armstrong, M. A. Berril, S. Schlüter, S. Berg, "A geometric state function for two-fluid flow in porous media," *Phys Rev Fluids* **3,** 084306 (2018)

2. W.G. Anderson, "Wettability literature survey-part4: Effects of wettability on capillary pressure," *J Pet Technol* **39,** 1283 (1987)

3. W.G. Anderson, "Wettability literature survey-part 5: The effects of wettability on relative permeability," *J Pet Technol* **39,** 1453 (1987)

4. Q. Lin, B. Bijeljic, S. Berg, R. Pini, M.J. Blunt, S. Krevor, "Minimal surfaces in porous media: Pore-scale imaging of multiphase flow in an altered wettability Bentheimer sandstone", *Phys. Rev. E*, **99,** 6 (2019).

5. M. Rücker , W.-B. Bartels , G. Garfi, M. Shams, T. Bultreys, M. Boone, S. Pieterse, G.C. Maitland, S. Krevor, V. Cnudde, H. Mahani, S. Berg, A. Georgiadis, P.F. Luckham, "Relationship between wetting and capillary pressure in a crude oil/brine/ rock system: From nano-scale to core-scale", *J Colloid Interface Sci.,* **562,** 159(2020)

6. M. Rücker, W.-B. Bartels, K. Singh, N. Brussee, A. Coorn, H. A. van der Linde, A. Bonnin, H. Ott, S. M. Hassanizadeh, M. J. Blunt, H. Mahani, A. Georgiadis, S. Berg, " The effect of mixed wettability on pore-scale flow regimes based on a flooding experiment in Ketton limestone," *Geophys. Res. Lett.* **46,** 3225-3234 (2019).

7. N.R. Morrow, "Wettability and its effect on oil recovery," J Pet Technol **42,** 1476 (1990)

8. G. Garfi, C.M. John, M. Rücker, Q. Lin, C. Spurin, S. Berg, S. Krevor, "Determination of the spatial distribution of wetting in the pore networks of rocks", *J Colloid Interface Sci.,* **613,** 786 (2022).

9. G. Garfi, C.M. John, Q. Lin, S. Berg, S. Krevor, "Fluid surface coverage showing the controls of rock mineralogy on the wetting state," *Geophys. Res. Lett.* **47** (8) (2020).

10. S. Foroughi, B. Bijeljic, Q. Lin, A.Q. Raeini, M.J. Blunt, "Pore-by-pore modeling, analysis, and prediction of two-phase flow in mixed-wet rocks," *Phys. Rev. E* **102,** 02330 (2020).

11. Q. Lin, B. Bijeljic, S. Berg, R. Pini, M.J. Blunt, S. Krevor, "Minimal surfaces in porous media: Pore-scale imaging of multiphase flow in an altered wettability Bentheimer sandstone", *Phys. Rev. E*, **99,** 6 (2019).

12. A. Scanziani, K. Singh, M.J. Blunt, A. Guadagnini, "Automatic method for estimation of in situ effective contact angle from X-ray micro tomography images of two-phase flow in porous media", *J. Colloid Interface Sci.* **496,** 51 (2017).

13. W.-B. Bartels, M. Rücker, M. Boone, T. Bultreys, H. Mahani, S. Berg, S.M. Hassanizadeh, V. Cnudde, "Imaging spontaneous imbibition in full Darcy-scale samples at pore-scale resolution by fast X-ray tomography", *Water Resour. Res.,* **55** 7072 (2019).

14. A.M. Alhammadi, A. AlRatrout, K. Singh, B. Bijeljic, M.J. Blunt, "In situ characterization of mixed-wettability in a reservoir rock at subsurface conditions", *Sci. Rep.,* **7,** 10753 (2017).

15. M. Andrew, B. Bijeljic, M.J. Blunt, "Pore-scale contact angle measurements at reservoir conditions using X-ray microtomography", *Adv. Water Resour.*, **68,** 24 (2014).

16. M. Prodanović, W.B. Lindquist, R.S. Seright, "Residual fluid blobs and contact angle measurements from X-ray images of fluid displacement," *XVI International Conference on Computational Methods in Water Resources,* (2006).

17. C.J. Radke, A.R. Kovscek, H. Wong, "A pore-level scenario for the development of mixed wettability in oil reservoirs," SPE 24880, *SPE Ann. Tech Conf. and Exhib.,* Washington, D.C. USA, Oct, 1992.

18. B.V. Derjaguin, E.V. Obukhov, "Anomalous properties of thin polymolecular films V.," Acta Physicochim. URSS, **10**(1), 25-44 (1939a).

19. B.V. Derjaguin, E.V. Obukhov, "Anomalous properties of thin polymolecular films V.," Acta Physicochim. URSS, **10**(2), 153-174 (1939b).

20. G.R. Jerauld, J. Fredrich, N. Lane, Q. Sheng, B. Crouse, D.M. Freed, A. Fager, and R. Xu, "Validation of a workflow for digitally measuring relative permeability," SPE 188688, *SPE Abu Dhabi Int. Pet. Exhib. & Conf.,* Abu Dhabi, U.A.E., Nov., 2017.

21. M. A. Al Ibrahim, N. F. Hurley, W. Zhao, D. Acero-Allard, "An automated petrographic image analysis system: capillary pressure curves using confocal microscopy, " SPE 159180-MS, *SPE Annual Technical Conference and Exhibition,* San Antonio, Texas, USA, Oct, 2012.

22. S. Chen and G. Doolen, "Lattice Boltzmann method for fluid flows," *Annu. Rev. Fluid Mech.*, 30, 329 (1998).

23. S. Chen, H. Chen, D. Martinez, and W. Mattheus,"Lattice Boltzmann model for simulation of magnetohydrodynamics," *Phys. Rev. Lett.*, **67**, 3776 (1991).

24. H. Chen, S. Chen, W. Matthaeus, "Recovery of the Navier - Stokes equations using a lattice-gas Boltzmann method," *Phys. Rev. A*, **45**, R5339 (1992).

25. Y. Qian, D. d'Humieres, P. Lallemand, "Lattice BGK models for Navier-Stokes equation," *Europhys. Lett.*, **17**, 479 (1992)

26. H. Chen, "Volumetric formulation of the lattice Boltzmann method for fluid dynamics: Basic concept," *Phys. Rev. E*, **58**, 3955(1998)

27. H. Otomo, H. Fan, R. Hazlett, Y. Li, I. Staroselsky, R. Zhang, R., and H. Chen, H., "Simulation of residual oil displacement in a sinusoidal channel with the Lattice Boltzmann method," *Comptes Rendus Mécanique*, **343**, 559 (2015).

28. H. Andrä, N. Combaret, J. Dvorkin, E. Glatt, J. Han, M. Kabel, Y. Keehm, F. Krzikalla, M. Lee, C. Madonna, M. Marsh, T. Mukerji, E. H. Saenger, R. Sain, N. Saxena, S. Ricker, A. Wiegmann, X. Zhan, "Digital rock physics benchmarks—part II: Computing effective properties," *Comp. Geosci.*, **50,** 33 (2013)

29. J. Wu, J.J. Huang, "Dynamic behaviors of liquid droplets on a gas diffusion layer surface: Hybrid lattice Boltzmann investigation," *J. App. Phys.*, **118,** 044902 (2015)

30. M.A. Safi, N.I. Prasianakis, J. Mantzaras, A. Lamibrac, F.N. Buchi, "Experimental and pore-level numerical investigation of water evaporation in gas diffusion layers of polymer electrolyte fuel cells," *Int J Heat and Mass Trans.*, **115**, 238 (2017).

31. L. Martin, P. Benedikt, K. Benjamin, W. Julius, D. Timo, S. Volker, L. Arnulf, "Understanding electrolyte filling of Lithium-ion battery electrodes on the pore scale using the Lattice Boltzmann Method," *Batteries and Supercaps*, **e202200090**, (to be published)

32. D.H. Jeon, "Wettability in electrodes and its impact on the performance of Lithium-ion batteries," *Energy Storage Materials*, **18**, 139 (2019)

33. X. Shan and H. Chen, "Lattice Boltzmann model for simulating flows with multiple phases and components," *Phys Rev E*, **47**, 1815 (1993).

34. H. Otomo, H. Fan, Y. Li, M. Dressler, I. Staroselsky, R. Zhang, and H. Chen, "Studies of accurate multicomponent Lattice Boltzmann models on benchmark cases required for engineering applications," *Jour. Comp. Sci.*, **17**, 334 (2016).

35. H. Otomo, B. Crouse, M. Dressler, D.M. Freed, I. Staroselsky, R. Zhang, H. Chen, "Multi-component Lattice Boltzmann models for accurate simulation of flows with wide viscosity variation," *Computer Fluids*, **172,** 674 (2018)

36. H. Chen, C. Teixeira, and K. Molvig, "Realization of fluid boundary conditions via discrete Boltzmann dynamics," *Int. J. Mod. Phys. C*, **9**, 1281 (1998).

37. B. Crouse, D. M. Freed, N. Koliha, G. Balasubramanian, R. Satti, D. Bale, and S. Zuklic, "A Lattice-Boltzmann based method applied to Digital Rock characterization of perforation tunnel damage,". SCA 2016-058, *Int. Symp. Soc. Core Analysts*, Snow Mass, CO, USA., (2016).

38. C. McPhee, J. Reed, I. Zubizarreta, *Core Analysis: A best Practice Guide*, **64,** 449 (2015).

39. N. Dyn, K. Hormann, S-J. Kim, D. Levin, "Optimizing 3D triangulations using discrete curvature analysis," *Mathematical methods for curves and surfaces*, **1,** 135 (2001)

40. *The VTK User's Guide*, (11th Edition, Kitware Inc., 2010)

41. W.B. Haines, "Studies in the physical properties of soil : the hysteresis effect in capillary properties and the modes of moisture distribution." *J Agric Sci*, **20,** 97 (1930)

42. R. T. Armstrong, N. Evseev, D. Koroteev, S. Berg, "Modeling the velocity field during Haines jumps in porous media," *Adv Water Resour,* **77**, 57-68, (2015).

43. J.G. Roof, "Snap-Off of oil droplets in water-wet pores," *SPE J.,* 10, 85 (1970)

44. R. T. Armstrong, S. Berg, O. Dinariev, N. Evseev, D. Klemin, D. Koroteev, S. Safonov, "Modeling of pore-scale two-phase phenomena using density functional hydrodynamics,"

45. V. Khosravi, S.M. Mahmood, D. Zivar, H. Sharifigaliuk, "Investigating the applicability of molecular dynamics simulation for estimating the wettability of sandstone hydrocarbon formations", *ACS Omega*, **5**, 22852 (2020).

46. M.P. Anderson, M.V. Bennetzen, A. Klamt, S.L.S Stipp, "First principles prediction of liquid/liquid interfacial tension," *J. Chem. Theory Comput.,* **10,** 3401 (2014).

47. M.P. Anderson, T. Hassenkam, J. Matthiesen, L.V. Nikolajsen, D.V. Okhrimenko, S. Dobberschutz, S.L.S Stipp, "First-principles prediction of surface wetting," *Langmuir,***36,** 12451 (2020).

 https://doi.org/10.1051/e3sconf/202336701006

Fast wettability assessment on small rock samples using a 3D, high-resolution, image-based Amott-like test

Maria Repina*, *Regis* Brugidou, *Alexandre* Dufour and *Richard* Rivenq

TotalEnergies OneTech, CSTJF 64000 Pau, France

Abstract. In this paper, we propose a new experimental set-up to carry out a full Amott-like cycle on small core samples (4 mm. diameter, 15 mm. length), imaged with a high-resolution micro-CT scanner. We conduct this experiment on two core samples from a carbonate reservoir. Each step of the cycle was followed by a 3D high-resolution scan along the entire sample length and analysed using digital image-processing algorithms. The characterization of wettability is based on multiple criteria: a) a pseudo-wettability index, analogous to the Amott-Harvey index, calculated from both resolved and unresolved pores; b) experimental observations during the spontaneous phases, and c) analysis of the spatial fluid distribution in the pore space. This new type of experiment has several advantages: a) it respects the experimental sequence of the industry standard Amott test, yet with a significantly reduced test duration; b) it provides complementary information describing the spatial distribution of the fluids in the visible pore network during spontaneous and forced phases; c) it is less sensitive to imaging resolution than existing approaches based on topological characteristics and contact angles.

1 Introduction

Wettability is a key factor controlling multi-phase flow in porous media. The standard and well-established laboratory procedures used to characterize the wettability of an oil/brine/rock system in the O&G industry are the Amott and USBM tests. They provide wettability indexes based on a combination of spontaneous and forced displacement properties of the oil/brine/rock system. They are very valuable for petrophysicists and reservoir engineers but take long to perform, up to several months, often even more than a year, and thus often come too late during reservoir pre-development evaluations.

In the last few years, there has been a growing interest in the assessment of wettability within the pore space, often associated with estimates of contact angles, interfacial curvature, or topological functions. These estimates have the advantage of considering millions of resolved pores. However, most of the calculated image-based wettability estimates often suffer from the lack of image resolution. Additionally, they might be influenced by the pore structure complexity.

In this paper, we explore the possibility of acquiring a qualitative wettability indicator, similar to that obtained using the conventional wettability characterization via Amott tests on centimeter-sized samples, using instead millimeter-sized samples imaged with a high-resolution micro-CT scanner. Following the idea presented in [6], also reported in [7] as a means of rapidly characterizing the wettability input required for Digital Rock Physics simulation, we used a similar experimental set-up to realize a full Amott-like cycle on small core samples (4mm diameter, 15mm long) with fluid saturations determined from micro-CT image analysis. It is also expected that pore scale imaging, at least in the part of pore network that can be fully resolved, will provide additional information on the wetting properties of the system. This approach was tested on two carbonate samples from the same reservoir. The duration of the proposed wettability assessment test may vary, in fact it depends strongly on the length of the ageing phase. In the presented examples, the wettability assessment was finished in nine weeks (wettability restauration, four weeks ageing and image analysis included).

2 Materials and methods

2.1 Rock samples and fluid properties

Two cylindrical samples of a reservoir carbonate rock were used in this study, their mineralogical composition is shown in Fig. 1.

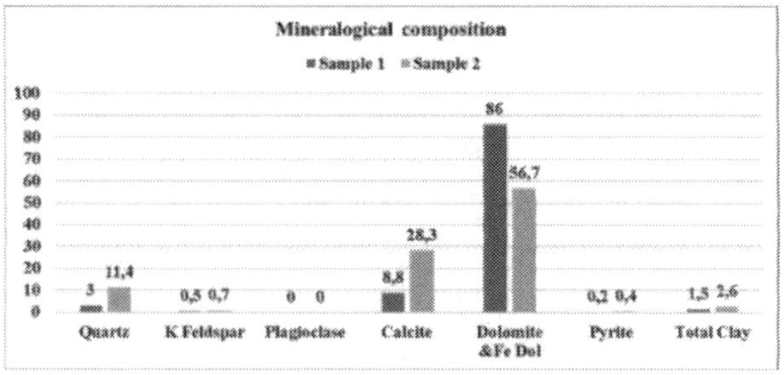

Fig. 1. Mini core samples mineralogical composition

* Corresponding author: repina.maria@totalenergies.com

Their main geometrical characteristics are provided in Table 1. Given the high degree of heterogeneity of the parent samples, the significant difference in the porosity values between the centimeter-sized and the milimeter-sized plugs is expected.

Table 1. Samples characteristics.

Sample	Length, mm	Diameter, mm	Porosity of the big parent plug, %	Permea-bility of the big parent plug, mD
1	13.95	3.91	20	20
2	14.50	3.90	8	16

The oil phase used was Marcol 52 (M52). The brine used for drainage and imbibition stages is a formation water dopped with KI, its composition is presented in Table 2.

Table 2. Brine composition.

Sel	C (g/L)
KI	100.0
NaCl	81.1
$CaCl_2$, 2H2O	14.5
$MgCl_2$, 6H2O	3.9
KCl	3.7
Na_2SO_4	1.5

For the characterization of the pore space with the Differential Imaging technique [3, 5], potassium iodide concentration was multiplied by three, following the example of [5] .

The dead oil used for the ageing comes from the field from which the carbonate reservoir samples were taken. This oil was centrifuged before being received; it was not filtered before the injection into the core.

2.2 Experimental program

The experimental program presented below, was applied to two samples.

Table 3 summarizes experimental temperature and pressure parameters.

Table 3. Experimental parameters.

Ageing temperature (°C)	95
Temperature during spontaneous and forced phases (°C)	28
Pore pressure during imaging (bar)	11
Confining pressure during imaging (bar)	50
Maximum confining pressure during forced phases (bar)	120
Maximum pore pressure during forced phases (bar)	100

2.2.1 Sample cleaning (carried out separately for both samples)

1. Sample was put in the Viton sleeve before being placed in a carbon flow cell
2. Confining pressure of 25 bars was applied and maintained within the cell to compress the Viton sleeve around the core sample to avoid fluid bypass. Temperature was raised to 65°C
3. 30 PV of 30 g/L NaCl/$CaCl_2$ (mass 90 %/10 %) brine was injected to wash out salts present in the sample
4. 95PV of toluene was injected at 5 µL/min
5. It was followed by the cyclic injection of toluene and mix of toluene and isopropanol in both directions. The cycles were composed of two phases: 30 PV at 5 µL/min toluene or toluene + isopropanol injection followed by maceration lasting between 2 and 8 hours. During the maceration period, the pore pressure was kept at 5 bars
6. Cleaning was finished with the injection of 78 PV of isopropanol at 10 µL/min
7. Sample was blown with the N_2
8. Temperature was lowered from 65°C to 28°C

2.2.2 Experimental protocol for porosity, saturation, and wettability characterization

For each sample, the experiment was performed as follows:

1. Core holder with the vertical flowing cell was installed into the micro-CT scanner
2. Four continuous tomograms "Dry scans" were taken for the whole sample with a voxel size 2.07 µm. The overlap of approximately 30% was used to stitch the four scans together
3. Vacuum was installed.
4. 25 PV of 300 g/L KI brine was injected into the rock sample to ensure that it was fully saturated with the highly contrasted brine. This step was necessary to characterize the sub-resolution pore space by applying the Differential Imaging method [3]. For more details, refer to the next section
5. Four continuous "KI_300" scans were taken for the whole sample with a voxel size 2.07 µm. Acquisition parameters were kept the same as for the dry scans.
6. High salinity brine was replaced by 48 PV of KI 100g/L brine. During the replacement brine injection, pressure was maintained close to 10 bars.
7. The primary drainage was effectuated by injecting 715 PV of the mineral oil M52 at 250µL/min. The direction was changed four times starting in a downwards direction. During the primary drainage, the confinement pressure was raised to 120 bars. SWi was fixed at 20-25%, with the most uniform SWi saturation profile. Fig. 2 shows the pressure recording during the primary drainage. The recovery pressure stayed equal at 11 bars.

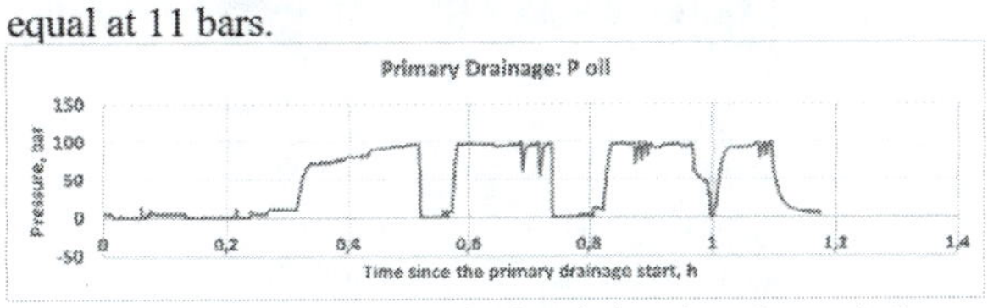

Fig. 2. Sample 1: Pressure monitoring during the primary drainage

8. Confinement pressure was decreased to 50 bars. The sample was moved to the oven.

9. 6 PV of toluene was injected to serve as a buffer between M52 and the dead oil.
10. M52 was replaced by the dead oil.
11. Four weeks ageing began. The temperature of the oven was kept at 95°C, pore pressure - at 6 bars. At the beginning of the second week, 6 PV of dead oil was injected at 0.6 μL/min in a downwards direction. One week after the direction of oil injection was changed, 700 μL of dead oil was injected at 0.6 μL/min. At the beginning of the last week of ageing, 6 PV of dead oil was injected at 0.6 μL/min one more time.
12. The temperature was progressively decreased to 65°C.
13. 6 PV of toluene was injected to serve as a buffer between dead oil and M52.
14. Dead oil was replaced by the mineral oil M52.
15. Four continuous "SWi_aged" scans were taken for the whole sample with a voxel size 2.07 μm. Acquisition parameters were kept the same as for the dry scans.
16. Flow cell was connected to the imbibition experimental set-up shown in Fig. 3.

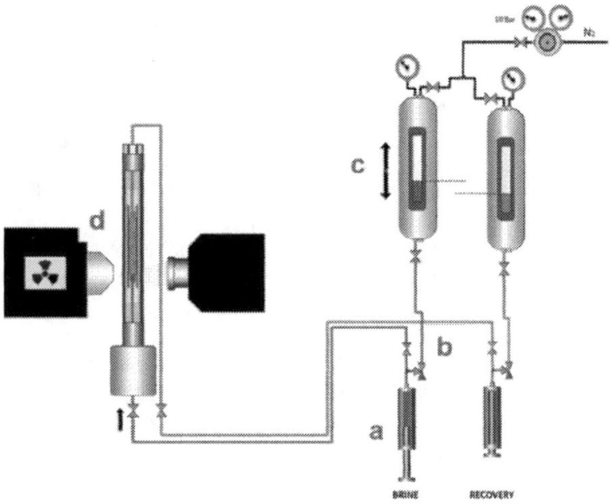

Fig. 3. Experimental set-up for the spontaneous imbibition: water level is controlled by the height of the water bottle. Steps indicated by a, b, b, d are detailed in the following paragraph.

17. The spontaneous imbibition was conducted as follows:
a) 5 PV of brine was injected at 0.6 μL/min to put the brine level at the inlet entrance, the brine arrival was controlled with the radio, as shown in Fig. 4

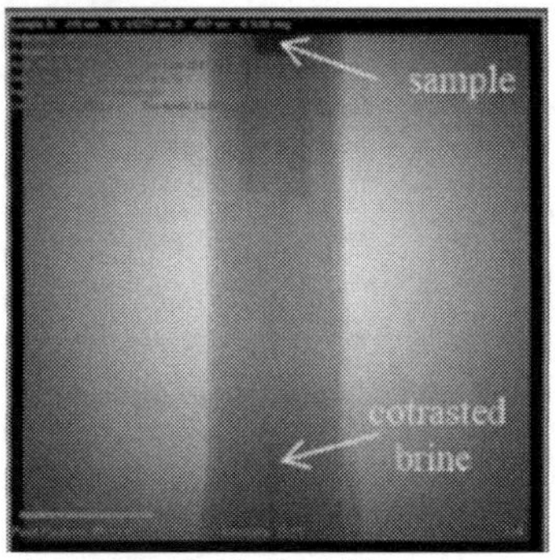

Fig. 4. Radio view: water rise in the tubing.

b) Injection was stopped, and the valve was switched. At that stage, the brine from the bottle started to interact with the sample and the recovery valve stayed opened.
c) Water bottle height was progressively adapted so that water pressure was sufficient to keep the brine level at the sample entrance, see Fig. 5.

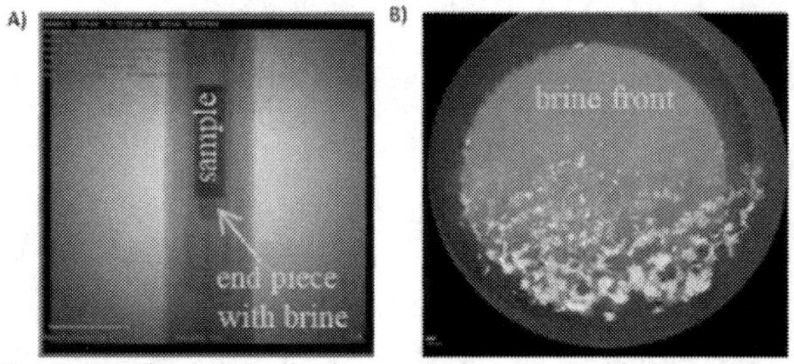

Fig. 5. Sample 1 A) Radio view: water in contact with the sample; B) image view: water in contact with the sample at the inlet

d) Over the next 6 days, the brine invasion was checked regularly with the radio and with the high-resolution acquisitions.
18. Four continuous "Spont_Imbib" scans were taken for the whole sample with a voxel size 2.07 μm. Acquisition parameters were kept the same as for the dry scans.

19. The line connected to the water bottle was closed. The forced imbibition began. It consisted of injecting:
a) 610 PV of the brine at the progressively increasing flow rate from 0.6 μL/min to 700 μL/min in the upward direction.
b) 350 PV of the brine at 700 μL/min in the downward direction.

The recovery and the confinement pressures were fixed at 11 and 120 bars respectively.

20. Confinement pressure was decreased to 50 bars. Brine pressure was regulated at 10 bars. The recovery valve was closed. Four continuous "Forced_Imbib" scans were taken for the whole sample with a voxel size 2.07 μm. Acquisition parameters were kept the same as for the dry scans.

21. The flow cell was connected to the drainage experimental set-up shown in Fig. 6

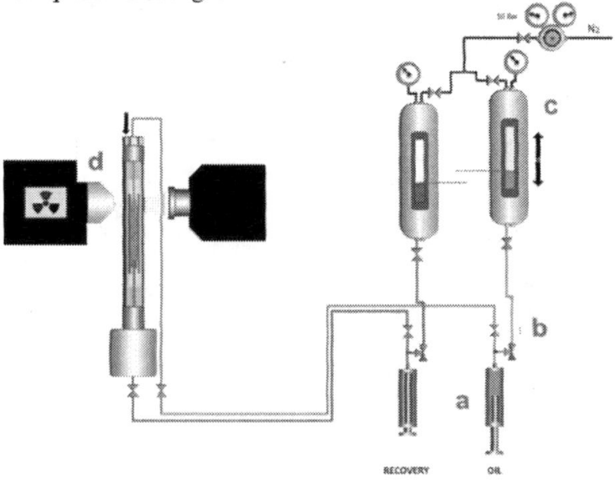

Fig. 6. Experimental set-up for the spontaneous drainage: M52 level is controlled by the height of the M52 bottle. Steps indicated by a, b, c, d are detailed in the following paragraph.

22. Spontaneous drainage was conducted as follows:
a) 15 PV of M52 was injected at 1 μL/min to fill the line connected to the sample and thus, put the M52 level at the inlet entrance. The arrival of the M52 was controlled with the X-ray monitoring in radio mode. Fig. 7 shows two stages: before and after M52 front arrival.

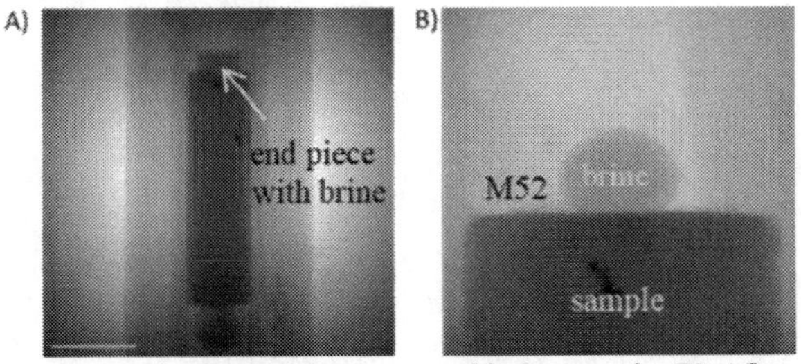

Fig. 7. A) Before: water in contact with the sample; B) After: zoom at the top of the sample: M52 pushed the water, however one water droplet stayed attached.

b) The injection was stopped, and the valve was switched. At that stage, M52 from the bottle started to interact with the sample and the recovery valve stayed opened.
c) Water bottle height was progressively adapted so that the M52 column pressure was sufficient to keep M52 level at the sample entrance.
d) Over the next 6 days, M52 spontaneous invasion was regularly checked in radio mode and with the high-resolution acquisitions.

23 Four continuous "Spont_Drain" scans were taken for the whole sample with a voxel size 2.07 μm. Acquisition parameters were kept the same as for the dry scans.

24. The line connected to the M52 bottle was closed. The forced drainage was started. It consisted of injecting:
a) 300 PV of M52 at the maximal possible flow rate in a downwards direction. The flow rate limitation was due to the max pressure limit equal to 100 bars. The max. constant injection flow rate that did not provoke the pump to shut down was equal to 60 μL/min
b) 300 PV at 60 μL/min in the upward direction
c) 325 PV at 55 μL/min in the downward direction.

The recovery and the confinement pressures were fixed at 10 and 120 bars respectively.

25. Confinement pressure was decreased to 50 bars. M52 pressure was regulated at 10 bars. The recovery valve was closed. Four continuous "Forced_Drain" scans were taken for the whole sample with a voxel size 2.07 μm. Acquisition parameters were kept the same as for the dry scans.

2.3 Imaging methods and processing

2.3.1 Imaging method

The images were taken by an Xradia Versa 520 high-resolution microCT. The X-ray energy was 90 kV and the power was 8W. The number of projections for each radio was 2001. The pixel size of all images was equal to 2.07 μm. All tomograms were reconstructed into 3D images using Zeiss reconstruction software.

Dry, KI_300, SWi_aged, Spont_Imbib., Forced_Imbib., Spont_Drain., Forced_Drain. images were first stitched together to cover the whole length of the sample. After, they were registered to the dry images to have the same orientation and to be easily comparable. A nonlocal means filter was used to remove noise. The image analysis was performed with the help of software packages: Mango [8] for the routine actions, ImageJ for the trainable Weka segmentation [2].

2.3.2 Characterization of the pore space and fluids inside

Differential imaging [3, 4] was used to quantify both resolvable and sub-resolution porosity. This method relies on the observation that the difference image between the KI saturated scan and the dry scan maximizes the phase contrast between grain, macro pores and zones containing sub-voxel porosity phases (Fig.8A – Fig.8C, zones 1-3 contain sub-resolution porosity) and consequently, provides a good input for image segmentation.

We clearly observe some sub-resolution pores with intermediate grey scale values in Fig. 8C. Results of the Weka trainable segmentation of the difference image can be found in Fig. 8D. Fig. 8E illustrates how good a trainable Weka segmentation is at capturing complex textures of the dual pore network.

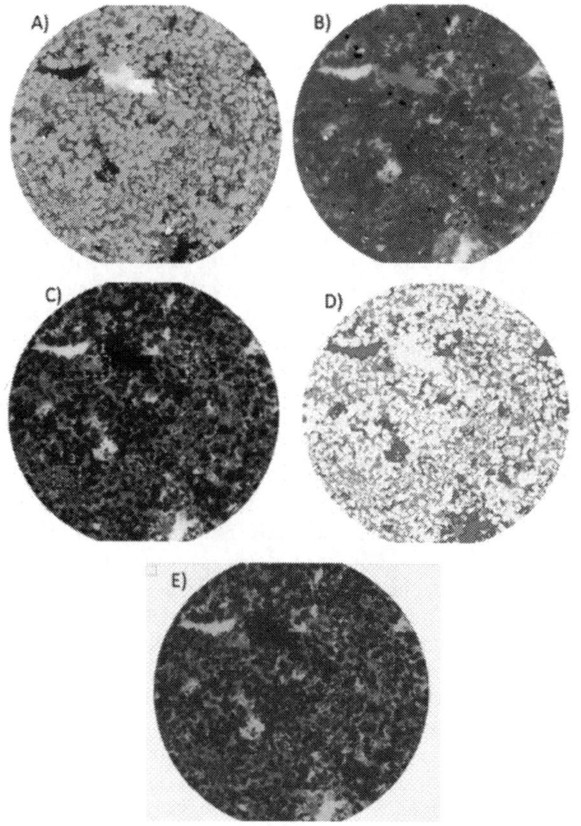

Fig. 8. Sample 2: two-dimensional cross-section of three-dimensional micro-CT images at the same location (13.18 mm from the sample inlet). A) dry scan; B) 300 g/L brine saturated scan; C) difference image between A and B; D) segmented difference image (resolved pores in red, unresolved porous zones in blue and tight grains in yellow); E) overlay of C and D images. Boxes 1-3 give examples of the zones containing sub-resolution porosity

Porosity of the sub-voxel zones was estimated based on the linear interpolation between two reference values (grey value of tight grains and grey value of brine in macro pores).

Fig.9 illustrates the profiles of total, macro and micro porosities for the second sample. As can be seen, the sample is heterogeneous in porosity. The mean total porosity slightly increases towards the outlet of the second examined mini core sample. As expected from the visualization of the pore network, a big part of the total porosity comes from the sub-voxel porosity.

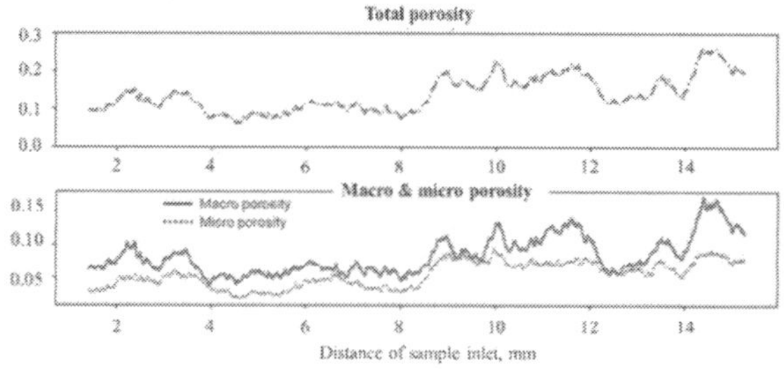

Fig. 9. Sample 2: porosity profiles along the sample length

Porosity quantification for samples 1 and 2 is given in Table 4.

Table 4. Porosity average values.

	Sample 1	Sample 2
Sub-resolution porosity	0.06 ± 0.01	0.06 ± 0.01
Resolved porosity	0.04	0.08
Total porosity	0.10 ± 0.01	0.14 ± 0.01

Two examples of the maps of the total (resolved and sub-voxel) porosity are illustrated in Fig.10.

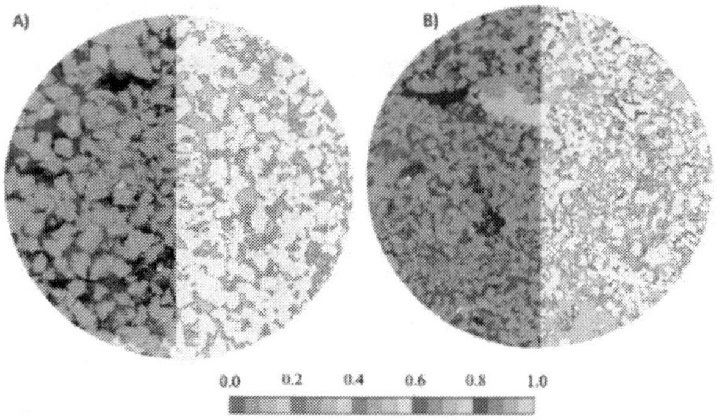

Fig. 10. Porosity map for A) Sample 1, slice is located at 12.68 mm from the inlet; B) Sample 2, slice is located at 13.18 mm from the inlet.

Grey-scale multiphase images for all stages of the second experiment are illustrated in Fig. 11 – Fig. 12. The black is mineral oil, the dark grey is grains, the light grey is doped brine.

https://doi.org/10.1051/e3sconf/202336701006

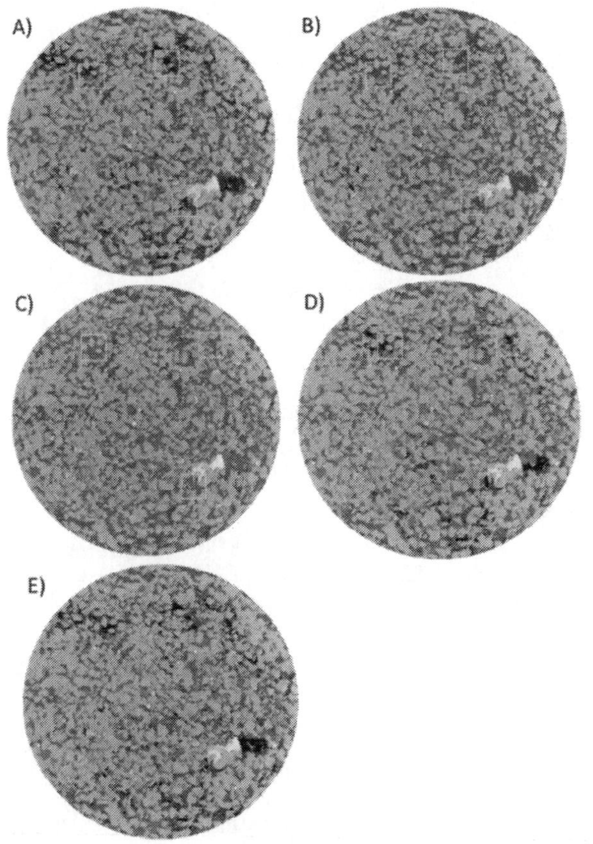

Fig. 11. Sample 1. Two-dimensional cross sections of three-dimentional micro-CT images. Slices are located at 1.56 mm from the bottom of the sample. Radius of each slice is equal to 1.53 mm. A) SWi_aged scan; B) Sp_Imbib scan; C) For_Imbib scan; D) Sp_drainage scan; E) For_Drainage scan. The yellow boxes indicate regions where fluid occupation was changed in comparison with the previous or the following saturation state

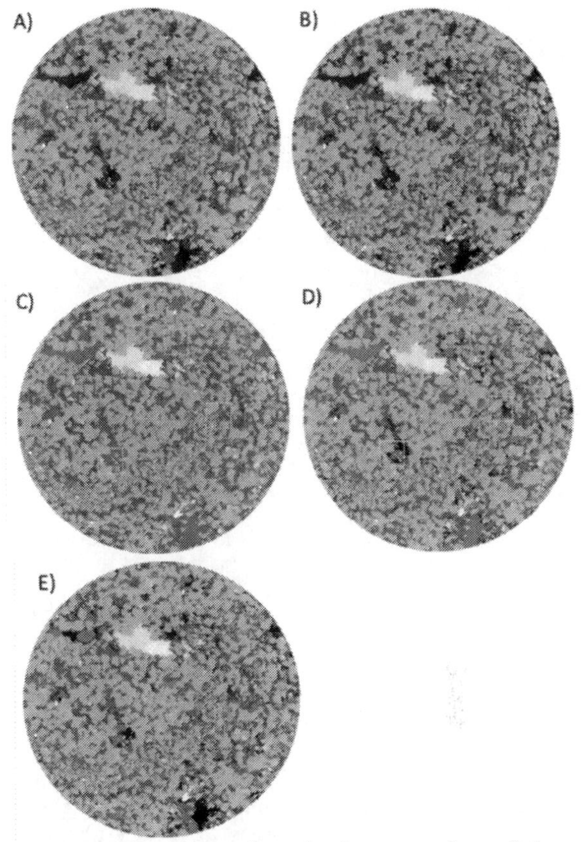

Fig. 12. Sample 2. Two-dimensional cross sections of three-dimensional micro-CT images. Slices are located at 13.18 mm from the bottom of the sample. Radius of each slice is equal to 1.53 mm. A) SWi_aged scan; B) Sp_Imbib scan; C) For_Imbib scan; D) Sp_drainage scan; E) For_Drainage scan. The yellow boxes indicate regions where fluid occupation was changed in comparison with the previous or the following saturation state

Fluid occupancy in the macro pores was identified by the direct two-phase segmentation using the converging active contour method with Mango software. Water saturation characterization in the sub-resolution pores was carried out following the principals of the differential imaging method.

The brine saturation profiles across the second sample are shown in Fig. 13.

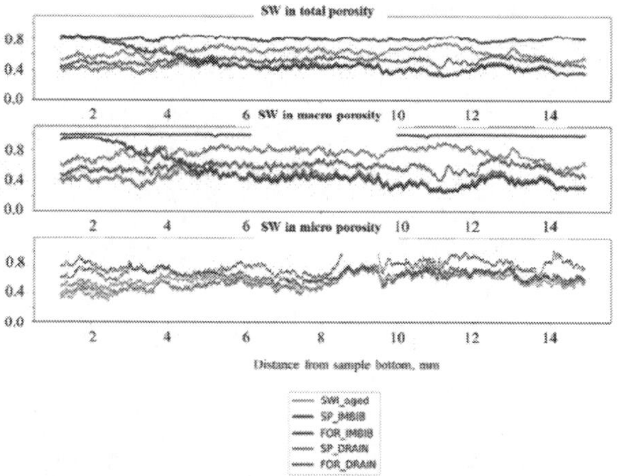

Fig. 13. Brine saturation profiles along the second sample from bottom to top: total; in macro pores; in sub-voxel zones. Brine access during the spontaneous imbibition was at the bottom, mineral oil access during the spontaneous drainage was at the top

It is apparent from Fig. 13 that the spontaneous imbibition of water happened only at the bottom, only one quarter of the sample had seen the increase of the water saturation coming principally from the macro pores. As for the spontaneous drainage, it was of higher intensity. Water saturation profiles suggest that it occurred along the entire sample, both macro and sub-resolution pores participated in the spontaneous drainage. Similar behaviour was seen for the first sample.

Average brine saturation values for both samples are presented in Table 5 – Table 6.

Table 5. Sample 1: water saturation average values.

	Ageing	Sp. Imb.	For. Imb.	Sp. Drn.	For. Drn.
Sub-resolution pores	0.60	0.48	1	0.86	0.80
Resolved pores	0.18	0.29	0.99	0.90	0.58
Total	0.46	0.48	0.99	0.87	0.72

Table 6. Sample 2: water saturation average values.

	Ageing	Sp. Imb.	For. Imb.	Sp. Drn.	For. Drn.
Sub-resolved pores	0.45	0.47	0.50	0.44	0.44
Resolved pores	0.42	0.50	0.99	0.75	0.57
Total	0.44	0.49	0.80	0.63	0.52

Table 5 and Table 6 illustrate that for both samples water saturation after the secondary drainage is higher than after the primary drainage. This increase of the water saturation was expected: during the secondary drainage it was much more difficult to inject M52 than during the first drainage. As a reminder, during the first and second drainage, the maximum accepted injection pressure was limited to 100 bars. During the second drainage, the maximum injection flow rate was divided by 4-5 vs. the primary drainage to respect the pressure limitation. This could suggest that during the four-week ageing, the initial wettability was shifted to a more oil-wet wettability, provoking more intensive water trapping that was noticed during the secondary drainage.

It should be noted that water saturation of the resolved porosity after the forced imbibition might be overestimated. In fact, the volume of oil films on the rock surface cannot be quantified with a high degree of precision, because it is below image resolution. The uncertainty of the saturation after this step is the highest, it is estimated at 3-5 %.

2.4 Wettability analysis results and discussions

To characterize the wettability behavior, we proposed to conduct the Amott–like test on two mini-core samples. This test includes a four-displacement sequence, similar to that of an Amott test, starting from the irreducible saturation state after ageing. It assumes that each sample is subjected to spontaneous and forced flows, first by water (imbibition) and afterwards by oil (drainage). The spontaneous flows are capillary driven. The forced flows are implemented by viscous displacement.

A pseudo-wettability index is proposed to be calculated by the following Eq. 1:

$$I_w - I_o = \frac{S_w^{sp.imb} - S_w^{Swi}}{S_w^{for.imb} - S_w^{Swi}} - \frac{S_o^{sp.dr} - S_o^{for.imb}}{1 - S_w^{for.dr} - S_o^{for.imb}} \quad (1)$$

Where I_w – is imbibition index, I_o – is drainage index, sp.imb and sp.dr are spontaneous imbibition and drainage, for.imb and for.dr are forced imbibition and drainage. This index is comprised between +1 and -1, it refers to the quantitative degree of wettability like the Amott-Harvey index [1].

Table 7 shows the results of pseudo-wettability index calculation for two samples.

Table 7. Result of pseudo-wettability index calculation.

	Sample 1			Sample 2		
	Iw	Io	Iw- Io	Iw	Io	Iw- Io
All pores	0.04	0.44	-0.41	0.14	0.61	-0.47

Results obtained from the pseudo-wettability index calculation suggest that the studied rock-fluids systems are most probably oil wet, for both samples the drainage index is systematically much higher than the imbibition index.

Moreover, this conclusion is supported by the following experimental observations:

1) Difficulty in setting Swi: for both samples, total initial water saturation remains high, suggesting capillary trapping of water

2) For both samples, the spontaneous imbibition remains limited to the lower part of the sample. The increase of water saturation after the spontaneous imbibition might be even attributed to the initial water advancement provoked under the initial non-stabilized pressure of the water bottle. For this reason, it was advised to adapt the experimental set-up to rule out, or at least substantially reduce the probability of the accidental forced imbibition at the sample inlet

3) Spontaneous drainage, on the other hand, happened over the entire length of the sample. We found a clear oil saturation increase in both resolved and the sub-resolution porosity across two samples. Moreover, we even recorded in the radio

https://doi.org/10.1051/e3sconf/202336701006

mode M52 production at the bottom of the second sample during the first 6 days, as shown in Fig. 14

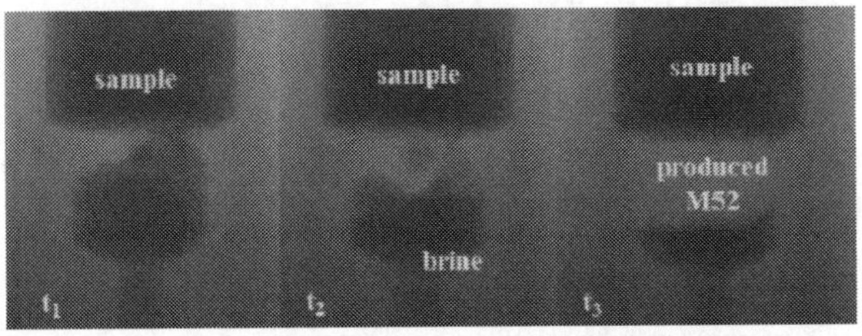

Fig. 14. Bottom of the second sample imaged in the radio mode during the spontaneous drainage: M52 production over time

4) Distribution of oil and brine visible interfaces within the resolved pores also suggests oil wettability, Fig. 15

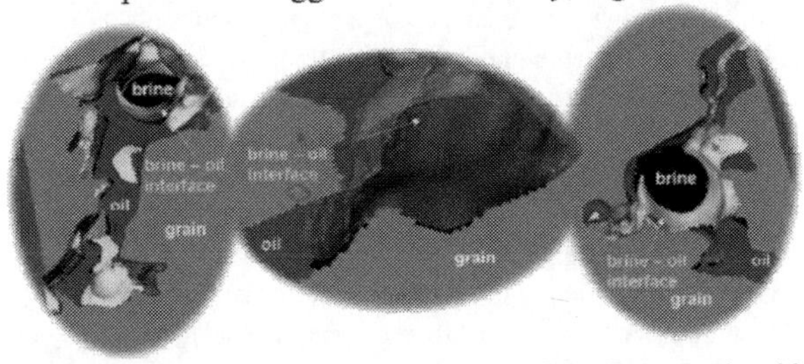

Fig. 15. Examples of some resolved pores of the first sample with brine-oil interfaces having negative curvature and occupying the large pore spaces

3 Conclusions

To assess the wettability of reservoir samples in a quick and physically meaningful manner, we designed a new experimental set-up allowing us to carry out a full Amott-like cycle on a small core sample imaged with a high-resolution micro-CT scanner.

The advantage of this kind of experiment is that it provides global wettability characterization based on multiple experimental observations, the analysis of the fluid distribution in the porous space and the result of the pseudo-wettability index calculation. Moreover, it respects the experimental sequence of the industry standard Amott test, yet with a significantly reduced test duration. Thanks to the high-quality 3D imaging, it provides complementary information describing the spatial distribution of the fluids in the visible pore network during spontaneous and forced phases. On top of that, it is less sensitive to imaging resolution than existing approaches based on topological characteristics and contact angles.

Comparison with the results of conventional Amott test on neighbour centimeter-sized core samples was planned and initiated in parallel to the Amott-like image-based tests on small core samples. These standard tests are still in progress. Future work should involve the comparison of the results of the wettability characterization on small core samples with the proposed experimental design to the results of the standard Amott ones.

References

[1] Amott, E. 1959. Observations Relating to the Wettability of Porous Rock. *Trans.* 216, 01, 156–162.

[2] Arganda-Carreras, I., Kaynig, V., Rueden, C., Eliceiri, K. W., Schindelin, J., Cardona, A., and Sebastian Seung, H. 2017. Trainable Weka Segmentation: a machine learning tool for microscopy pixel classification. *Bioinformatics (Oxford, England)* 33, 15, 2424–2426.

[3] Gao, Y., Lin, Q., Bijeljic, B., and Blunt, M. J. 2017. X-ray Microtomography of Intermittency in Multiphase Flow at Steady State Using a Differential Imaging Method. *Water resources research* 53, 12, 10274–10292.

[4] Gao, Y., Raeini, A. Q., Selem, A. M., Bondino, I., Blunt, M. J., and Bijeljic, B. 2020. Pore-scale imaging with measurement of relative permeability and capillary pressure on the same reservoir sandstone sample under water-wet and mixed-wet conditions. *Advances in Water Resources* 146, 2, 103786.

[5] Lin, Q., Al-Khulaifi, Y., Blunt, M. J., and Bijeljic, B. 2016. Quantification of sub-resolution porosity in carbonate rocks by applying high-salinity contrast brine using X-ray microtomography differential imaging. *Advances in Water Resources* 96, 306–322.

[6] Regaieg M, Brugidou R and Nono, F. 2022. *A porous sample wettability parameter determining method and related system,* US Patent No. 63/342,923.

[7] Regaieg M, Nono F, Farhana Faisal T, Varloteaux C and Rivenq R. 2022. Pore network simulations coupled with innovative wettability anchoring experiment to predict relative permeability of a mixed-wet rock. *Symposium of the Society of Core Analysis* (2022).

[8] Sheppard, A. *MANGO*. ANU.

 https://doi.org/10.1051/e3sconf/202336701007

Water-Gas Imbibition Relative Permeability: Literature Review, Direct versus Indirect Methods and Experimental Recommendations

Fabrice Pairoys[1*], *Cyril* Caubit

[1]TOTALENERGIES CSTJF, Pau, France

Abstract. When an active aquifer encroaches into a gas bearing reservoir or when an oil rim sweeps gas during late depletion of the gas cap, gas displacement by liquid is important for estimating the gas recovery. In the water displacing gas condition, the viscosity ratio is extremely favorable, resulting in a sharp waterfront in the reservoir matrix: it results that changing the relative permeability Kr shape has negligible effect, while endpoints water relative permeability Krw Max and residual gas saturation Sgr are much more important to understand gas flow performance for estimation of gas recovery with active aquifer or productivity decline after water breakthrough. Three main methods are used to determine water/gas relative permeability curves: imbibition unsteady-state, imbibition steady-state or indirect approaches such as co-current spontaneous imbibition if transient data are available. One of the other popular indirect methods is called Brooks-Corey approach: by measuring the drainage Pc curve using centrifuge or porous plate methods, it is possible to calculate a pore size distribution index λ. This coefficient is used in a Brooks-Corey model to determine the drainage Kr curve. It is also required to measure and determine the relationship between the residual gas saturation Sgr and the initial gas saturation Sgi relationship. Finally, it is accepted that there is no hysteresis on the water relative permeability Krw curve, as water is always the wetting phase in the gas/water couple. As non-wetting phase, gas exhibits strong hysteresis between drainage and imbibition curves: it is therefore necessary to apply a correction on the drainage Krg curve to build the imbibition one using correcting models. The aim of this paper is to compare gas/water relative permeability of clastic rocks using direct waterflooding information and indirect approach using Brooks-Corey model. It is shown that using the indirect approach leads to results like those experimentally obtained. Also, additional numerical simulations are proposed to discuss the relevance of measuring the entire water-gas imbibition relative permeability curve using the steady-state approach.

1 Introduction

Trapping mechanisms at pore scale of non-wetting fluids such as gas (versus liquids) have been mainly investigated through observations using glass micromodels or transparent glass bead systems. Micro-CT recently started 3D imaging of trapped gas in rocks to better understand this trapping phenomenon. Both techniques have shown that gas phase is trapped as discontinuous phase at the middle of the large pores while water occupies smaller pores as shown by [1].

Different types of displacement mechanisms in imbibition cycle have been put in evidence on glass micromodels by [2]: it is shown that, in liquid-wet domain, the imbibition is mainly dependent on pore size distribution and contact angle. Later, [3], [4] and [5] showed that:

- The ratio between pore throat and pore diameters (aspect ratio) matters, with more trapping at larger aspect ratio
- The number of pore throats connected to a given pore is also very important on trapping, with large coordination numbers tending to lead to less trapping
- An increase in contact angle suppresses snap-off, leading to less trapping

If these results have been observed on glass micromodels and pore network models, micro-CT should confirm the above-mentioned trends. Today, the resolution of micro-CT systems limits these investigations to fair to very good quality rocks.

At the macroscopic scale, in most cases, the displacement of gas by liquid occurs at favorable to extremely favorable viscosity ratios. During laboratory experiments, there is no or negligible variation of gas saturation after the water breakthrough. In the field, flood is almost piston-like in the reservoir matrix, with no mobile gas behind the waterfront. It is finally concluded that attention is devoted to Sgr and maximum Krw value at Sgr rather than the shape of the Kr curve. Nevertheless, imbibition relative permeability bounding curve, measured or derived, is still required for reservoir numerical simulations.

In this paper, we made a literature review on the methods to estimate the residual gas saturation, looking at the parameters that can affect Sgr, providing tips for the waterflooding tests. Also, the indirect Brooks-Corey approach, coupled with the Land's hysteresis model, was tested and compared to the experimental data results. It confirmed the findings and conclusions from [6].

2 Background

2.1 Experimental Methods to estimate Sgr

Several gas/water imbibition experimental methods can be used to determine Sgr:

* Corresponding author: fabrice.pairoys@totalenergies.com

- Porous plate method: after a primary drainage of water displaced by humidified nitrogen at highest capillary pressure, Pc is then gradually decreased in a stepwise manner allowing controlled imbibition of gas-saturated water. Residual gas is obtained by material balance when the capillary pressure Pc=0. This test is time-consuming, especially on low permeability rocks. On very low permeability samples, mercury imbibition can be attempted to obtain trapped non-wetting phase saturation, but it is not recommended
- Unsteady-state method; it is a forced displacement of gas by water at constant low rate Qw or low differential pressure dP. Residual gas saturation Sgr is obtained at the end of the flood when the gas production ceases (generally right after the water breakthrough). In situ saturation monitoring is recommended: it generally confirms the sharp front with no variation of the gas saturation behind the front
- Steady-state method: it is a co-injection of gas and water, but it is not the best appropriate approach to determine Sgr, except maybe on heterogeneous rocks and gas/condensate studies. It can be used to obtain the full Kr curve
- Spontaneous imbibition by water or capillary rise method: it is the most popular technique to determine Sgr. The spontaneous imbibition test requires a strongly wetting liquid to achieve a non-wetting residual saturation. Two configurations exist: co-current and counter-current displacements, respectively with the liquid touching the bottom of the rock (capillary rise test) and with the rock fully immersed in the liquid

Based on the past literature, there is no ideal technique and decision of choosing such or such method: it is case dependent. [7] and [8] reported very similar trapped gas saturations using steady-state, unsteady-state and spontaneous imbibition. More recently, some studies from [9] put in evidence some significant differences on Sgr computed from the different methods. Based on the above observations, it is recommended to make a combination of techniques during a study. Important to note that large uncertainty in results is anticipated for very tight gas reservoirs with porosity ϕ below 5% and permeability K below 0.01 mD. In this specific case we won't develop in this work, non-standard approach such as evaporation method may be tested.

2.2 Various Potential Effects on Sgr (Literature review and in house experience)

Regarding the types of liquid, [10] showed that same trapped gas saturation is found using different imbibing fluids, provided the liquids strongly wet the rock.

For the forced displacements, it is recommended to perform the liquid injection at low rate despite most of the past studies show no Sgr dependency on displacement rate ([7], [8], [11]: only [9] showed significant differences in Sgrw values between porous plate method and unsteady-state coreflooding tests.

Fluid pressure or temperature do not affect residual gas saturation ([11], [12]), [13], [14]).

Regarding the effect of wettability, it is always assumed that gas is the non-wetting phase and liquid the wetting phase. However, it was early observed by [7] that some cores could imbibe oil but not water. It was confirmed by in house tests, showing that water is not always strongly wetting the rock in respect to gas. When water is not imbibing the rock, other fluids can be tested, such as toluene or refined oil: in house experience and literature have showed that these liquids are sometimes more wetting the rock than water ([15], [16]). When poor water wetness is observed, it can either result from true original reservoir wettability or from core contamination.

Effect of Sgi on Sgr:

Residual gas saturation Sgr strongly depends on initial gas saturation, both on sandstones ([7], [10], [14], [17]) and on carbonates ([18]). Several formulas can be used to capture this effect: the most used is the Land's formula ([17]):

$$\frac{1}{S_{gr}^*} - \frac{1}{S_{gi}^*} = C = \frac{1}{S_{gr(Swi)}} - \frac{1}{1-S_{wi}} \qquad \text{Eq 1}$$

$$S_{gi}^* = \frac{S_{gi}}{1-S_{wi}} \text{ and } S_{gr}^* = \frac{S_{gr}}{1-S_{wi}} \qquad \text{Eq 2}$$

Other ones like [18] or [19] are available.

[21] measured the residual gas saturation Sgr as function of initial gas saturation Sgi on 60 sandstone samples. Their porosity and permeability ranged from 0.06 to 0.25 and from 0.1 to 2000 mD. Sgrm values ranged from 0.04 to 0.65. It was confirmed that the piecewise linear relationship from [19] is the best relationship to describe Sgr-Sgi curves. None of the hyperbolic laws can correctly describe the observed experimental behavior. The relationship from [20] gave the poorest estimates of Sgr as a function of Sgi.

The existence of Sgr plateau as Sgi decreases is linked to the absence of gas trapping inside micropores. Assuming pores are drained in order of size while Sgi increases, the lowest Sg value of the piecewise linear and horizontal relationship between residual and initial gas saturation may be defined as the limit between drained pores that trap gas and pores too small to trap gas. It indirectly confirms that there is no gas trapping in micropores.

Despite it was observed that model from [19] best captures experimental relationships between Sgr and Sgi, the industry and particularly numerical codes continue using Land's model.

Effect of rock properties, Sgr versus porosity with or without clays:

It has been observed large Sgr variations in a given field with decreasing Sgr trend as porosity increases ([14]) but several in house and external studies have shown more complex trends ([9]). [22] has collected data results from the literature showing wide dispersion of the maximum residual gas saturation Sgrm/porosity cross plot:

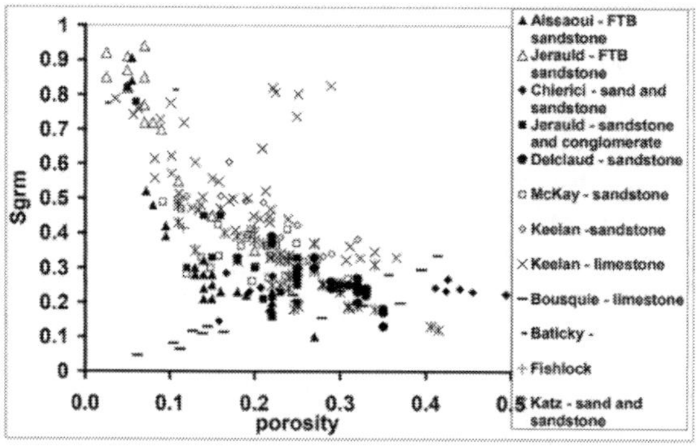

Fig. 1. Literature review 1 of maximum trapped gas saturation versus porosity, in absence of irreducible water saturation

Despite some studies concluded that Sgr increases as porosity decreases, Figure 1 shows the difficulty that encountered [13] to find, in vain, a general correlation between porosity (permeability and Sgi too) and Sgr. The largest investigation was performed by [22] and [23] with more than 300 Sgr measurements on sandstones:

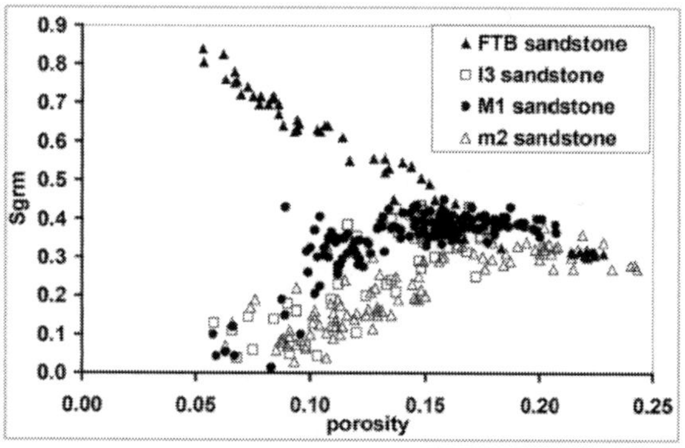

Fig. 2. Literature review 2 of maximum trapped gas saturation versus porosity in absence of irreducible water saturation

For porosity values below 15%, Figure 2 shows that Sgrm decreases for I3, M1, M2 sandstones and increases for FTB (Fontainebleau outcrops) as porosity decreases. Above 15% of porosity, the data results from all reservoirs converge, with Sgrm values varying from 25% to 35%.

The difference between the Fontainebleau sandstone and the other ones was explained by the presence of micro-porosity or clays, the Fontainebleau sandstone being mesoporous and clay-free.

Later, [24] provided the following plot of Sgrm versus porosity, adding the clay %. It shows that greater microporosity in the clay mineral assemblage enhances water imbibition and limits gas trapping.

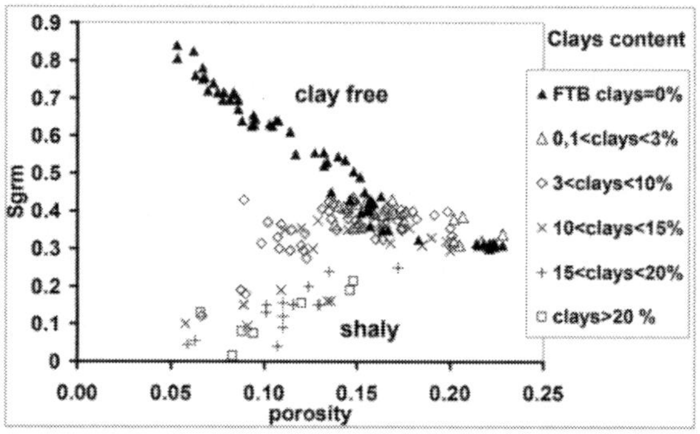

Fig. 3. Literature review 3 of maximum trapped gas saturation versus porosity with clay %, in absence of irreducible water saturation

The observations show that assessing the variability of Sgr within a reservoir and finding clues to explain it from a sedimentological point of view are key objectives.

Based on the literature review, it is obvious that laboratory work is required to guess a trapped gas saturation. Any estimation should be given ±10 s.u. at minimum, with larger uncertainty when dealing with carbonate rocks and sandstone rocks with less than 15 p.u. porosity.

2.3 Waterflood tips:

In water displacing gas tests, most of the studies show no additional gas production post breakthrough, whatever the displacement rate as expected from a liquid-wet system. During laboratory waterflood, 1 to 2 PVs injected are enough. Moreover, as the viscosity ratio is extremely favorable, resulting in a very sharp waterfront in conventional matrix, the endpoints Sgr and Krw at Sgr are more important than the Kr shape by itself.

Some cautions are required when performing waterflood, such as equilibrating the phases to avoid any mass transfer, avoiding compressing gas in the pore network during the test. Coreflooding tests can be performed with a low solubility gas such as nitrogen, at ambient temperature and ambient or reduced pore pressure.

2.4 Deriving water/gas imbibition Kr curves from Sgr/Sgi experiments:

In strongly water wetting conditions, it is stated that the shape of the imbibition relative permeability is not important while the Sgr/Sgi relationship and Krw Max value are. Nevertheless, imbibition Kr curves are still needed as input in reservoir simulations.

They are few developed methods to obtain this Kr curve in water-wet conditions. The approach proposed by [17] is generally used when drainage gas/water Kr and Sgr/Sgi relationship are available. It assumes that:

- The water wetting phase does not exhibit any hysteresis, meaning that drainage and imbibition Krw overlay
- The gas non-wetting phase exhibits a strong hysteresis between drainage and imbibition, but imbibition Krg can be derived from drainage Krg

- The gas non-wetting phase can be treated as two separate saturations, one trapped that does not contribute to the flow, another one mobile used to calculate Krg(Sg)

2.5 Calculation Krg(Sg):

1- Imbibition starts from the maximum gas saturation post drainage

2- Sgr is derived from the experimental Sgr/Sgi relationship using Land's formula (Equation 1 and Equation 2)

3- Mobile non-wetting phase saturations S_{gF}^* are calculated for various gas saturations using the following equation:

$$S_{gF}^* = \frac{1}{2}\left(S_g^* - S_{gr}^* + \sqrt{\left(S_g^* - S_{gr}^*\right)^2 + \frac{4}{C}\left(S_g^* - S_{gr}^*\right)}\right) \quad \text{Eq 3}$$

Where Sg* is the effective total gas saturation, including trapped gas.

4- Imbibition relative permeability to gas is then calculated using:

$$K_{rg} = S_{gF}^{*2}\left(1 - \left(1 - S_{gF}^*\right)^{\varepsilon-2}\right) \quad \text{Eq 4}$$

Where the exponent ε corresponds to the drainage exponent of the wetting phase Brooks-Corey exponent detailed later ($K_{rw} = S_w^{*\varepsilon} = S_w^{*\frac{(2+3\lambda)}{\lambda}}$).

The above method implies that drainage relative permeability curve has been measured, which is, unfortunately, not always the case. Nevertheless, if primary drainage capillary pressure has been measured, drainage relative permeability can be inferred from capillary pressure.

2.6 Deriving water/gas imbibition Kr curves from water/gas drainage capillary pressure:

Assuming that the rock is strongly wetted by the liquid in the gas/liquid system, the effect of different wettability on relative permeability curves is nil. Therefore, several methods of estimation can be used to estimate gas/liquid Kr, such as [25], [26], [27], [28]. In this study, the Brooks-Corey approach ([28]) is used to determine the drainage relative permeability from the drainage capillary pressure.

The tested procedure is made several steps:

1- Obtaining drainage capillary pressure either from MICP, centrifuge or porous plate tests. It first allows to determine the various rock types and saturation function per rock type. Capillary pressure averaging is necessary to provide one single Pc curve, using Leverett J-function for instance ([29])

2- For each rock type, plot of Pc versus saturation on a log-log scale and fit the curve using a Brooks-Corey equation:

$$P_c = P_{ce}\left(S_w^*\right)^{\frac{-1}{\lambda}} \quad \text{Eq 5}$$

Where Pce is the entry capillary pressure, Sw* the reduced water saturation and λ the pore size distribution index

3- Build the drainage Brooks-Corey relative permeability curves using the above λ fitting parameter and the below equations:

$$K_{rw} = S_w^{*\frac{(2+3\lambda)}{\lambda}} \quad \text{Eq 6}$$

$$K_{rg} = (1 - S_w^*)^2\left[1 - (S_w^*)^{\frac{(2+\lambda)}{\lambda}}\right] \quad \text{Eq 7}$$

4- The resulting Brooks-Corey Krg curve must be fitted with a simple Corey function to determine the water Corey exponent n_w for K_{rw}, the water Corey exponent n_w is directly calculated $n_w = \frac{2+3\lambda}{\lambda}$: the curve is now following a Corey model

5- The indirect calculated drainage Corey coefficients can be compared to the cloud of Corey coefficients from drainage Kr experimental results stored in our database

6- In parallel, plot the Sgr-Sgi relationship: in our example, the Land's method was tested, obtaining a Land's coefficient for each rock type

7- Using the drainage Krg, the Krg correction is performed to calculate Krg imbibition using Equation 4 (Krg=f[SgF*]), imbibition Krw being similar to the drainage one, ranging from Sgi to Sgr

3 Experimental Program

We present in this paper an example taken from a real gas field, with an active aquifer encroaching into the reservoir. The reservoir rock is a sandstone with low to medium clay content. Only data results from the best rock type RT1, containing no clay, are presented as example.

To validate the indirect Brooks-Corey method for inferring the imbibition relative permeability curve for a specific rock type, the experimental program consisted in:

- Measuring nitrogen/brine primary drainage porous plate capillary pressure with measurement of the maximum gas permeability KgMax at Swi
- Low imbibition waterflooding rate tests to establish Sgr/Sgi relationship with measurement of the maximum brine permeability KwMax at Sgr
- Final Dean-Stark quality check

The experimental program was performed at 3145 psi of net confining pressure and ambient temperature. To avoid gas compression, the design of the waterflooding test was done such a way the differential pressure was less than 2% of the 10 bars fluid pressure. Nitrogen and brine of 130 kppm NaCl equivalent salinity were equilibrated at the test pressure to avoid any mass transfer.

4 Experimental Results

Table 1 represents all data results collected during the program.

Table 1. RT1 core analysis properties

Id	ϕ (%.)	Kkl (m D)	Swi (%)	Kg (Swi) (mD)	Sgr (%)	Kw (Sgr) (mD)	Krw (Sgr)
1	23.0	238	14.6	202	20.6	22,8	0,113
2	22.4	263	18.9	204	19.3	29,9	0,146
3	23.4	413	15.0	323	19.8	34,9	0,108
4	22.9	491	13.3	388	21.6	43,3	0,112
5	23.7	143	16.7	128	19.9	23,5	0,184
6	23.0	602	12.1	474	19.4	95,2	0,201
7	23.7	466	17.7	418	20.8	63,5	0,152

Important to note that the rock has no complex pore structure and negligible clay content.

For the rock type RT1, an averaged capillary pressure curve was obtained using the J-Leverett approach. Figure 4 represents the J-function versus water saturation Sw on a log-log plot with two "framing/bounding" functions, pessimistic and optimistic respectively:

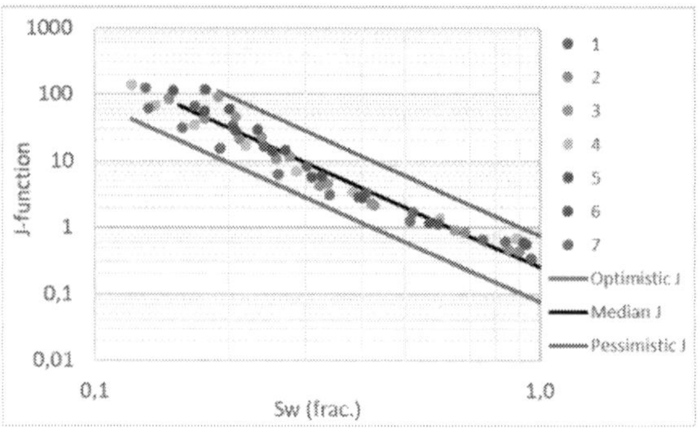

Fig. 4. Median J-function versus Sw with envelope

A power law type function $J=aSw^{-b}$ is used to fit the J-data points and to frame the data point cloud. In addition to the mean function plot, a pessimistic and optimistic cases were positioned to frame the J-data points. This approach results in finding associated Pc envelope in addition to the averaged Pc curve (Figure 5).

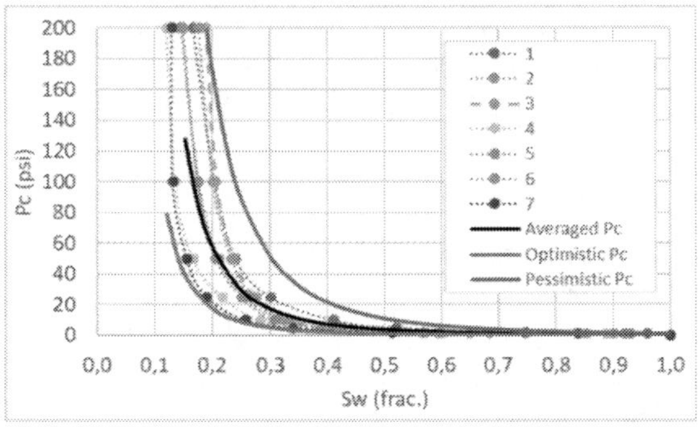

Fig. 5. Average capillary pressure Pc curve (in black) with its envelope (optimistic Pc in red, pessimistic Pc in blue)

Note that the transformation from the mean J-function to the averaged capillary pressure Pc was obtained using arithmetic average of the porosity and permeability. The Pc envelope helps the reservoir engineer to understand the uncertainty on the measured Pc and adjust its saturation's law for static reservoir modeling with observed logging information if needed.

Then, by plotting the log-log plot Pc versus effective water saturation Sw* and by using a power regression $Y=aX^b$, the respective values of entry pressure Pce (Pce=a) and pore size distribution index λ (λ=-1/b) from Equation 5 can be determined. Note that low Sw values may be discarded to obtain an acceptable correlation coefficient R^2.

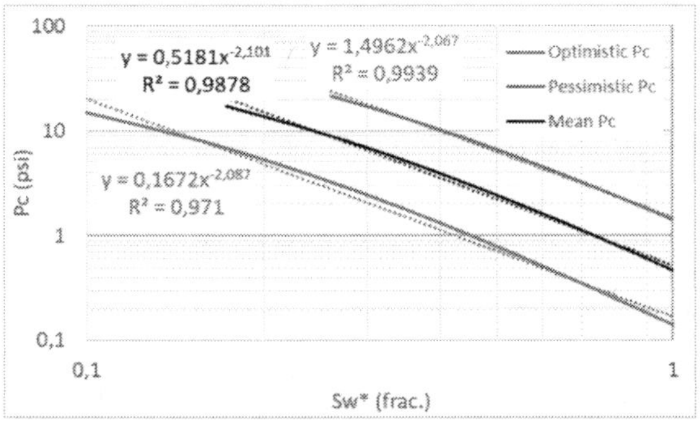

Fig. 6. Log-log plot of averaged capillary pressure Pc

The Brooks-Corey parameters can be computed from the regressions, for the averaged Pc and its envelope.

Table 2. Brooks-Corey parameters for Pc

	Optimistic Pc	Pessimistic Pc	Averaged Pc
Pce (psi)	0.17	1.50	0,52
λ	0.479	0.483	0,476
Swi (%.)	12.1	18.9	15.4

The obtained low λ values result from the good grain sorting and size, high permeability/porosity of this sandstone reservoir rock type. λ values from Table 2 are then substituted in Equations 6 and 7 to build the Kr curves. It is shown that the three λ values obtained from low case, high case and mean case are very close together. The resulting Kr curves have same curvature. The only difference is the value of Swi, according to the case.

The Sgr versus Sgi plot is then built to determine the C Land parameter that will be later used in the model.

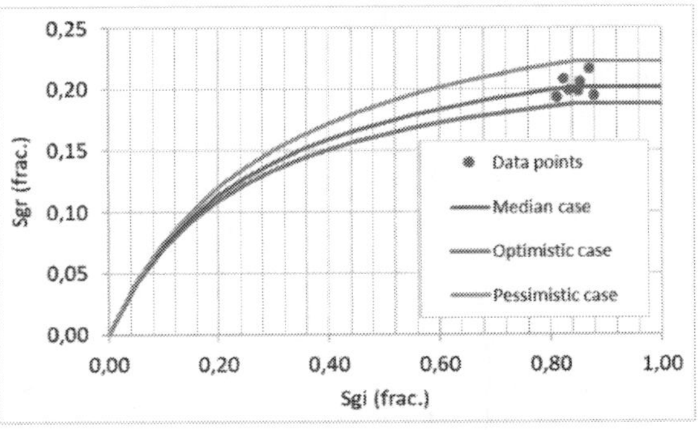

Fig. 7. Log-log plot of averaged capillary pressure Pc

The uncertainty on Sgr values may be captured by bounding the data points cloud. For RT1, the Sgr value may vary from 18.8% to 22.3% (Figure 7). The equivalent and median C parameter was found to be equal to 3.2.

The following step shows only the median case. Using Equations 3 and 4, with the C and λ parameters determined earlier (respectively C=3.2 and λ=0.476), the imbibition Krg is then calculated. The Sgr value was taken as the average of the seven measured Sgr from Table 1. Also, the Krw Max values measured at the end of the respective waterflooding tests are plotted in Figure 8.

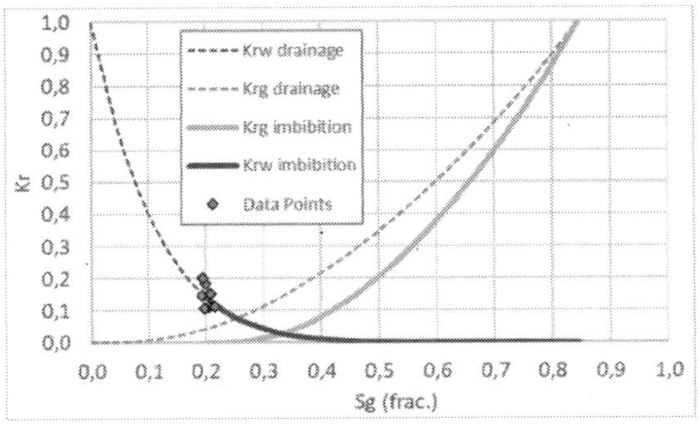

Fig. 8. Imbibition relative permeability curves from the median Pc curve and measured Krw Max at Sgr

As explained earlier, the shape of Kr curves and respective equivalent imbibition Corey exponents (Nw=7.2 and Ng=2.1) are not of high importance, only Krw Max and Sgi/Sgr relationship are. In Figure 8, it is observed that the indirect Brooks-Corey method falls in the cloud of Krw Max experimental data points. In our example, the indirect Brooks-Corey Krw Max value was equal to 0.142 while the averaged Krw Max values measured on the seven samples was 0.145.

For the indirect Sgr value, all can understand that it must match because the value is calculated from the average of the individual and experimental Sgr values.

Note that, in our example, the nitrogen Corey exponents for drainage and imbibition modes were found to be equal, though there is a hysteresis on the Krg.

At the end of the imbibition cycle, the water saturations measured by material balance on the seven samples during the waterflooding tests are compared with the saturation measured by Dean-Stark.

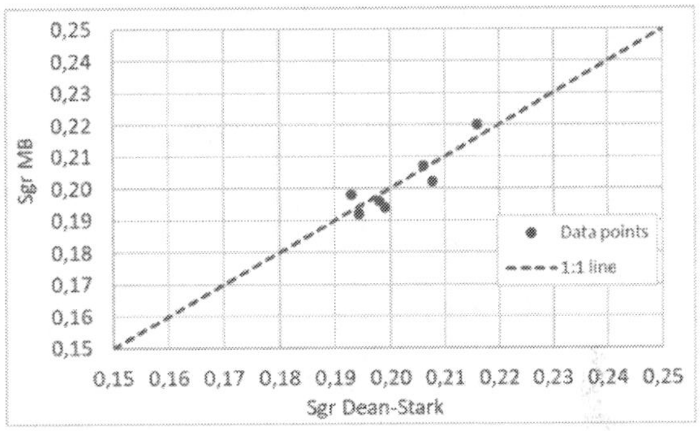

Fig. 9. Comparison between Dean-Stark Sgr and Sgr determined by material balance (Sgr MB)

Both methods lead to close Sgr values (+/- 1% S.U. max.), validating the measured production by material balance during the waterfloods.

5 Discussion

In gas-liquid imbibition cycle, it is often attempted to measure the relative permeability curve using steady-state approach. It consists in co-injecting water and gas at a total constant flowrate but varying ratios, increasing the water fractional flow.

In the case of water-gas test, where the viscosity ratio is extremely favorable and where the water is strongly wetting, we can question the relevance of performing this kind of experiment. As a matter of fact, the imbibition water-gas relative permeability curves are defined on a small range of saturation. This is because it requires extremely low water ratio to obtain data points at low water saturation. Due the experimental limitations, it may be impossible for the laboratories to design and perform such test. This statement can be proven by using numerical simulations.

CYDAR 1D numerical simulations were performed to design a steady-state relative permeability test. In our example, we used the same parameters than those presented earlier on RT1 (rock geometry and properties, fluid properties, Brooks-Corey Kr parameters, initial and final conditions).

https://doi.org/10.1051/e3sconf/202336701007

Table 3. Simulation parameters

Rock		
L	6,60	cm
D	3,81	cm
ϕ	23,2	%
Kkl	388	mD
Fluids		
μw	1,38	cP
μg	0,017	cP
ρw	1,1	g/cc
ρo	0,001	g/cc
Kr parameters		
Nw	7,2	
Ng	2,1	
Sgi	84,6	%
Sgr	20,2	%
Krg Max	1	
Krw Max	0,142	

Note that:

- The rock properties come from the average of core plug dimensions, porosity and Klinkenberg permeability
- The fluid properties were measured for the brine, calculated for the nitrogen
- The Kr parameters come from the experimental program and Brooks-Corey Kr coefficients discussed above

A first design was tested, with no imbibition capillary pressure (Pc=0), a total constant flow rate of 100cc/h with 8 ratio and an additional bump rate at 450cc/h.

Table 4. Design for simulation

Time (day)	Qw (cc/h)	Qg (cc/h)
25	0,01	99,99
2	0,1	99,9
2	1	99
2	10	90
2	50	50
2	80	20
2	100	0
2	450	0

The minimum flow rate for a high precision pump is around 0,01 cc/h. This rate value corresponds to the first fractional water rate applied in the simulation.

Even on the large permeability sample, the first ratio requires 25 days before obtaining equilibrium but:

- More than 50% of the produced gas is produced during this first ratio
- The dP stabilization is attained at 5 mbar, which may be readable with the resolution of the dP transducers but the confidence in the measurement is poor (high uncertainty)
- The transient saturation profiles simulated are vertical, showing no additional production after breakthrough

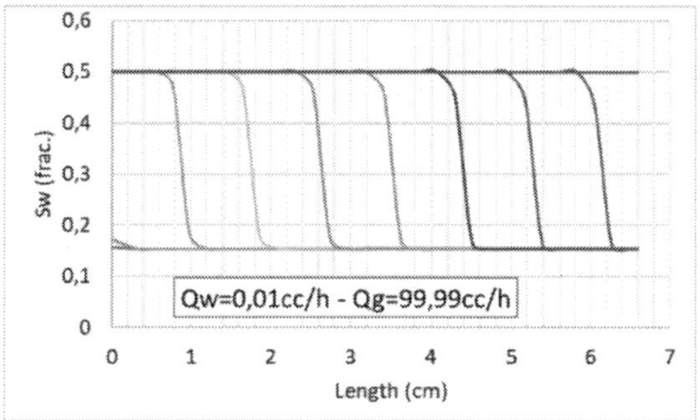

Fig. 10. Saturation profiles during the first ratio (Qt=100 cc/h)

The simulated data results with the above experimental design are presented in Figure 11.

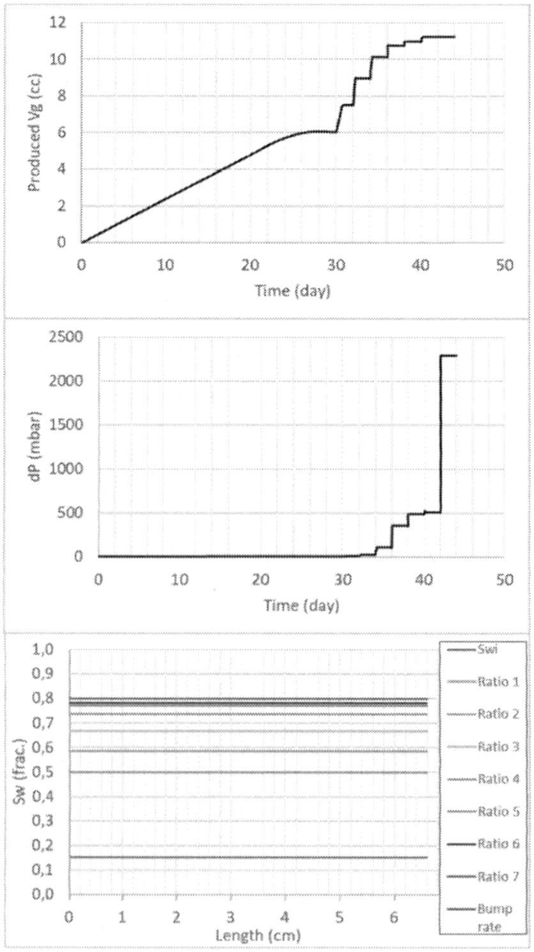

Fig. 11. Gas production, dP signal and saturation profiles at equilibrium

The simulated results presented in Figure 11 show that it is extremely difficult to measure the imbibition water-gas relative permeability. Even at the lowest water rate, the first ratio leads to more than 50% of the gas production. It is impossible to get Kr data points between Swi (=15.4%) and Sw=50%.

It is also extremely difficult to ensure good dP reading before the third ratio because of dP values at equilibrium below 30 mbars, resulting in high kr data points uncertainty too.

The Kr data points can only be defined in a reduced water saturation range 0.585<Sw<0.798, equivalent to 0.415<Sg<Sgr=0.202.

3 additional simulations were run to observe the impact of capillary pressure, considering 3 different cases:

- Case 1 is considered as an extreme case with a strongly water-wet rock, using the positive primary drainage curve. In this case, the corresponding Amott water index Iw is equal to 1

- Case 2 is a median case, less water-wet, using a log(beta) function to build a curve with a positive and negative parts. The choice was to impose a value of Sw(Pc=0) equal to 0.5. It results in Iw close to 0.5

- Case 3 is another extreme case with a significant and non-realistic negative Pc (chosen for the sensitivity study), considering the rock is not water-wet. The Sw(Pc=0) value is then equal to Swi, with no spontaneous imbibition of water. The resulting Iw is nil.

For relative permeability, same parameters than those from Table 3 were taken. Obviously, there is a problem of consistency between Kr parameters and Pc parameters for case 3 mainly (water-wet Kr vs non-water-wet Pc), but it helps in understanding the impact of Pc on the simulations thus on the steady-state experimental design.

The averaged primary drainage curve was taken to build the case 1, rescaled in the imbibition cycle saturation range. As explained previously, it is an extreme case.

A log(beta) function was used to build case 2 and case 3. It is a 3 input-parameter function [Po, β, Sw(Pc=0)] as shown in Equation 8:

$$Pc = c.Po.ln\left(\frac{\left(1-S_w^{*\beta}\right)}{S_w^{*\beta}}\right) - b \qquad \text{Eq 8}$$

With Po a pressure coefficient to control the magnitude of the Pc curve, Sw* the reduced/normalized water saturation, β coefficient to control the asymmetry of the function, b, a function dependent on the water saturation at Pc=0, and c parameter calculated as a function of β to impose a slope equal to Po at the middle of the Pc curve (Sw*=0.5).

The Kr simulation parameters detailed in Table 3 were taken, in addition to the Pc parameters for case 1 from Table 2 (Brooks-Corey Pc) and Pc parameters for case 2 and 3 indicated in Table 5.

Table 5. Simulation parameters for Pc

CASE 1, Brooks-Corey Pc parameters		
Pce	0.52	psi
λ	0.476	
Iw	1,0	
CASE 2, log(beta) Pc parameters		
Po	11.6	psi
β	5	
Sw(Pc=0)	0,50	frac.
Iw	0,5	
CASE 2, log(beta) Pc parameters		
Po	11.6	psi
β	5	
Sw(Pc=0)	0,16	frac.
Iw	0,0	

The resulting Pc curves are shown in Figure 12 below:

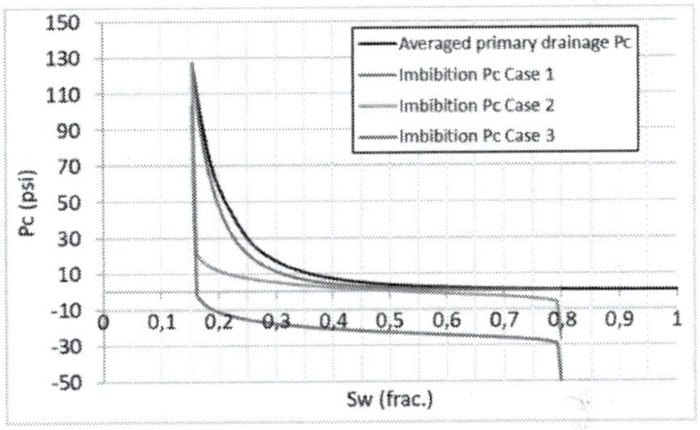

Fig. 12. Primary drainage (black) and imbibition capillary pressure curves for case 1 (blue), case 2 (orange) and case 3 (red)

The results from the three simulations are presented in Figure 13.

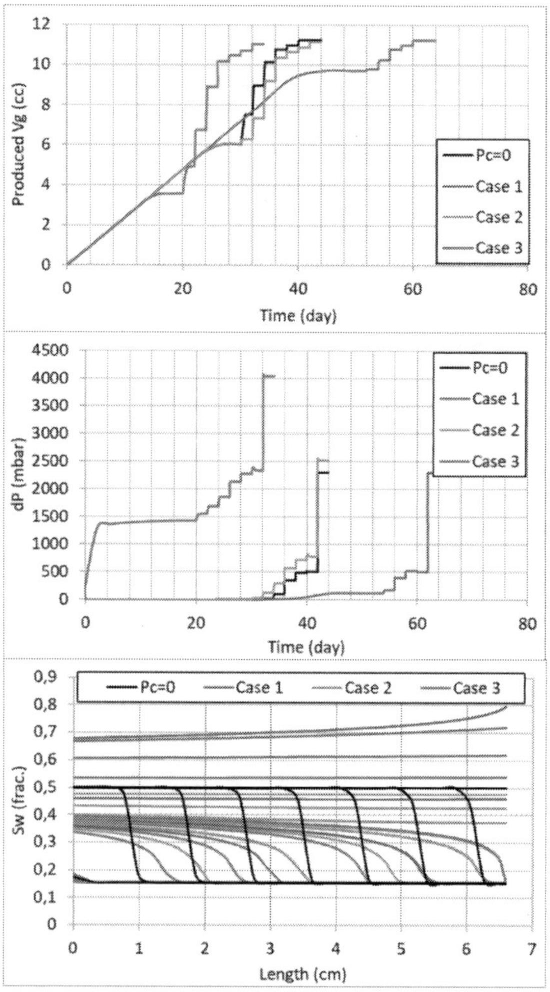

Fig. 13. Gas production, dP signal and saturation profiles during the first ratio (Qw=0.01cc/h – Qg=99.99cc/h)

The results from the simulations show that:

- The Pc effect is visible on the extreme cases in term of experimental durations, with shorter experiment with the least water-wetness condition and longer experiment with strongly water-wet condition
- More water-wet condition, longer and higher gas production at the first rate, but also lowest level of dP making the experiment almost not useable and uncertain, confirming that the steady-state Kr curve cannot be described over the full saturation range
- Case 2 and case with Pc=0 are very close regarding dP and produced Vg. Even with Iw=0.5, case 2 Pc curve remains flat, close to Pc=0 over a large range of saturation. Nevertheless, the saturation profiles are different (case 2 profiles tilted compared to "front like displacement" profiles at Pc=0)
- Case 3 results in short experimental duration, with readable high dP value and less produced Vg at the first ratio, allowing to potentially design an acceptable test. But the non-water wet Pc curve is not consistent with the measured Kr curve on the specific presented experiment. But it shows that, if the rock is not strongly water-wet compared to gas (as observed in the literature on some CO_2-brine coreflooding studies), it may be possible to measure the relative permeability curve

As explained in the last bullet point, imbibition gas-water Kr could be measured. But additional limitations can be encountered. As steady-state approach requires dP and produced Vg stabilization to calculate the relative permeability point at equilibrium, a lot of PVs must be injected through the sample: it may lead to some gas diffusion in water. Moreover, at high measured dP, gas compressibility and expansion may affect the in-situ saturation at ambient conditions: it is then preferable to perform more complex experiments at high pore pressure, with live fluids and specific HPHT separators and so on. Again, it only makes sense if the rock is not strongly water-wet.

6 Conclusions and Recommendations

In addition to a bibliography review on gas-water relative permeability and trapping relationship, experimental results from a consolidated sandstone reservoir study were compared with an indirect method to determine imbibition relative permeability. It is concluded that, after properly defining rock types using the measured primary drainage capillary pressure, the Brooks-Corey approach coupled with Land's hysteresis model is a good proxy for providing both drainage and imbibition gas-water relative permeability when water is strongly wetting the rock compared to gas. Due to the strong water wettability to the rock, it is extremely difficult or even impossible to experimentally determine imbibition relative permeability with unsteady-steady or steady-state coreflooding tests. The key parameters for the imbibition cycles are Krw Max and the Sgr-Sgi relationship: the shape of the Kr curve has negligible influence on the microscopic efficiency in imbibition cycle at strongly water-wet conditions, with very favorable viscosity ratio resulting in piston like displacement (no mobile gas behind the front).

It is not required to perform steady-state experimental test to attempt obtaining the imbibition gas-water Kr curve on water-wet rock. Numerical simulations confirmed this statement using the presented experimental case. Now, considering a non-water-wet system, as observed in some CO_2-brine studies at HPHT conditions, unsteady or steady-state coreflooding tests may be performed to capture the relative permeability curves.

All these observations confirm the findings in [6]: they will help in defining experimental protocols and interpretation of petrophysical parameters Kr and Pc in gas reservoirs with aquifer encroachment but also in saline aquifers that are identified as suitable storage formations for carbon dioxide (CO_2): in this last example, as the CO_2 plume rises upward in the formation due to the density contrast, the re-imbibition of water in the CO_2 flooded area will trap CO_2 as discontinuous non-wetting phase: assessing the capacity of the aquifer to trap the CO_2 is obviously of high importance.

The authors would like to thank TOTALENERGIES for permission to publish this work and Gerald Hamon for his knowledge sharing throughout his career.

Nomenclature

C: Land's coefficient
CT: computed tomography
D: diameter, cm
dP: differential pressure, mBar
FTB: Fontainebleau outcrops
K: permeability, mD
Kg: gas permeability, mD
Kkl: Klinkenberg gas permeability, mD
Kw: water permeability, mD
Kw Max: maximum water permeability, mD
Kr: relative permeability
Krg: gas relative permeability
Krg Max: maximum gas relative permeability
Krw: water relative permeability
Krw Max: maximum water relative permeability
L: length, cm
MB: material balance
MICP: mercury injection capillary pressure
Ng: gas Corey exponent
Nw: water Corey exponent
Pc: capillary pressure
Pce: entry capillary pressure
p.u.: porosity unit
Qg: gas flow rate
Qw: water flow rate
Sg: gas saturation

Sg: normalized/reduced gas saturation*
SgF: free and normalized/reduced water gas saturation*
Sgi: initial gas saturation
Sgi: normalized/reduced initial gas saturation*
Sgr: residual gas saturation
Sgr: normalized/reduced residual gas saturation*
Sgrw: residual gas saturation post waterflood
Sgrm: maximum residual gas saturation with Sgi=1
Sw: water saturation
Sw: normalized/reduced water saturation*
Swi: irreducible water saturation
s.u.: saturation unit
Vg: produced gas volume

ε: *function of* λ, *with* $\varepsilon=(2+3\lambda)/\lambda$
ϕ: *porosity*
λ: *pore size distribution index*
ρg: *gas density*
μg: *gas viscosity*
ρw: *water density*
μw: *water viscosity*

References

1. Kumar, M. et al.: "Imaging of pore scale distribution of fluids and wettability", SCA 2008-16.
2. Lenormand, R. et al.: "Mechanisms of the Displacement of One Fluid by Another in a Network of Capillary Ducts", (1983), J. Fluid Mech., 135, 337-353.
3. Wardlaw, N.C.: "Fluid topology, pore size and aspect ratio during imbibition", Transport in Porous Media, 1988, V3,17-37.
4. Jerauld, G. et al.: "The effect of pore-structure on hysteresis in relative permeability and capillary pressure: Pore-level modeling", Transport in Porous Media, vol. 5, no. 2, pp. 103-151, 1990.
5. Blunt, M. et al.: "Pore-level modeling of wetting", Physical Review E, vol. 52, no. 6, pp. 6387- 6403, 1995.
6. Cense, A. et al.: "SCAL for gas reservoirs: a contribution for better experiments", SCA 2016-023.
7. Geffen, T. et al.: "Efficiency of gas displacement from porous media by liquid flooding", Transactions AIME, vol 195, 1952, p 29-38.
8. Legatski, M. et al.: "Displacement of gas from porous media by water", SPE 899, 1964.
9. Bull, O. et al.: "The quest for the true residual gas saturation- An experimental Approach", SCA 2011-03.
10. Pickell, J. et al.: "Application of air-Mercury and oil-air capillary pressure data in the study of pore structure and fluid distribution", SPEJ, March 1966, p 55-61, also SPE 1227.
11. MacKay, B.: "Laboratory studies of gas displacement from sandstone reservoir having strong water drive", APEA Journal 1974.
12. Handy, L.: "Determination of effective capillary pressure for porous media from imbibition data", Petroleum Transactions AIME, 1960, vol 219, p 75-80.
13. Chierici, G. et al.: "Experimental research of gas saturation behind the waterfront in gas reservoirs subjected to water drive", World Petroleum congress Frankfort, 19-26 June 1963.
14. Delclaud, J.: "Laboratory measurements of the residual gas saturation", SCA 1991-23 EURO, 1991.
15. Keelan, D.: "A critical review of core analysis techniques", JCPT, April-June 1972, p 42-55.
16. Fahes, M. et al.: "Wettability alteration to intermediate gas-wetting in gas-condensate reservoirs at high temperatures", SPEJ, December 2007, p 397-407.
17. Land, C. S.: "Calculation of Imbibition Relative Permeability for Two- and Three-Phase Flow from Rock Properties", SPEJ, (June 1968), 149.
18. Keelan, D.K. et al.: "Trapped-gas saturation in carbonate formations", SPE 4535, (1975).

19. Aissaoui, A.: "Etude théorique et expérimentale de l'hystérésis des pressions capillaires et des perméabilités relatives en vue du stockage souterrain de gaz", PhD thesis, Ecole des Mines de Paris, 1983.
20. Jerauld, G.: "General three phase relative permeability model for Prudhoe Bay", SPERE 12(4) p 255-263, 1997
21. Suzanne, K. et al.: "Residual gas saturation of sample originally at residual water saturation in heterogeneous sandstone reservoirs", SCA 2003-14.
22. Suzanne, K. et al.: "Distribution of trapped gas saturation in heterogeneous sandstone reservoirs", SCA 2001-14.
23. Hamon, G. et al.: "Field-wide variations of trapped gas saturation in heterogeneous sandstone reservoirs", SPE 71524, 2001.
24. Suzanne, K. et al.: "Experimental relationships between residual gas saturation in heterogeneous sandstone reservoirs", SPE 84038, 2003.
25. Purcell, W. R.: "Capillary Pressures – Their Measurement Using Mercury and the Calculation of Permeability" Trans. *AIME*, (1949), 186, 39.
26. Burdine, N. T.: "Relative Permeability Calculations from Pore Size Distribution Data" *Trans. AIME*, (1953), 198, 71.
27. Corey, A. T.: "The Interrelation between Gas and Oil Relative Permeabilities", Prod. Mon. (1954), 19, 38.
28. Brooks, R. H. and Corey, A. T.: "Properties of Porous Media Affecting Fluid Flow", *J. Irrig. Drain. Div.*, (1966), 6, 61.
29. Leverett, M.C.: "Capillary behavior in porous solids". *Transactions of the AIME* (142): 159–172, 1941.

https://doi.org/10.1051/e3sconf/202336701008

Carbonated Smart Water Injection for Optimized Oil Recovery in Chalk at High Temperature

Md Ashraful Islam Khan[1*], *Sander Haaland Kleiberg*[1], *Ivan Dario Pinerez Torrijos*[1], *Tina Puntervold*[1] *and Skule Strand*[1]

[1]Department of Energy Resources, University of Stavanger, 4036 Stavanger, Norway

Abstract. Finding cost-efficient ways of increasing oil production with a low carbon footprint is the new challenge for the petroleum industry that wants to meet the net-zero emission goals by 2050. Smart water injection is an EOR process that increases oil production and delays water breakthrough by wettability alteration. Seawater is a smart water in chalk reservoirs, being especially effective at high temperatures. Different studies have shown that the effectiveness of seawater can be further improved by modifying the ion composition before injection.

Carbonated water (CW) has been proposed as a potential EOR fluid. In addition to producing extra oil, the reduction of greenhouse gas (CO_2) in the atmosphere can be achieved by using carbonated smart water as an injection fluid. The main mechanism behind increased oil recovery by injecting carbonated water is believed to be oil viscosity reduction and swelling, as the CO_2 is transferred from the aqueous phase to the oil phase. Wettability alteration has also been proposed as a possible mechanism, and this hypothesis is further investigated in this study along with other proposed mechanisms.

Stevns Klint outcrop chalk was used in this study, this material is recognized as an excellent analogue for North Sea chalk reservoirs. Optimized oil recovery by carbonated water in chalk was investigated at a high temperature (130°C) by flooding carbonated formation water (CFW) and carbonated seawater (CSW), to be compared with high saline formation water (FW) and seawater (SW) flooding. The oil/brine/rock/CO_2 interactions were tracked by measuring the pH of the produced water (PW) and by identifying any mineralogical changes by SEM (Scanning Electron Microscope) and EDX (Energy Dispersive X-Ray) analyses. The solubility of CO_2 in different brines was measured and compared with simulation data performed by PHREEQC. The diffusion of CO_2 from the aqueous phase to the oil phase was analysed to check if enough CO_2 can be diffused from the carbonated water into the oil phase.

By flooding CSW in both secondary and tertiary modes, a slight increase in the oil recovery was observed and was found to be the best performing brine. The oil recovery was also slightly increased using CFW in tertiary mode after FW which does not behave like smart water for carbonates.

The solubility of CO_2 was low and increased by increasing pressure and decreasing brine salinity. The acidity of CW did not increase by increasing pressure. No changes in pore surface minerals were observed after CW flooding, confirming limited mineral dissolution. A mass transfer of CO_2 from the brine phase to the oil phase was confirmed in the experimental work, but a significant amount of CO_2 remained in the brine phase.

The main mechanism behind this extra oil observed using CW is most likely not linked to oil swelling and viscosity reduction or mineral dissolution which could affect the porosity and the permeability of the rock system. Wettability alteration is a more likely explanation but needs to be looked further into for confirmation.

1 Introduction

The global energy demand is increasing by more than 1% every year because of the growth of the population and the increase in per capita energy consumption in developing countries. Although energy production from renewable sources is showing strong growth in recent years, oil is still the biggest source of energy production [1]. Therefore, to meet the net zero carbon emission goal by 2050, it is important to find a better way to produce oil in an environmentally friendly manner. Carbon capture, utilization, and storage (CCUS) might be a solution where CO_2 will be captured and used as a product to produce more oil, and therefore, CO_2 will be stored permanently in the reservoir after permanent plug and abandonment.

Previous lab experiments and field trials have shown that injection of CO_2 can increase oil recovery by reducing oil viscosity and increasing sweep efficiency [2]. The injected CO_2 was not fully recovered and thus shows storage possibility. However, direct CO_2 injection shows gravity segregation and viscous fingering effect that causes early CO_2 breakthrough along with technical difficulties in handling CO_2. These problems can be minimized by injecting carbonated water (CW) without sacrificing CO_2 storage potential [3].

* Corresponding author: sarjilsust09@gmail.com

Several mechanisms for CW EOR have been proposed in different literature [4]. The main mechanism behind extra oil recovery is believed to be the mass transfer of CO_2 from the water phase to the oil phase as the solubility of CO_2 is much higher in oil than in water [5-7]. This leads to oil swelling, reduction of interfacial tension (IFT) between oil and rock, and reduction of oil viscosity resulting in reconnection of isolated oil droplets, increases relative permeability of oil, and enhances the oil mobilization, thus increases sweep efficiency [3]. Wettability alteration due to CO_2 mass transfer was also proposed by Sohrabi et al. (2015). CO_2 dissolution into the oil phase destabilises the polar organic components of the oil attached to the rock surface and helps them to move. Decreases in the aqueous phase pH (in CW) also change the surface charges on the water/oil and water/rock interfaces which leads to subsequent changes in the wettability of the system and thus produces more oil [8].

The evolution of solution gas from the heavy oil caused by CO_2 dissolution is also considered one of the mechanisms for oil recovery by injecting CW [9]. The reaction between rock and CW resulting in rock dissolution and increase permeability by creating new a path to mobilize oil might also help to increase oil recovery [10].

Previously, our research has shown seawater (SW) works as smart water at high temperatures by changing the wettability of the reservoir. The effectiveness of SW can be further improved by removing NaCl from SW and spiking it with additional sulphate ions [11]. However, the preparation of this water demands additional energy to treat the water and obtain the desired compositions. Moreover, adding extra chemicals makes the injection less cost efficient. At high temperatures, non-stable sulphate concentration can lead to precipitation, affecting the porosity and permeability of the reservoir. Additionally, both the injection and the production facilities can experience severe scaling problems.

Therefore, there is a need for more environmentally friendly injection water that not only increases the oil recovery but also reduces the risk of precipitation. Injection of carbonated water can be a good alternative for the carbonate reservoir [4]. Many researchers reported extra oil recovery in carbonates by injecting carbonated water along with CO_2 storage possibilities [2].

Initial research done on carbonated water injection (CWI) in the 1940s by the Oil Recovery Corporation reported an extra 15% oil production when CW was injected after conventional water flooding [12]. Lake et al. (1984) have reported additional oil recovery of 26% in the tertiary mode by injecting CW in case of light oil. Recently, extensive research was conducted by Herriot Watt Institute of Petroleum Engineering Centre on CWI for oil recovery by core flooding and micro model which showed extra oil recovery both in the secondary and the tertiary mode [8]. Kilybay et al. (2016) performed a comparative oil recovery experiment by flooding different smart waters and carbonated smart water in carbonate reservoir core plugs and experienced extra oil recovery (~14%) for carbonated smart water injection. This extra oil recovery was attributed to the impact of CO_2 mass transfer from brine to oil inducing a viscosity drop, local flow diversion and trapped oil swelling. Carbonate dissolution and pore enlargement were also proven through NMR porosity and ICP-MS studies [14]. Sand pack flooding experiments conducted by Mosavat (2014) showed it is possible to store CO_2 by injecting CW. Kechut et al. (2011) showed about 47 - 51% of total CO_2 injected can be stored by carbonated water injection through numerical simulations and laboratory experiments.

The objective of this work is to determine if CW can induce EOR effects in chalk and to determine which mechanisms may be behind such effects. Thus, we compared the oil recovery performance of different core floods by injecting different carbonated brines at high temperatures in equally restored chalk cores. The reasons behind the extra oil observed were discussed. The solubility of CO_2 in different brine was studied by experimental work and simulations. The mass transfer of CO_2 from the brine phase to the oil phase was also confirmed through experimental work.

2 Experimental sections

2.1. Core Material

Outcrop Chalk collected from Stevns Klint (SK) quarry, near Copenhagen, Denmark, was used in this experiment. Cores were drilled from the same chalk block in the same direction and cut and shaped to the desired diameter of 3.8 cm and the desired length of 7 cm. All cores were inspected visually, and no visible fractures and distinct heterogeneities were found. These outcrop chalk cores consisted of 98% pure biogenic $CaCO_3$ and are similar to North Sea chalk reservoir cores. SEM (Scanning Electron Microscope) and EDX (Energy Dispersive X-Ray) analyses were done to see the changes in the minerals before and after the exposure to different carbonated brines. The specific surface area, (BET), porosity, and permeability of the cores were measured, and the values measured were found to be similar to previously published data [11, 16-21]. The properties of the cores are given in Table 1.

Table 1. Physical properties of the SK cores.

Core #	BET m^2/g	Porosity (%)	Water Permeability k_w (mD)	Pore volume (mL)
SK-3	2.0	48.5	3.7	38.7
SK-4		49.5	3.8	40.9
SK-6		45.4	4.1	36.3
SK-11		47.2	3.8	38.3

2.2 Crude Oil

A low asphaltenic stock tank oil with an acid number (AN) of 2.90 mg KOH/g and a base number (BN) of 0.95 mg KOH/g was used as base oil. This base oil was diluted with 40 weights % of heptane, centrifuged, and filtrated through a 5 μm Millipore filter. The AN of that diluted oil is found to be

 https://doi.org/10.1051/e3sconf/202336701008

~2.1 mg KOH/g and no precipitation of asphaltenic material was observed during storage after filtration. The surface active polar organic components were removed from a batch of the base oil using silica gel that provided an oil of ~0 AN. When the diluted oil and silica-treated oil were mixed, an oil (Oil A) with AN of 0.58 mg KOH/g and BN of 0.30 mg KOH/g was obtained which was used in the experiments. AN and BN of the oil samples were analysed by potentiometric titration. The density and viscosity of the prepared oil were determined to be 0.81 g/cm^3 and 2.4 cP, respectively.

2.3 Brines

Brines used in this experiment were made by mixing deionized water (DI) and reagent-grade salts. All the brines were mixed overnight by magnetic rotation and filtered through a 0.22 µm Millipore filter. SW composition is based on the seawater composition from the North Sea and the FW composition is based on the formation water from a North Sea Chalk reservoir. The properties of the brines are given in table 2.

Table 2. Properties of brines.

Ions(mM)/ Properties	SW	FW
[Na^+]	450.1	997.0
[K^+]	10.1	5.0
[Ca^+]	13.0	29.0
[Mg^{2+}]	44.5	8.0
[Cl^-]	525.1	1066.0
[HCO_3^-]	2.0	9.0
[SO_4^{2-}]	24.0	0.0
TDS (g/L)	33.34	62.83
Density (g/cm^3)	1.02	1.04
Bulk-pH	7.8	7.3

Carbonated seawater (CSW) and carbonated formation water (CFW) were prepared by equilibrating access CO_2 (g) with the respective brine in a pressure cylinder at 6-7 bar and 23 °C. The equilibrated carbonated brine was then moved to a separate cylinder at a higher pressure, so no gas cap was formed.

2.4 Core restoration

2.4.1 Establishing initial water saturation (S_{wi})

All cores were initially cleaned by flooding 5 PV of deionized (DI) water at room temperature to remove easily dissolvable salts, especially sulphate salts as described by Puntervold et al. (2007). The cores were then dried at 90°C to a constant weight. The initial formation water saturation (S_{wi}) of 10% was established by using the desiccator technique [22]. After they had reached 10% initial water saturation with formation water, the cores were stored in a sealed container for 3 days to allow an even ion distribution within the cores.

2.4.2 Oil exposure

The cores were then flooded with 4 PV of Oil (1.5 PV in each direction) at 50°C. Finally, the cores were wrapped in Teflon tape to avoid unrepresentative wetting on the outer surface and aged in the same oil for 2 weeks at 90°C to achieve a more homogeneous core wetting.

2.5 Oil Recovery by Spontaneous Imbibition

Spontaneous imbibition (SI) experiment was performed on a restored core to evaluate the degree of water wetness after core restoration. The experiment was performed at 110°C and 10 bar using formation water as the imbibing brine to avoid any chemically induced wettability alteration. The produced oil was collected in a glass burette that has a resolution of 0.1 ml. The volume of oil produced was calculated as % original oil in place (OOIP) versus time. The experimental setup is depicted in the following figure 1.

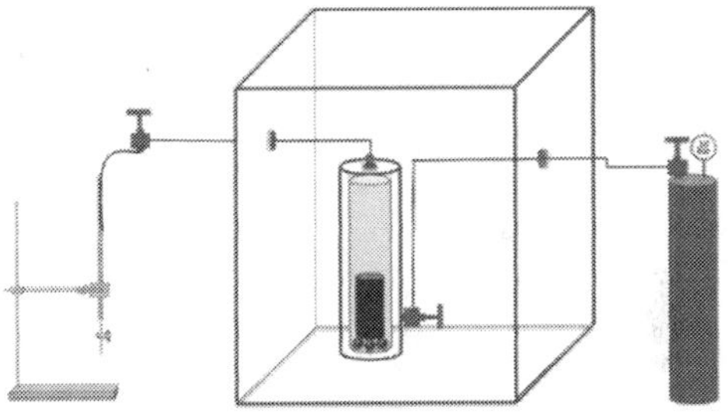

Fig. 1. Schematic of the setup used for spontaneous imbibition experiment.

2.6 Oil Recovery by Viscous Flooding (Forced Imbibition)

All experiments were performed at 130°C. The restored cores were mounted in the Hassler core holder with a confining pressure of 20 bar and a back pressure of 10 bar to prevent boiling. The cores were then successively flooded with different injection brines in secondary and tertiary mode at a constant injection rate of 1 PV/Day. When the recovery plateau was reached, the injection rate of the tertiary injection fluid was increased 4 times (4 PV/Day) to observe any end effects. The pressure drop was recorded during the experiment. Samples of produced water were collected during the experiment, and the pH of produced water were measured to catch any chemical reactions during the brine injections. The oil recovery, pressure drop, and the pH of the produced water was plotted against the PV brine injected. The experimental setup is illustrated in figure 2.

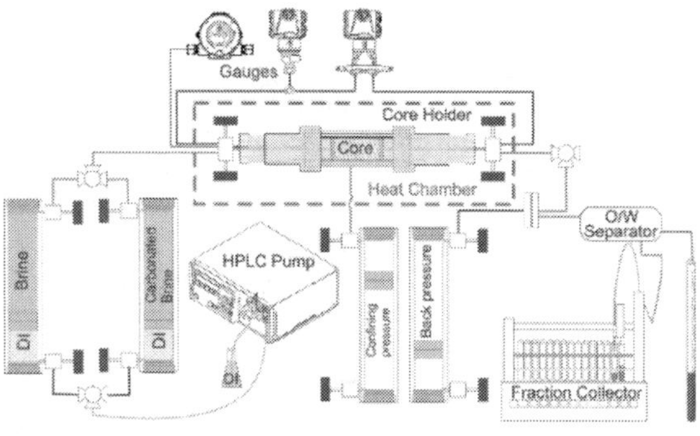

Fig. 2. Schematic of flooding setup.

2.7 Solubility of CO_2 in brines

The solubility of CO_2 in brines at 60°C was measured in the pressure range from 10 to 200 bar. 150 ml of a specific brine (DI, SW, FW) was mixed with 150 ml of CO_2 (g) at 10 bar in a pressured cylinder. The 2-phase system with a CO_2 gas cap above the brine was then temperature equilibrated overnight at 60°C. At each pressure stage the CO_2 – brine systems were mixed and allowed to equilibrate into 2 phases before a single flash of the equilibrated carbonated brine phase was performed. A small volume of the carbonated brine was flashed to standard conditions (SC). The pressure was maintained in the sample cylinder during the flash experiment. The mass of brine collected in the flash apparatus and liberated gas volume at SC was measured by a gasometer. The GOR as mg CO_2 per gram of brine was calculated based on the average of 3 flash experiments at each pressure stage. The pH of the carbonated brine was also measured in freshly taken samples after each flash experiment at ambient conditions.

2.8 Simulation of CO_2 solubility using PHREEQC

PHREEQC (version 3) software was used to simulate the solubility of CO_2 in different brines. The pH of the brines equilibrated with CO_2 was also simulated at specific pressure points at 60°C. The simulated data were then compared with the experimentally obtained data.

3 Results and discussion

3.1. Solubility of CO_2

To understand the potential of carbonated brine in EOR and the storage capacity of CO_2 in aquifers, it is important to understand the solubility of CO_2 in brines at pressure and temperature. Therefore, flash experiments were conducted to measure how much CO_2 is soluble in different brines at 60°C at a wide range of pressures. The amount of dissolved CO_2 in FW, SW, and DI water is presented in figure 3, as an average of the result from 3 flash experiments. The experimental data is also compared with simulated data obtained from PHREEQC.

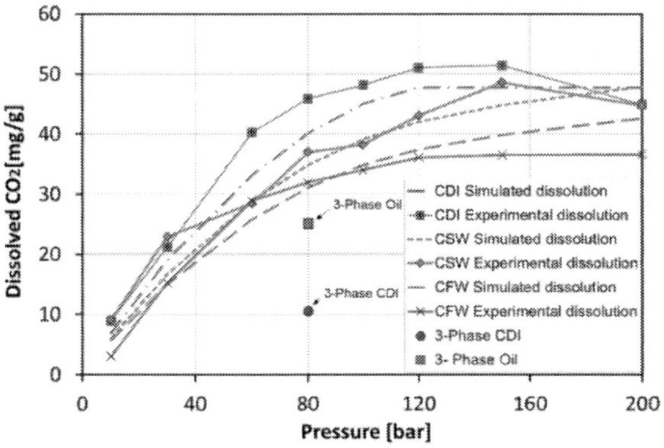

Fig. 3. Solubility of CO_2 in different brines at 60°C was plotted against pressure. Continuous lines are representing experimental data and dotted lines are representing the simulated data. 2 single points are representing the solubility of CO_2 in the water phase and the oil phase when both phases are present.

The experimental results show that the solubility of CO_2 gradually increases with the increase in pressure, but no significant increases are observed at pressures above 100 bar. CO_2 becomes a supercritical fluid when temperature and pressure exceed 31°C and 73 bar, without affecting the solubility of CO_2 in the brine phase. The solubility of CO_2 also depends on the salinity. The highest solubility is observed in DI water, and it decreases with increasing salinity. Thus, the formation water having the highest salinity (62830 ppm) showed the lowest solubility. However, the solubility difference was not more than 40% for FW and DI water at any pressure stages above the critical CO_2 pressure. The solubility of CO_2 in brines is low with maximum values of 50 mg/g in DI water. The solubility is less than 10 mg/g in all the brines at 10 bar which is the injection pressure for all the flooding experiments that will be presented in the next section.

The solubility of CO_2 in DI, SW, and FW has also been simulated using PHREEQC. The results from the simulations are presented in figure 3 and are in line with the experimental data, confirming the low CO_2 solubility at any pressure. This confirms that PHREEQC could be used to estimate CO_2 solubility in other brines or at other process temperatures.

The experimental data is aligned with simulated data and previously published experimental data by Duan et al. (2006) and Spycher and Pruess (2004).

The main mechanism behind extra oil recovery by carbonated water is the mass transfer of CO_2 from the water phase to the oil phase as the solubility of CO_2 is much higher in oil than in water. This phenomenon has been investigated in a 2-phase flash experiment at 80 bar and 60°C. CDI water at 80 bar and 60°C was equilibrated with the same amount of stabilized Oil A. Single flash experiments were performed from both equilibrated phases to SC.

At 80 bar and 60 °C, the CDI is able to dissolve 45 mg CO_2/g DI. After equilibrating with the same amount of crude oil, the amount of CO_2 in DI water was reduced 4 times down to 11 mg/g, while the amount of CO_2 transferred to the Crude oil Phase was 25 mg/g, giving a gas/oil ratio about 2.5 times higher than the gas/water ratio. Thus, this experiment confirms that CO_2 in carbonated water could be transferred to the oil phase as claimed by several researchers [5-7]. This transportation of CO_2 is because of the chemical potential difference between the two phases. But the amount of CO_2 that could be transported from carbonated water is very limited, so it is hard to believe that swelling and viscosity reduction of the oil phase could be the main mechanisms for enhancing oil recovery.

The brine phase collected from the single flash of carbonated water was collected, and the measured pH values for the individual brines at ambient conditions are presented in figure 4.

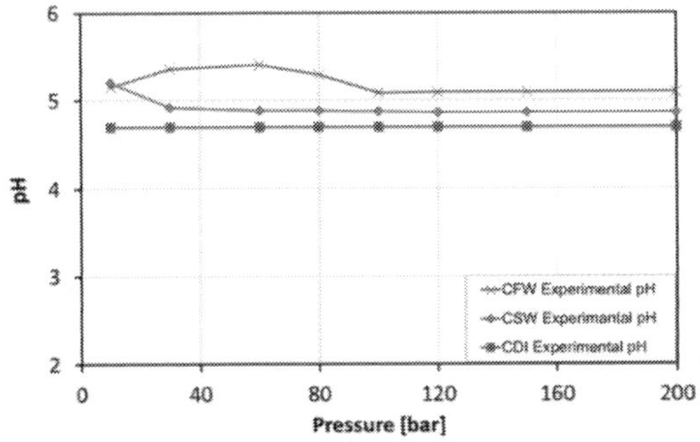

Fig. 4. The pH of different carbonated brines at different pressure and constant temperature (60°C). pH was measured at ambient conditions.

The pH was in the range of 4.7 to 5.2 for all brines and independent on the equilibrating pressure. CFW showed a slightly higher pH (~5) while CDI showed the lowest pH (~4.7).

Carbonic acid (H_2CO_3) is formed by dissolving CO_2 into different brine which is a weak acid (eq. 1). Carbonic acid is then rapidly separated into carbonate (CO_3^{2-}) and bicarbonate (HCO_3^-) (eq. 2 and 3) which provide hydrogen ions that give lower pH values (~5).

$$CO_2\,(aq) + H_2O \rightleftharpoons H_2CO_3 \quad (1)$$

$$H_2CO_3 \rightleftharpoons H^+ + HCO_3^- \quad (2)$$

$$HCO_3^- \rightleftharpoons H^+ + CO_3^{2-} \quad (3)$$

The constant pH values at different pressure points prove that the concentration of protons (H^+) cannot be increased by increasing the amount of CO_2 in the brine phase as we observed by increased pressure. Simulation of the pH of different carbonated brine using PHREEQC confirmed the same trend, figure 5. The simulated pH is lower than the measured because the pH was measured in ambient conditions.

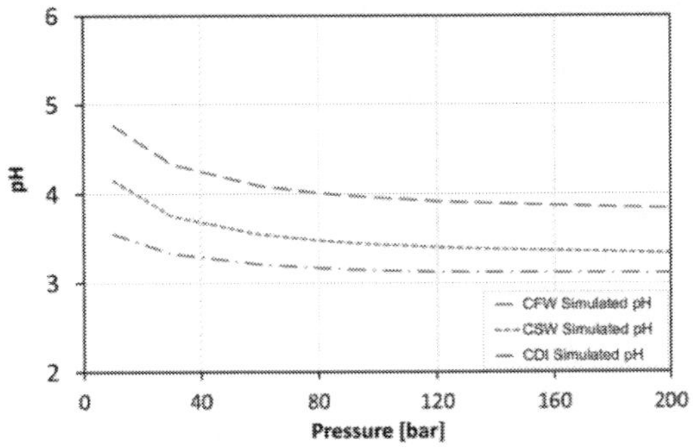

Fig. 5. Simulated pH of different carbonated brines at different pressure and constant temperature (60°C).

As the concentration of protons (H^+) cannot be increased with an increase in pressure, the chemical effect of carbonated water in EOR cannot be increased by increasing pressure.

3.2. Initial wetting of the restored cores

All oil recovery studies performed on equally restored SK chalk cores were taken from the same block to minimize the variation in physical core properties. Spontaneous imbibition (SI) is a practical way to approach and quantify the wettability of oil/brine/rock systems [25]. Both the speed of imbibition and the ultimate recovery gives valuable information to describe the core wettability after the core restoration. SI experiments could easily be performed at reservoir temperature, to reduce uncertainties regarding physical fluid properties and temperature effects. A spontaneous imbibition test was performed on the restored core SK-3 at 10 bar at 110°C. FW was used as the imbibing brine to exclude the effect of chemical induced wettability alteration during the test. The result is presented in figure 6.

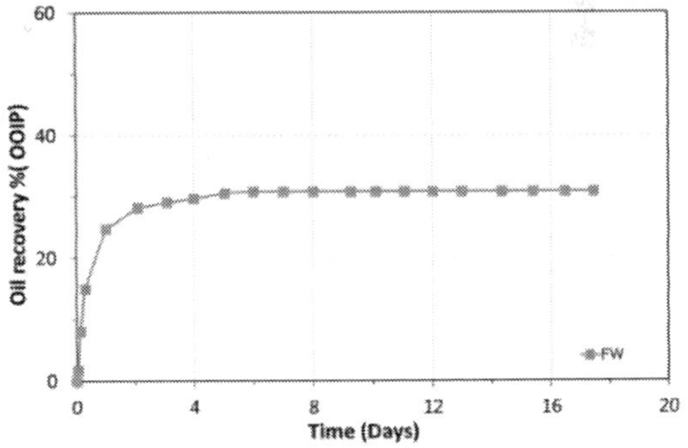

Fig. 6. Spontaneous imbibition into oil (AN=0.58 mgKOH/g) saturated (Swi=10%) chalk core at 110°C using FW as imbibing brine with a constant pressure of 10 bar. Oil recovery (% OOIP) was plotted against time (Days).

The Oil recovery reached its plateau at around 31% of OOIP on the 7th day, confirming positive capillary forces and that the core wettability is clearly on the water wet side.

A simplified wetting index only based on SI experiments can be used to calculate the wettability [26]. The degree of water-wetness can be quantified by a modified Amott water index (I^*_{W-SI}) (eq. 4) using a very water-wet core as a reference core.

$$I^*_{W-SI} = \frac{SI_C}{SI_{WWC}} \tag{4}$$

SI_{WWC} is the oil recovery (% OOIP) by spontaneous imbibition from a strongly water-wet SK reference core which have not been exposed to any crude oil. The value of the modified Amott water index is 1 for a strongly water wet core and 0 for a neutral to strongly oil wet core.

To calculate the modified Amott water index for SK cores, we have used a strongly water-wet SK core as the reference core described by Piñerez Torrijos et al. (2019). The reference core was restored with FW, S_{wi}=10%, heptane as oil phase, and FW was used as imbibing fluid at 23°C. A rapid oil production occurred, reaching a recovery of 68% OOIP after 40 minutes and an ultimate recovery of 71% OOIP after 4 hours. Therefore, the value of SI_{WWC} is 71. SI_C for the assessed SK-3 core is 31. I^*_{W-SI} for core SK-3 calculated by eq. 4 is then equal to 0.44 which suggests that the core has intermediate water wetness.

3.3. Flooding experiment

To evaluate the effectiveness of different types of injection brines on oil production, we have used equally restored sister cores of core SK-11, SK-6, and SK-4.

After core restoration, the cores were flooded at 130°C with different injection brines in both secondary and tertiary modes. The experiments were maintained by 20 bar confining pressure and 10 bar back pressure. The injection rate was constant and was 1 PV/D in all experiments performed. In the end, the injection rate was increased four times to evaluate any end effects. 4 different injection brines were used, formation water (FW), seawater (SW), carbonated formation water (CFW), and carbonated seawater (CSW). The carbonated brines were equilibrated with CO_2 gas at 6-7 bar, before the carbonated brine phase was transferred to a separate cylinder. The amount of CO_2 dissolved in the carbonated brines used for injection was then close to 5 mg CO_2/g of liquid. All injection brines were pressurized to 10 bar to match the back pressure of the system before injection.

In the first flooding experiment on core SK-11, FW was injected in secondary mode at a rate of 1 PV/D. The pressure difference between the inlet and outlet of the core was monitored. An ultimate oil recovery plateau of 68 %OOIP was reached after 2 PV was injected, figure 7a. The pressure drop (ΔP) gradually increased and reached the maximum value of 450 mbar after 0.5 PV was injected before it declined and stabilized at 150 mbar.

The pH in produced water samples was measured and compared with %OOIP in figure 7b. The first produced water was observed after 0.5 PV was injected (60 %OOIP). The produced water pH was close to 6.5, slightly lower than the bulk FW pH of 7.2.

After 3.5 PV injection, the injection brine was switched to CFW, figure 7. During the next 5 days (5 PV), 5% extra oil was mobilized. The ΔP immediately increased from 150 mbar to 800 mbar which could not be explained by a change in the viscosity of CFW. With 4 times increase in ΔP it is difficult to pinpoint if the extra oil mobilized is a result of the CFW or increased viscous forces.

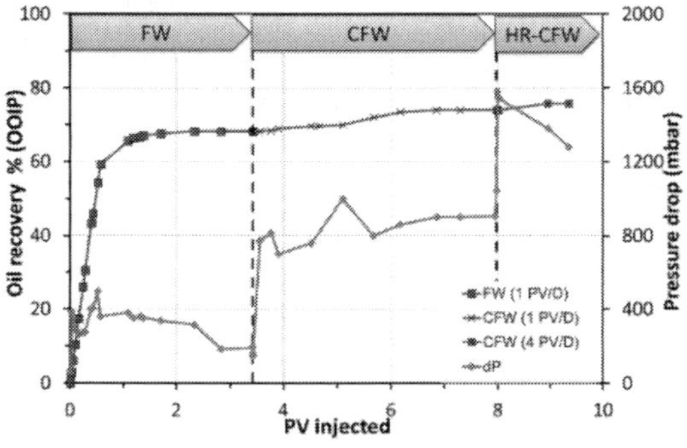

(a)

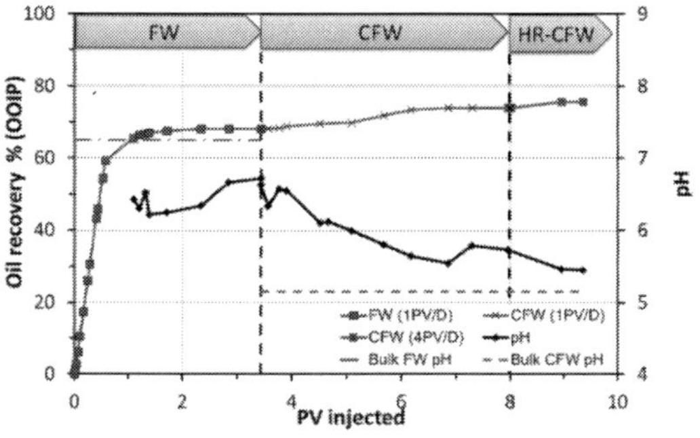

(b)

Fig. 7. Oil recovery tests on SK-11 core at 130°C. The flooding sequence was FW-CFW at a flooding rate of 1 PV/D. Oil recovery (%OOIP) was plotted against the PV injected and (a) average ΔP over the core, (b) pH of the produced water samples. At the end the injection rate was increased 4 times to 4 PV/D. Black dotted vertical lines are representing the change of injection fluid or change of injection rate. Coloured dotted horizontal lines represents the bulk pH of the injected fluids.

During CFW injection the pH in produced water samples went down below 6, confirming the presence of CFW which have a bulk pH of 5.2.

Since the chalk cores used in this experiment are mainly composed of calcite minerals, $CaCO_3$ dissolution could happen in presence of water, and $CaCO_3$ breaks into Ca^{2+} and CO_3^{2-} ions. CO_3^{2-} that take up a proton (H^+) from water and becomes HCO_3^- leaving OH^- behind. On the other hand, CO_2 dissolved in water will form a weak Carbonic acid, H_2CO_3 which turns to H^+ and HCO_3^-. Then Ca^{2+} can take two ions of HCO_3^- and becomes soluble $Ca(HCO_3^-)_2$ while OH^- and H^+ combine and become H_2O. Thus, the weak carbonic acid system is buffered in the porous media. This chemical process is illustrated simply in figure 8. The chemical equilibrium depicted in the figure does not change too much regardless of which brine (FW, SW, or DI) is being used.

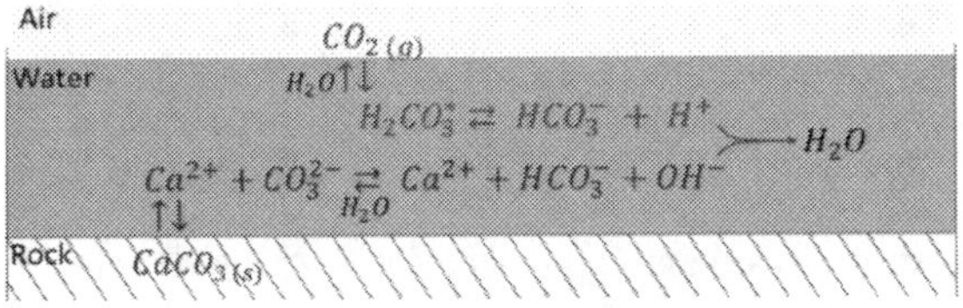

Fig. 8. Simple illustration of chemical processes happening inside the core in presence of both $CaCO_3$ (s) and CO_2.

Therefore, the carbonate is working as a buffer to the injected CFW and increases its pH. Nevertheless, a change in pH can change the reactivity of the polar organic components (POC) towards the positively charged calcite rock surface as seen in the equations below.

$$\mathbf{R-COOH} + H_2O \rightleftharpoons R-COO^- + H_3O^+ \quad (5)$$

$$R_3N: + H_2O \rightleftharpoons \mathbf{R_3NH^+} + OH^- \quad (6)$$

Negative charged POC (R-COO⁻) has the strongest affinity towards the positively charged calcite surface [28]. A reduction in pH could theoretically reduce the amount of R-COO⁻ species and affect wettability, but the pK_a values for Naphthenic Acids described in Eq. 5 are typically below 4.5 [29] and will not significantly change the amount of non-protonated carboxylic acids.

However, the pressure difference during CFW was four times higher than FW. If viscous forces are important in the oil mobilization this could also cause higher oil recovery.

According to Wiebe and Gaddy (1940), the solubility of CO_2 reduces with an increase in temperature. In our experiment, the injection fluid was prepared at room temperature and then injected into the core at high temperature. The significant increase in pressure drop development is most likely linked to gas liberation and a third fluid phase in the pore system.

Many researchers have claimed that when carbonated water (CW) is injected into the carbonate reservoir, the acidic nature increases the dissolution of carbonate minerals which could promote an increase in porosity and permeability [31-34]. These changes in petrophysical properties help to increase oil recovery by creating a new flow path. $CaCO_3$ and $MgCO_3$ are carbonate minerals that can easily react with carbonated water. FW has a high concentration of both Ca^{2+} and Mg^{2+} ions which will limit the dissolution of carbonates even though the pH is somewhat reduced, as illustrated in figure 8. If there was a new flow path created during the oil recovery experiment in figure 7, the pressure drop should have been reduced.

After the oil production reached its plateau during CFW injection, the injection rate was increased from 1 PV/D to 4 PV/D to observe the effect of viscous forces and any end effects. A marginal extra oil mobilization of 1.5 %OOIP. The pH of the produced water was somewhat lower than at 1 PV/day, which could be explained by that the injection fluid has less time to interact with the minerals.

In the second oil recovery experiment on core SK-6, CSW was flooded in secondary mode followed by SW in tertiary mode. Injection of CSW gave an ultimate oil recovery of 79 %OOIP which was reached after 2.5 PV injected, figure 9.

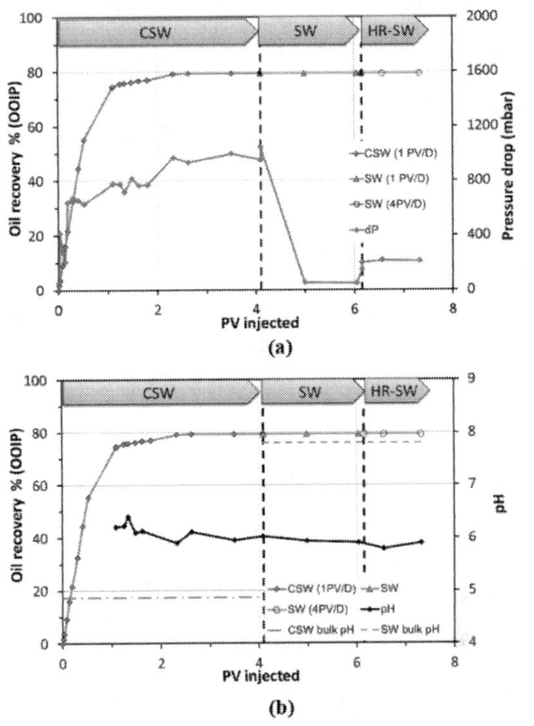

Fig. 9. Oil recovery tests on SK-6 core at 130°C with a back pressure of 10 bar. The flooding sequence was CSW-SW-HRSW. The flooding rate was 1 PV/D followed by a higher rate (HR) of 4 PV/D at the end. Oil recovery % (OOIP) was plotted against the PV injected of respective brine and (a) the average ΔP around the inlet and the outlet of the core, (b) the pH of the produced water. Black dotted lines are representing the change of injection fluid or change of injection rate. Colored dotted lines are representing the bulk pH of the injected fluid.

The ΔP was gradually increased during the hole CSW injection of 4 PV, even though water breakthrough took place after 0.5 PV injected. As previously discussed, the solubility of CO_2 reduces in higher temperatures and free CO_2 gas may promote the increase in ΔP.

After switching to SW, no extra oil mobilization was observed, but a significant decline in ΔP developed. This clearly indicates that free CO_2 liberated during CSW injection is taken up by SW and a new 2-phase flow system of Oil A and brine reestablished a more normal ΔP of 80 mbar. The results clearly indicate that CO_2 liberated from the brine phase trapped in the pore system is not easily diffusing into the residual oil but is being taken up in SW when that is passing through.

Increasing the injection rate 4 times did not mobilize any extra oil, but the ΔP increased to 200 mbar confirming typical pressure drop performance for a 2-phase fluid system in a heterogenous core. The pH of produced water is slightly acidic with a pH close to 6 during injection of both CSW and SW injection. This is slightly above the bulk pH of CSW and significantly below the bulk pH of SW.

 https://doi.org/10.1051/e3sconf/202336701008

Comparing the recovery results from FW-CFW injection (figure 7), and CSW- SW injection (figure 9), the CSW is significantly more efficient than FW in secondary mode. We know that SW behaves as a Smart Water and is able to change wettability and improve the sweep by inducing more positive capillary forces [11, 35]. To be able to evaluate if the significant improvement in the oil recovery during CSW injection is a result of wettability alteration by the ions present in the SW, or if the CO_2 present in the aqueous phase also contributes to the extra mobilized oil, a third core flooding experiment was performed on core SK-4 by injecting SW followed by CSW.

SW injection in secondary mode gave an ultimate oil recovery of 74 %OOIP after 2 PV injected. The water breakthrough was observed after 0.5 PV injected when 65 %OOIP was recovered. The experimental result is presented in figure 10.

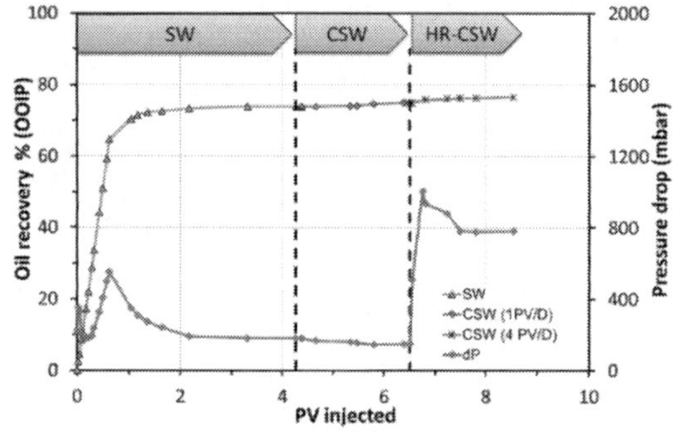

(a)

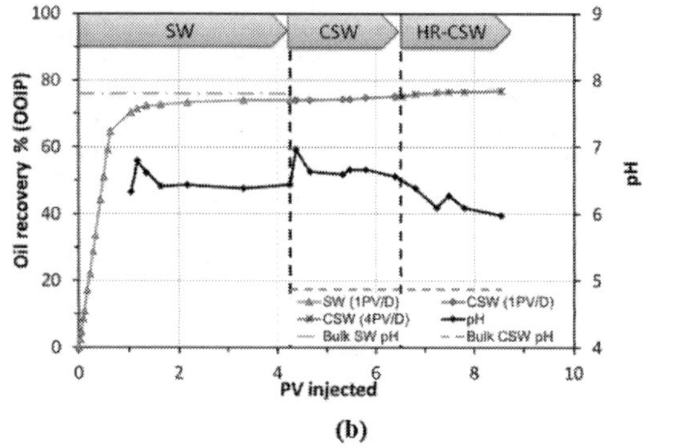

(b)

Fig. 10. Oil recovery tests on SK-4 core at 130°C with a back pressure of 10 bar. The flooding sequence was SW-CSW-HR CSW. The flooding rate was 1 PV/D followed by a higher rate (HR) of 4 PV/D at the end. Oil recovery % (OOIP) was plotted against the PV injected of respective brine, and (a) the average pressure drops around the inlet and the outlet of the core, (b) the pH of the produced water. Black dotted lines are representing the change of injection fluid or change of injection rate. Coloured dotted lines are representing the bulk pH of the injected fluid.

The ΔP profile is more normal and follows the same trend as observed during secondary FW injection when no gas phase is present, figure 7. The highest ΔP of 500 mbar was observed after 0.6 PV, gradually declining to 170 mbar when the oil recovery plateau was reached. The PW pH was almost constant and was around 6.5 during the SW injection.

After 4 PV, the injection brine was changed to CSW. During the next 3 PV, a slight increase of 1% OOIP extra oil was observed, indicating a tiny effect of the CO_2 present, supported by a slight decline in the PW pH towards 6. The ΔP during CSW injection indicated no free gas phase developing, and a low and stable ΔP of 150 mbar was observed, peaking at 1000 mbar during high-rate injection, before stabilizing again at 800 mbar. The significant increase in injection rate and ΔP had only a minor effect on oil mobilization, improving the recovery by only 2 %OOIP.

The overall recovery results confirm that by changing the chemistry of the injection brines, significant changes in ultimate oil recoveries could be observed. SW behaves as a Smart Water in mixed wet chalk, and the SW injection improved the ultimate oil recovery of 74% OOIP compared to baseline oil recovery using FW giving 68% OOIP. By adding CO_2 to the brine phase, carbonated brines could be formed. By injecting CSW in secondary mode an ultimate oil recovery of 79% OOIP was reached which is significant above SW recovery. Tertiary injection of CFW after FW and CSW after SW injection indicates smaller EOR effects.

The experimental results clearly demonstrate that the amount of CO_2 that could be dissolved in the brine phase is very limited. A significant amount of the CO_2 will still remain in the brine phase after equilibrium with the crude oil. The classical explanation linked to swelling and viscosity reduction of the oil phase is not likely the main mechanism for the extra oil mobilized during CW injection.

To evaluate the EOR mechanism of carbonated brines, we need to understand what could happen at the mineral surfaces linked to mineral dissolution processes and/or wettability alteration which we know could improve the sweep efficiency in water displacement processes in the presence of CO_2. We clearly observe a pH reduction during CW injection which due to mineral dissolution which was described in figure 8.

3.4. SEM and EDX analysis

Carbonated water will affect the mineral dissolution of calcite as described in figure 8. The effect of carbonate water exposure on mineral composition and pore surface minerals have been investigated by performing SEM and EDX analyses. In table 3, the mineral composition of SK Chalk in atomic weight % retrieved from EDX analyses before carbonated brine exposure is given in Table 3.

Table 3. Atomic weight % of SK cuttings retrieved from EDX before carbonated brine exposure.

Element	Na	Mg	Al	Si	S	K	Ca
Atomic %	0.02	0.13	0.19	0.81	0.54	0.23	98.06

The result confirms that SK chalk is very pure, consisting of more than 98% $CaCO_3$. The rest of the minerals are silicate minerals, Quartz, Clay, and/or Feldspars.

https://doi.org/10.1051/e3sconf/202336701008

A Scanning Electron Microscopy (SEM) picture of the sample is given in figure 11.

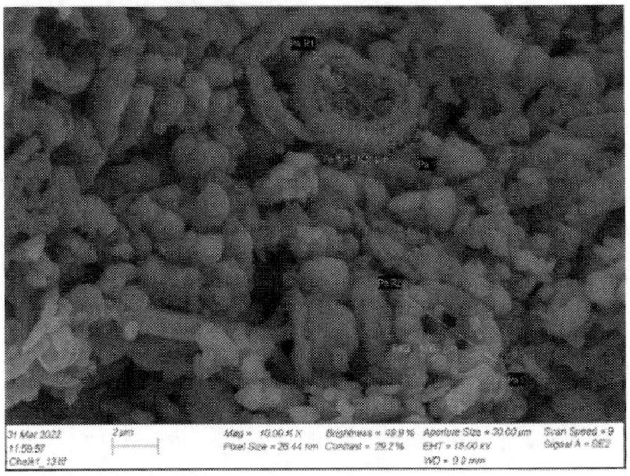

Fig. 11. SEM photo of the core sample before oil recovery test.

The presence of the intact coccolith ring confirms the biogenic origin of the Chalk.

After the core flooding experiments with carbonated brines, rock samples from the flooded cores were taken, dried, and analysed by SEM and EDX to track any mineralogical or visual changes in the mineral surfaces. Figure 12, 13, and 14 is representing the SEM photos from core SK-11, SK-4, and SK-6. Compared with figure 11, no visual changes on the mineral surfaces after carbonated brine exposure were detected. The coccolith rings are still intact and the grain sizes of the rock fragments are similar. No significant changes were found in the elements after the exposure by EDX.

The unchanged cation composition confirms negligible mineral dissolution by carbonated water. A few PV with carbonated brine is not enough to significantly affect the mineral surfaces when we have 2 m^2 of mineral surfaces in one gram of Chalk. At the same time, the calcite dissolution in brines with a high Ca^{2+} concentration as we have in FW and SW should be low due to the common ion effect.

However, Kono et al. (2014) observed significant dissolution of carbonate minerals by SEM. They reported smaller and smoother grains after carbonated water exposure and observed a significant increase of Ca^{2+} in the effluent brine. But in these core experiments, the total pore volume of fluid injected was several hundred, and without reporting injection rates. It is apparent that high injection volumes and rates could cause carbonate dissolution. Riazi (2011) found even sandstone is corroded by carbonated brine due to prolonged exposure to CWI for 2 weeks at the process conditions of 2000 psi and 38°C, which was a static exposure test, not a flooding experiment.

In our experiments, we have flooded SK cores with only about 4 PV which is more realistic to what could happen in the main part of the reservoir. Exposure with several PV with carbonated brines is more realistic to near wellbore effects for injection wells.

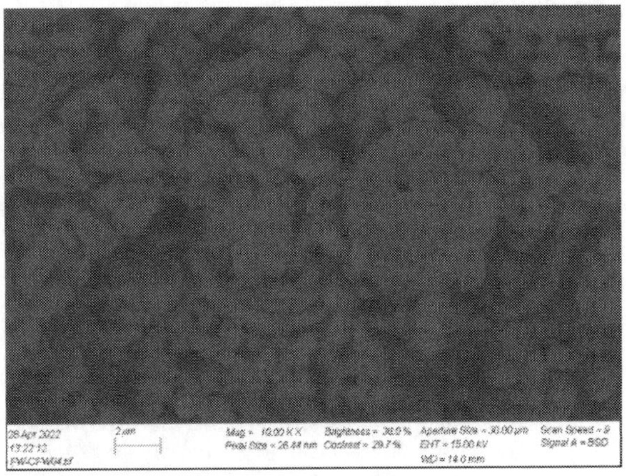

Fig. 12. SEM photo of chalk sample from core SK-11 after the oil recovery test by brine flooding (FW-CFW-HR CFW).

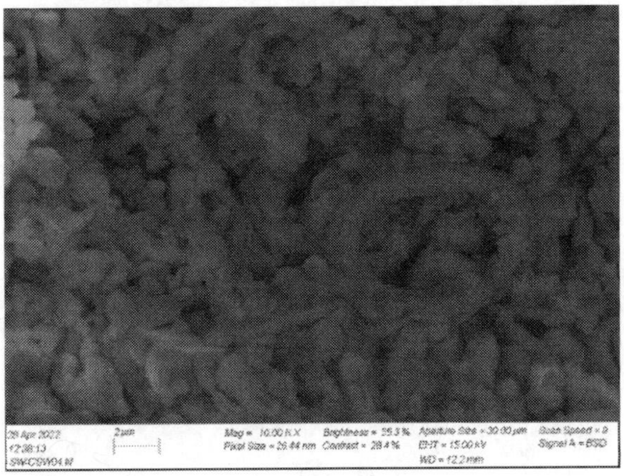

Fig. 13. SEM photo of chalk sample from core SK-4 after the oil recovery test by brine flooding (SW-CSW-HR CSW)

Fig. 14. SEM photo of chalk sample from core SK-6 after the oil recovery test by brine flooding (CSW-SW-HR SW)

4 Conclusions

Stevns Klint chalk was exposed to different carbonated brines at 130 °C to evaluate the effect of carbonated brines as an EOR fluid. The studies were coupled with CO_2 solubility experiments with 2 and 3 phases (CO_2, brines, and oil), as well as PHREEQC simulations. The main findings of this study were:

- The solubility of CO_2 in brine was low and depends on brine salinity. After reaching the super critical conditions of CO_2 the solubility did not increase significantly. The experimental solubility studies validated a relatively good performance of the PHREEQC simulator.
- Oil recoveries were slightly increased in chalk at high temperature both in secondary and tertiary mode. A mechanism linked to swelling and viscosity reduction of the oil phase is not likely be a main mechanism for the extra oil.
- Dissolution of $CaCO_3$ in presence of carbonated brines can also contribute to change in petrophysical properties and thus produce extra oil. However, no changes on pore surface minerals after flooding with carbonated brines were observed by SEM and EDX analyses.
- Wettability alteration might play a vital role in increasing oil recovery but need further investigation.

The author gratefully acknowledges the financial support from the University of Stavanger and the laboratory work support from master's student Mahmood Fani and Saja H. A. Algazban.

References

1. BP. *Statistical Review of World Energy*. 2020 [cited 2022 20 May]; Available from: https://www.bp.com/en/global/corporate/energy-economics/statistical-review-of-world-energy/oil.html.
2. Esene, C., et al., *Comprehensive review of carbonated water injection for enhanced oil recovery*. Fuel, 2019. **237**: p. 1086-1107.
3. Riazi, M., et al. *Oil Recovery Improvement Using CO2-Enriched Water Injection*. in *EUROPEC/EAGE Conference and Exhibition*. 2009.
4. Bisweswar, G., A. Al-Hamairi, and S. Jin, *Carbonated water injection: an efficient EOR approach. A review of fundamentals and prospects*. Journal of Petroleum Exploration and Production Technology, 2020. **10**(2): p. 673-685.
5. D. W. Green, a.G.P.W., *Enhanced oil recovery*. SPE Textbook Series,. 1998: Society of Petroleum Engineers, Richardson, Texas.
6. Abedini, A. and F. Torabi, *Parametric Study of the Cyclic CO2 Injection Process in Light Oil Systems*. Industrial & Engineering Chemistry Research, 2013. **52**(43): p. 15211-15223.
7. Dong, Y., et al. *An Experimental Investigation of Carbonated Water Flooding*. in *SPE Annual Technical Conference and Exhibition*. 2011.
8. Sohrabi, M., et al. *A Thorough Investigation of Mechanisms of Enhanced Oil Recovery by Carbonated Water Injection*. in *SPE Annual Technical Conference and Exhibition*. 2015.
9. Mosavat, N., *Utilization of Carbonated Water Injection (CWI) as a Means of Improved Oil Recovery in Light Oil Systems: Pore-Scale Mechanisms and Recovery Evaluation*, in *Faculty of Graduate Studies and Research*. 2014, University of Regina.
10. Kono, F., et al. *Laboratory Measurements on Changes in Carbonate Rock Properties due to CO2-Saturated Water Injection*. in *Abu Dhabi International Petroleum Exhibition and Conference*. 2014.
11. Puntervold, T., et al., *Modified seawater as a smart EOR fluid in chalk*. Journal of Petroleum Science and Engineering, 2015. **133**: p. 440-443.
12. Adiputra, E., L. Mucharam, and S.D. Rahmawati. *Experimental Evaluation of Carbonated Water Injection to Increase Oil Recovery Using Spontaneous Imbibition*. in *Selected Topics on Improved Oil Recovery*. 2018. Singapore: Springer Singapore.
13. Lake, L.W., et al., *Isothermal, multiphase, multicomponent fluid flow in permeable media*. In Situ; (United States), 1984. **8:1**: p. Medium: X; Size: Pages: 1-40 2009-12-16.
14. Kilybay, A., et al. *Hybrid EOR Technology: Carbonated Water and Smart Water Improved Recovery in Oil Wet Carbonate Formation*. in *SPE Annual Caspian Technical Conference & Exhibition*. 2016.
15. Kechut, N.I., M. Jamiolahmady, and M. Sohrabi, *Numerical simulation of experimental carbonated water injection (CWI) for improved oil recovery and CO2 storage*. Journal of Petroleum Science and Engineering, 2011. **77**(1): p. 111-120.
16. Parkhurst, D.A., C.A.J. , *User's Guide to PHREEQC (Version 2)--A Computer Program for Speciation, Batch-Reaction, One-Dimensional Transport, and Inverse Geochemical Calculations*, U.S.G. SURVEY, Editor. 1999, U.S. DEPARTMENT OF THE INTERIOR.
17. Piñerez, I., et al., *Core wettability reproduction: A new solvent cleaning and core restoration strategy for chalk cores*. Journal of Petroleum Science and Engineering, 2020. **195**: p. 107654.
18. Hopkins, P., T. Puntervold, and S. Strand, *Preserving initial core wettability during core restoration of carbonate cores*, in *The International Symposium of the Society of Core Analysts*. 2015: St. John's Newfoundland and Labrador, Canada.
19. Shariatpanahi, S.F., S. Strand, and T. Austad, *Initial wetting properties of carbonate oil reservoirs: effect of the temperature and presence of sulfate in formation water*. Energy & Fuels, 2011. **25**(7): p. 3021-3028.
20. Fathi, S.J., T. Austad, and S. Strand, *Water-based enhanced oil recovery (EOR) by "Smart Water": optimal ionic composition for EOR in carbonates*. Energy & Fuels, 2011. **25**(11): p. 5173-5179.

21. Puntervold, T., S. Strand, and T. Austad, *New method to prepare outcrop chalk cores for wettability and oil recovery studies at low initial water saturation.* Energy & Fuels 2007. **21**(6): p. 3425-3430.
22. Springer, N., U. Korsbech, and H.K. Aage, *Resistivity index measurement without the porous plate: A desaturation technique based on evaporation produces uniform water saturation profiles and more reliable results for tight North Sea chalk,* in *Paper presented at the International Symposium of the Society of Core Analysts* 2003: Pau, France.
23. Duan, Z., et al., *An improved model for the calculation of CO2 solubility in aqueous solutions containing Na+, K+, Ca2+, Mg2+, Cl−, and SO42−.* Marine Chemistry, 2006. **98**(2): p. 131-139.
24. Spycher, N. and K. Pruess, *CO2-H2O Mixtures in the Geological Sequestration of CO2. II. Partitioning in Chloride Brines at 12-100oC and up to 600 bar.* 2004.
25. Anderson, W.G., *Wettability literature survey- part 1: rock/oil/brine interactions and the effects of core handling on wettability.* Journal of Petroleum Technology, 1986. **38**(10): p. 1125-1144.
26. Austad, T., et al., *Smart Water EOR: Beneficios y Aplicaciones en Reservorios Clásticos y Carbonatos.* 2017.
27. Piñerez Torrijos, I.D., et al., *Impact of Temperature on Wettability Alteration by Smart Water in Chalk.* 2019. **2019**(1): p. 1-11.
28. Standnes, D.C. and T. Austad, *Wettability alteration in chalk. 1. Preparation of core material and oil properties.* Journal of Petroleum Science and Engineering, 2000. **28**(3): p. 111-121.
29. Standal, S., et al., *Effect of polar organic components on wettability as studied by adsorption and contact angles.* Journal of Petroleum Science and Engineering, 1999. **24**: p. 131-144.
30. Wiebe, R. and V.L. Gaddy, *The Solubility of Carbon Dioxide in Water at Various Temperatures from 12 to 40° and at Pressures to 500 Atmospheres. Critical Phenomena*.* Journal of the American Chemical Society, 1940. **62**(4): p. 815-817.
31. Bowker, K.A. and P.J. Shuler. *Carbon Dioxide Injection and Resultant Alteration of Weber Sandstone (Pennsylvania-Permian), Rangely Field, Colorado.* in *SPE Annual Technical Conference and Exhibition*. 1989.
32. Shiraki, R. and T.L. Dunn, *Experimental study on water–rock interactions during CO2 flooding in the Tensleep Formation, Wyoming, USA.* Applied Geochemistry, 2000. **15**(3): p. 265-279.
33. Wellman, T.P., et al. *Evaluation of CO2-Brine-Reservoir Rock Interaction with Laboratory Flow Tests and Reactive Transport Modeling.* in *International Symposium on Oilfield Chemistry*. 2003.
34. Grigg, R.B. and R.K. Svec. *CO2 Transport Mechanisms in CO2/Brine Coreflooding.* in *SPE Annual Technical Conference and Exhibition*. 2006.
35. Austad, T., et al., *Seawater as IOR fluid in fractured chalk,* in *Paper SPE93000 presented at the 2005 SPE International Symposium on Oilfield Chemistry*. 2005: Houston, TX, USA.
36. Riazi, M., *Pore scale mechanisms of carbonated water injection in oil reservoirs,* in *Petroleum Engineering*. 2011, Heriot-Watt University.

Nano-colloid based suspensions and emulsions used as means for enhanced oil recovery

Anastasia Strekla[1,2], *Christina* Ntente[1,3], *Maria* Theodoropoulou[1] and *Christos* Tsakiroglou[1,*]

[1]Foundation for Research and Technology Hellas - Institute of Chemical Engineering Sciences, 26504 Patras, Greece
[2]University of Patras, Department of Physics, 26504 Patras, Greece
[3]University of Patras, Department of Chemistry, 26504 Patras, Greece

Abstract. Aqueous solutions of polyphenols extracted from parsley leaves are mixed with solution of ferric chloride hexahydrate to synthesize suspensions of polyphenol-coated iron oxide nanoparticles, which are then used to prepare Pickering oil (n-decane)-in water emulsions. The performance of the nano-colloid suspensions and emulsions as agents of enhanced oil recovery (EOR) is assessed with secondary imbibition tests performed under constant flow rate on a glass-etched pore network, after a drainage / primary imbibition cycle for paraffin oil / brine (0.25M NaCl solution) system. The transient changes of the fluid saturation profile along with the pressure drop across the porous medium are recorded. Comparative analysis of the oil recovery efficiency attained with nano-colloid suspensions of varying iron concentration, and Pickering emulsion, enables us to classify them, and select the most efficient ones for further studies.

1 Introduction

The use of inorganic nanoparticles in petroleum engineering applications including enhanced oil recovery (EOR) has attracted the scientific interest in recent years [1]. An approach to improve the dispersion ability of nanoparticles and tailor their properties for a particular application is to covalently attach polymers to the nanoparticle surface, resulting in polymer-coated nanoparticles (PNPs). While less work has been carried out with PNPs for EOR, recent work suggests they may be superior to unmodified nanoparticles for EOR [2]. PNPs can be tailored for a specific interface and application. PNPs can decrease the interfacial tension of oil and water or water and air, which can lead to more stable emulsions [3-6]. For example, silica nanoparticles coated with a polyelectrolyte were used to stabilize oil-in-water emulsions [7]. More recently, stable oil-in-water emulsions were generated by using silica nanoparticles coated with a pH responsive polymer, and it was found that the most stable emulsions were formed at lower polymer chain grafting densities [8,9]. Related studies on star polymers [10], bottlebrush polymers [11] and paramagnetic particles with adsorbed amphiphilic polymers found stable emulsions [12] and reductions in the oil-water interfacial tension at relatively low (0.1 wt %) particle contents [13]. PNPs can respond to temperature, pH, and light through a change in surface properties [6].

Moreover, significant progress has been done on the development of PNPs for the in-situ remediation of contaminated porous media (e.g. soils, sediments, and aquifers) by injecting zero-valent iron nanoparticles (nZVI) [14-16], and know-how obtained in the context of such environmental studies might be helpful in preparing stable PNPs and designing efficient PNP-based EOR processes [17]. During the last years, green synthesis of NPs has been gaining importance in view of the advantages such as economic benefits, environment friendliness, and recycling. Such 'greener' processes for obtaining various nanomaterials have been widely reported and are gathering attention in several nanotechnology-based applications [18]. As one of the green synthetic approaches, plant extracts were considered for the synthesis of nanoparticles through biosynthetic routes which involve primarily the extraction of polyphenols, which are natural polymers acting as reducing agents and stabilizers [19-21].

In the present work, aqueous solutions of polyphenols extracted from the leaves of parsley are mixed with solution of ferric chloride hexahydrate to produce suspensions of iron oxide nanoparticles, and an ultrasound probe is utilized to generate stable oil-in-water emulsions by mixing the suspensions with n-decane. The performance of the nano-colloid suspensions and emulsions as agents of enhanced oil recovery (EOR) is assessed with secondary imbibition tests performed after a drainage / imbibition cycle for paraffin oil / brine (NaCl solution) system. Displacement tests are conducted under constant flow rate in a transparent glass-etched pore network. The transient changes of the fluid saturation profile along with the pressure drop across the porous medium are recorded. The oil saturation is measured with image analysis of successive snap-shots captured by a CCD camera [22]. Comparative analysis of the oil displacement efficiency attained with nano-colloid suspensions and Pickering emulsions enables us to classify them and pre-select the most efficient ones for further studies.

* Corresponding author: ctsakir@iceht.forth.gr

2 Materials and Methods

2.1 Synthesis of nanoparticles suspensions and emulsions

Initially, 1000 mL of tri-distilled (3D) water were heated to 80 ° C on a hot plate. Then, 50 g of chopped parsley were added in the water and left for 1 hr under stirring at constant temperature. Afterwards, the liquid extract was separated from the solid residue by vacuum filtration with the use of filters of pore size 4 μm, and 2.5 μm (Whatman) and centrifugation at 12000 rpm for 20 min (Heraeus Megafuge-16). The total polyphenol concentration (TPC) of extract was measured in terms of Gallic Acid Equivalent (g/L GAE), by using the Folin-Ciocalteu method [23] and UV-Vis spectroscopy (HITACHI U-3000).

The preparation of the iron oxide nanoparticles (IONP) was accomplished by adding, under stirring, the polyphenol extract (PPH) to a 0.1M $FeCl_3$ solution at a volume ratio 2:1 [24]. The reaction progress was monitored by recording continuously the pH and Redox Potential (ORP), Eh, of the solution with two probes (Vernier pH and ORP electrodes) connected with a data acquisition card (LabQuest Steam) (Fig.1a,b). When pH and Eh were stabilized, the solution colour changed to dark brown signifying the completion of the reaction. After having adjusted its pH to 6.0, the suspension was placed in a refrigerator and kept there until being used.

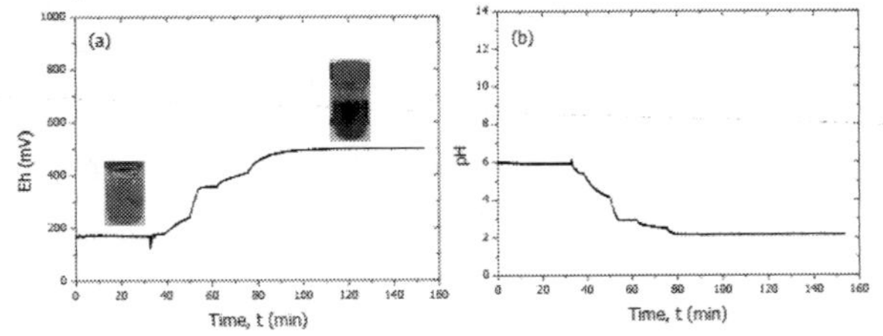

Fig.1. Transient response of the (a) Oxidation Reduction Potential, Eh, and (b) pH of solution, during the synthesis of iron oxide nanoparticles.

The preparation of emulsions was done by mixing 20 mL of aqueous phase (IONP suspension) with 20 mL of n-decane at a volume ratio 1:1 for 10 minutes inside an ultrasonic probe (Hielscher, UP400St).

2.2 Characterization of suspensions

Suspensions of PPH-coated IONPs were prepared at three iron (Fe) concentrations, 0.25, 0.5 and 0.75 g/L. The particle size distribution, and ζ-potential of all IONP suspensions were determined with Dynamic Light Scattering -DLS (Zetasizer Nano System, Malvern) after having diluted each suspension at ratio 1:50 with 3D water.

The surface tension of each solution/suspension, as well its interfacial tension with paraffin oil were measured at 25°C on a Sigma 702 (KSV Instruments) tensiometer by using the DuNouy Ring method.

The air/water and oil/water static contact angles were measured from images of drops of aqueous phase surrounded by air or oil on a glass plate, using a HY-2307 CMOS camera (Hayear) connected via USB with PC and equipped with image capturing software. The images were analyzed by the OpenDrop software [25].

2.3. Visualization tests

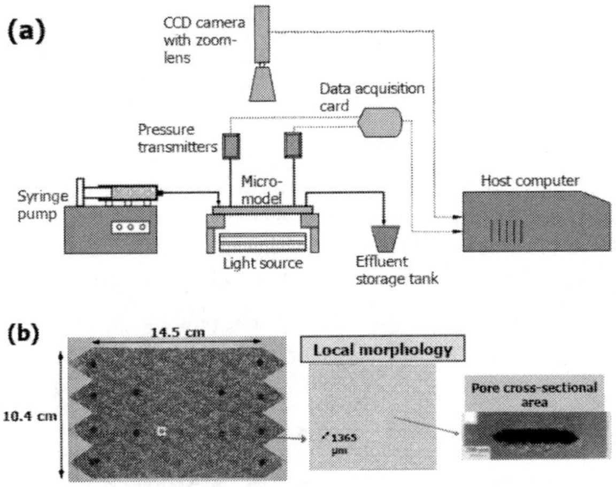

Fig. 2. (a) Schematic diagram of experimental setup. (b) Morphology of glass-etched pore network model.

Immiscible displacement tests were conducted on a glass-etched pore network model (Fig.2a,b). The visualization tests (Fig.2a) enable us to assess the capacity of the various types of NPs to reduce the residual oil remaining in the porous medium after primary drainage (flow rate, q_o=0.08 mL/min) primary imbibition (flow rate, q_w=0.2 mL/min), and secondary imbibition (flow rate, q_w=0.2 mL/min) cycles, where each displacement step started at the point that the previous step stopped, so that a continuous saturation history was recorded. Paraffin oil (Labkem, Spain) colored with oil red (concentration=1500 ppm) was selected as the model viscous oil phase. Aqueous solution of NaCl 0.25M was used as the brine in drainage and primary imbibition tests, and suspension of PPH-coated IONPs, along with its Pickering emulsion with n-decane were used as the displacing fluid in secondary imbibition tests. At each series of displacement tests, images were captured every 10-20 s by a CCD video camera (Panasonic AW-E300E). The pressures were measured continuously on two holes placed on the boundaries of visualized area (Fig.2a) through two pressure transmitters SGM (Electrosystems) properly connected with a data acquisition system (ADAMS-4561 & ADAMS-4117, Advantech). Recorded snap-shots of the flow pattern were processed with image analysis software (ImageJ) to determine the gradual variation of residual oil saturation. (Fig.2a,b).

The capillary number, Ca, and viscosity ratio, κ, are defined by

$$Ca = \frac{u_0 \mu_{inj}}{\gamma_{ow}} \tag{1}$$

$$\kappa = \frac{\mu_{inj}}{\mu_{disp}} \tag{2}$$

where u_0 is the superficial flow velocity (=q/A), μ_{inj} and μ_{disp} are the viscosities of injected and displaced fluid, and γ_{ow} is the oil/water interfacial tension. Moreover, the pore-scale, Ca_{LI} and network-scale, Ca_{LN}, capillary numbers [26] are defined by

$$Ca_{L1} = fCa \quad (2)$$
$$Ca_{LN} = Ca_{L1}(L_N/L_P) \quad (3)$$
$$f = \frac{L_P r_H}{k} \quad (4)$$

where L_N is the network length (L_N=0.14 m), L_P is the pore length (L_P=1.365 mm), r_H is the equivalent hydraulic pore radius (r_H=60 μm) and k the absolute permeability of the porous medium (k=20.5 Da).

3 Results and discussion

The polyphenol concentration in parsley extract, measured by the Folin-Ciocalteu method, was found equal to 2.26 g/L GAE. The physicochemical and interfacial properties of all fluid systems are shown in Tables 1 and 2, respectively.

Table 1. Physicochemical properties of fluid systems

Solution / Suspension	Particle size distribution		ζ-potential (mV)	Density, ρ (g/cm³)
	$<D_p>$ (nm)	σ_p (nm)		
PPH-coated IONPs C_{Fe}=0.25g/L	22.8	6.2	-22.1	0.993
PPH-coated IONPs C_{Fe}=0.25g/L with 0.25 M NaCl	40.2	12.2	-	0.981
PPH-coated IONPs C_{Fe}=0.5 g/	55.0	21.7	-24.7	0.993
PPH-coated IONPs C_{Fe}=0.75 g/L	19.6	7.3	-10.3	0.990

Table 2. Interfacial properties of fluid systems

Solution / Suspension	Surface tension γ_s (mN/m)	Interfacial tension γ_{ow} (mN/m)	Air/water contact angle θ_a (°)	Oil/water contact angle θ_{ow} (°)
PPH solution	55.1	19.4	-	-
0.25M NaCl solution	72.5±0.5	44.0	 53.8±0.4	 54.5±1.3
PPH-coated IONPs C_{Fe}=0.25g/L	62.8±0.5	27.9	 53.1±0.2	 39.2±5.9
PPH-coated IONPs with 0.25 M NaCl	57.4±0.25	20.7	 55±0.1	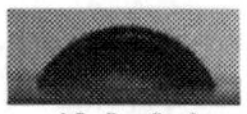 43.9±0.4
PPH-coated IONPs C_{Fe}=0.5 g/	57.5±0.3	26.1	 51.2±1.2	 44.8±0.7
PPH-coated IONPs C_{Fe}=0.75 g/L	53.7±0.1	20.7	 35.2±2.7	 11.5±3.2

It is evident that the interfacial properties along with the contact angle are decreasing functions of the iron concentration (Table 1), and the oil displacement efficiency is expected to be favoured when using IONP suspensions of progressively higher Fe concentration. On the other hand, the dynamic surface tension, measured by the pendant drop method is a decreasing function of time (Fig. 3a) revealing the capacity of the system to generate stable foams.

The relative volumes of the various phases as functions of time for a Pickering emulsion synthesized from IONP suspension with C_{Fe}=0.25 g/L and C_{NaCl}=0.25M is shown in Fig.3b.

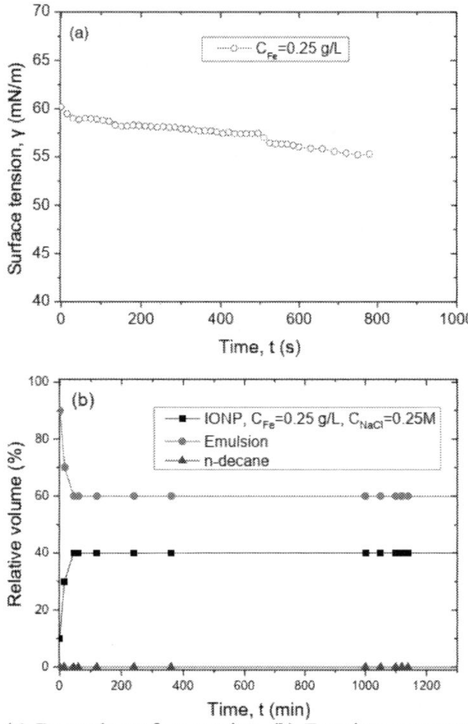

Fig.3. (a) Dynamic surface tension. (b) Transient response of the relative volume of the various phases of emulsion.

Table 3. Experimental conditions

Injected fluid	A	B	C	D
q (mL/min)	0.08	0.2	0.2	0.2
u_0 x 10^5 (m/s)	0.94	2.35	2.35	2.35
μ (Pa s)	0.026	0.001	0.001	0.001
Ca x 10^6	8.0	0.77	0.84	1.13
κ	26.0	0.038	0.038	0.038
Ca_{L1}	0.032	0.0031	0.0034	0.0046
Ca_{LN}	3.32	0.32	0.34	0.47

Injected fluid	E	F	G
q (mL/min)	0.2	0.2	0.2
u_0 x 10^5 (m/s)	2.35	2.35	2.35
μ (Pa s)	0.0125	0.001	0.001
Ca x 10^6	14.1	0.9	1.13
κ	0.48	0.038	0.038
Ca_{L1}	0.057	0.0036	0.0046
Ca_{LN}	5.86	0.38	0.46

*__A__=Paraffin oil; **B**=Aqueous solution C_{NaCl}=0.25M; **C**=IONP solution C_{Fe}=0.25 g/L; **D**=IONP solution C_{Fe}=0.25 g/L C_{NaCl}=0.25 M; **E**=Pickering emulsion; **F**=IONP solution C_{Fe}=0.5 g/L; **G**=IONP solution C_{Fe}=0.7 g/L.

Successive images of the drainage/ primary imbibition/ secondary imbibition steps are shown in Figs. 4-8. Drainage follows a frontal flow pattern (Fig.4a,5a,6a,7a,8a) due to the high viscosity ratio and high Ca_{LN} (Table 3, [26]). The primary imbibition flow pattern is dominated by capillary fingering (Fig.4b, 5b, 6b, 7b, 8b) due to the relatively low value of capillary number at pore, Ca_{L1}, and network, Ca_{LN}, scale (Table 3). Secondary imbibition by IONP suspensions is still dominated by capillary fingering (Fig.4c,5c,6c,7c. Table 3). However, the secondary imbibition by the emulsion transits to a frontal drive (Fig.8c) which is consistent with the increase of the viscosity ratio along with the pore & network scale capillary number by one order of magnitude (Table 3).

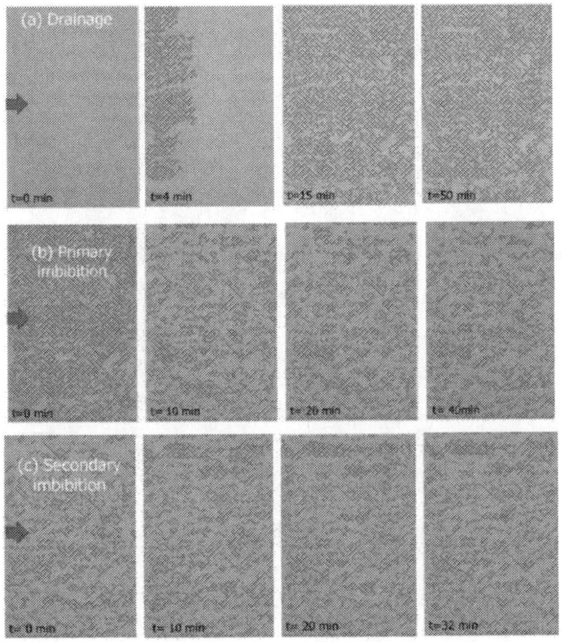

Fig.4. (a) Displacement of brine (C_{NaCl}=0.25M) by paraffin oil (μ=0.026 Pa s). (b) Displacement of paraffin oil by brine (C_{NaCl}=0.25M). (c) Displacement of residual paraffin oil by IONP suspension (C_{Fe}=0.25 g/L, C_{NaCl}=0.25M).

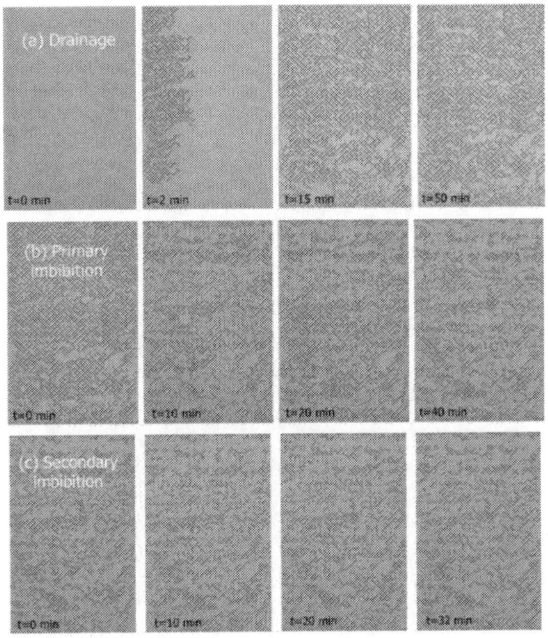

Fig.5. (a) Displacement of brine (C_{NaCl}=0.25M) by paraffin oil (μ=0.026 Pa s). (b) Displacement of paraffin oil by brine (C_{NaCl}=0.25M). (c) Displacement of residual paraffin oil by IONP suspension (C_{Fe}=0.25 g/L).

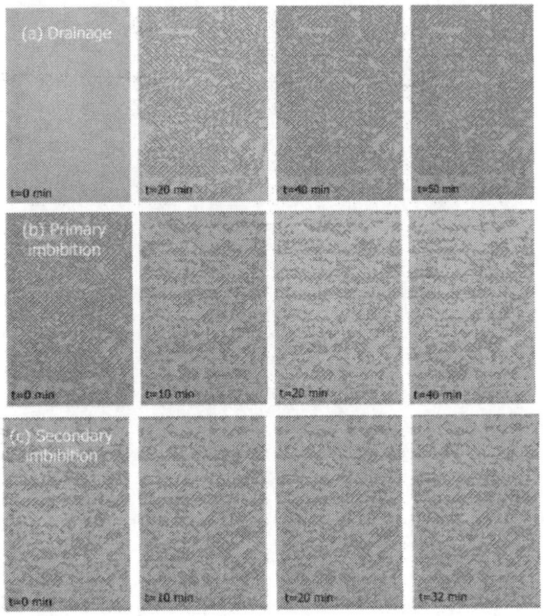

Fig.6. (a) Displacement of brine (C_{NaCl}=0.25M) by paraffin oil (μ=0.026 Pa s). (b) Displacement of paraffin oil by brine (C_{NaCl}=0.25M). (c) Displacement of residual paraffin oil by IONP suspension (C_{Fe}=0.50 g/L).

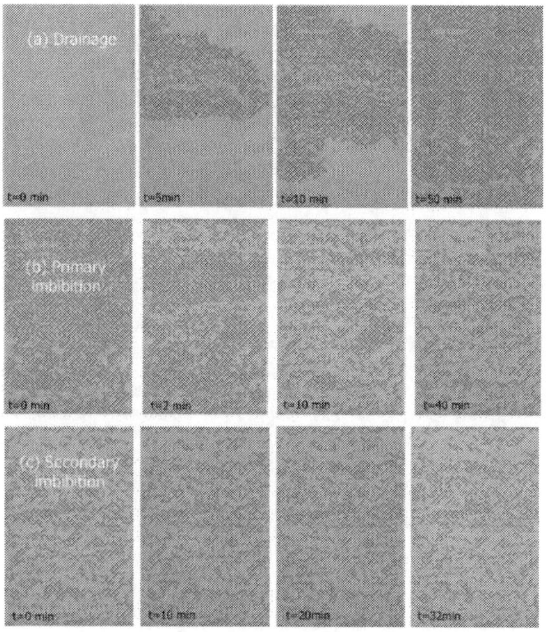

Fig.7. (a) Displacement of brine (C_{NaCl}=0.25M) by paraffin oil (μ=0.026 Pa s). (b) Displacement of paraffin oil by brine (C_{NaCl}=0.25M). (c) Displacement of residual paraffin oil by IONP suspension (C_{Fe}=0.75 g/L).

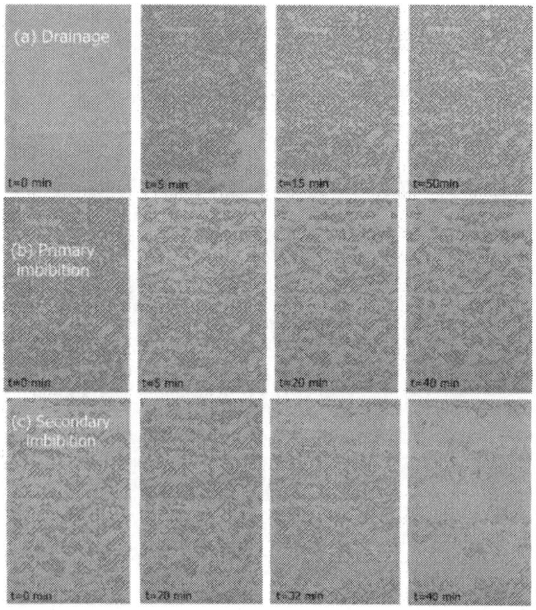

Fig.8. (a) Displacement of brine (C_{NaCl}=0.25M) by paraffin oil (μ=0.026 Pa s). (b) Displacement of paraffin oil by brine (C_{NaCl}=0.25M). (c) Displacement of residual paraffin oil by Pickering emulsion.

During the injection of PPH-coated IONP suspensions, oil trapped at the left side of the visualized area (near the inlet ports) was mobilized and transferred towards the central area, leading to a temporary or permanent increase of oil saturation (Fig.9a-blue circles, Fig. 9b-blue circles, Fig. 9c-green circles, Fig.9d-blue circles). However, on average, the local oil saturation decreases over the majority of pore clusters, at the final stages of the process (Fig.9a,b,c,d). It is worth mentioning that during primary or secondary imbibition, oil mobilization is carried out via a mechanism of successive and locally occurring drainage / imbibition events. On the other hand, the displacement of residual oil by the emulsion follows a clearly frontal-like pattern and is a very efficient displacement, in terms of the percentage of residual oil recovery (Fig.8c).

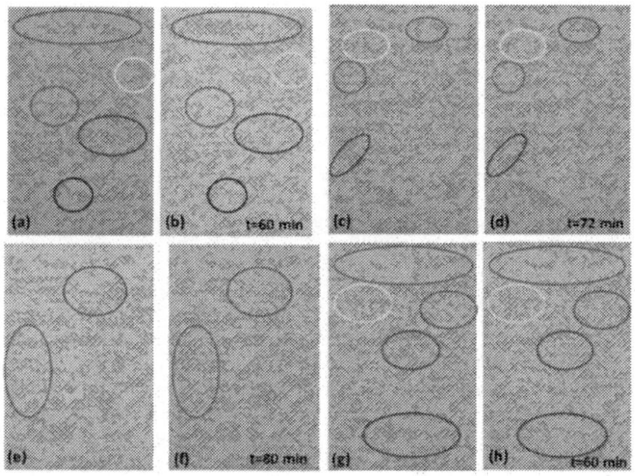

Fig. 9. Comparison between the final snap-shots of primary and secondary imbibition during the displacement of residual oil by (a,b) IONP suspension in brine with C_{Fe}=0.25 g/L, C_{NaCl}=0.25M), (c,d) IONP suspension with C_{Fe}=0.25 g/L, (e,f) IONP suspension with C_{Fe}=0.50 g/L, (g,h) IONP suspension with C_{Fe}=0.75 g/L.

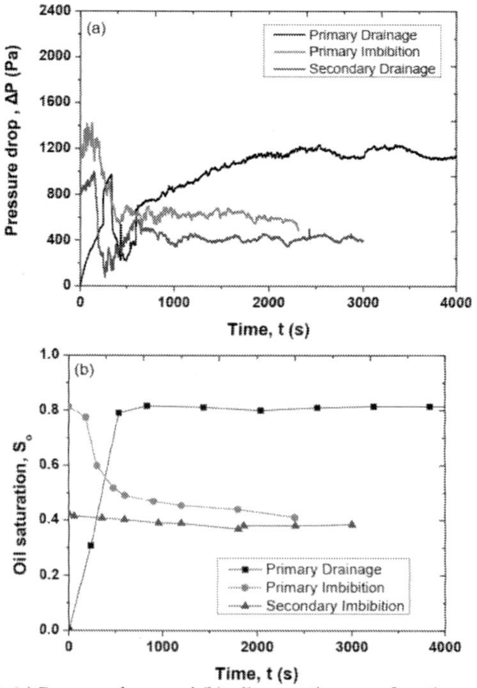

Fig.10. (a) Pressure drop, and (b) oil saturation, as a function of time. IONP suspension with C_{Fe}=0.25 g/L and C_{NaCl}=0.25M was injected at the secondary imbibition step.

The transient response of the pressure drop across the central region of the glass micromodel along with the gradual change of oil saturation are shown in Figs.10-14.

 https://doi.org/10.1051/e3sconf/202336701009

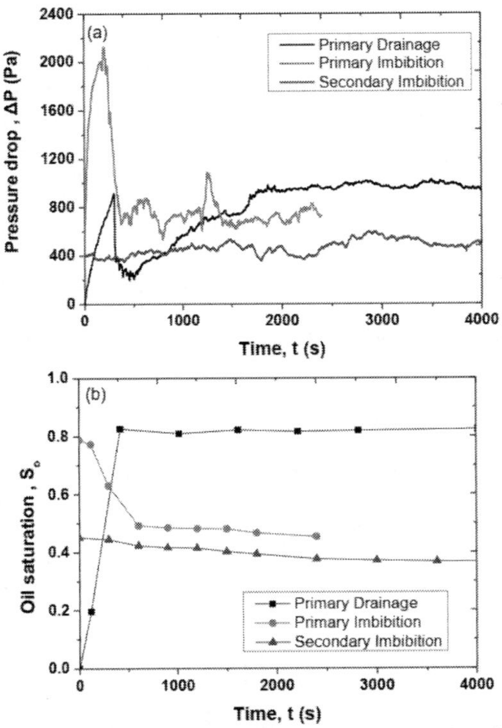

Fig.11. (a) Pressure drop, and (b) oil saturation, as a function of time. IONP suspension with C_{Fe}=0.25 g/L was injected at the secondary imbibition step.

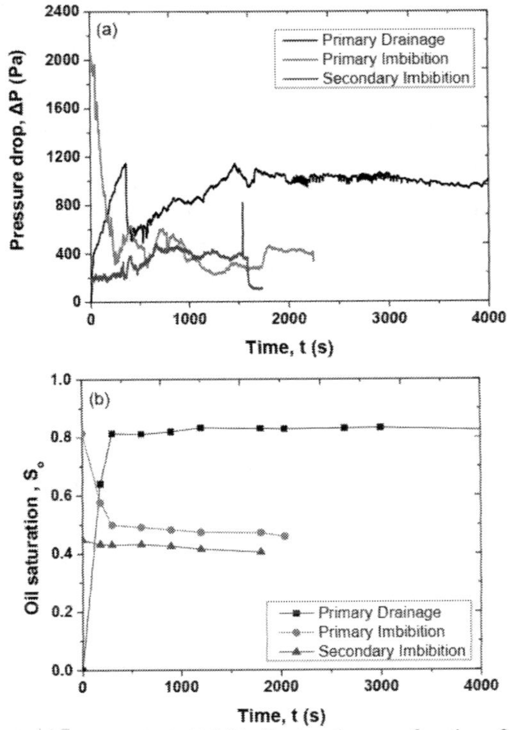

Fig.12. (a) Pressure drop, and (b) oil saturation, as a function of time IONP suspension with C_{Fe}=0.5 g/L was injected at the secondary imbibition step.

The decrease of the residual oil saturation during the secondary imbibition step by IONP suspensions (Fig.10b,11b,12b,13b) is reflected in the relatively low pressure drop across the porous medium compared to the corresponding one of primary imbibition step (Fig.10a,11a,12a,13a). It is worth mentioning that during secondary imbibition with IONP suspensions, the residual oil displacement efficiency seems to be higher without the presence of NaCl (Fig.11b, 12b, 13b, Table 4). In general, various mechanisms have been proposed in the literature to explain the higher oil recovery with low-salinity water-flooding [27]. Perhaps, oil recovery increases because of changes in the wettability of the porous medium toward more water-wet conditions [27]. On the other hand, it has been observed that at long term, the presence of NaCl favours the destabilization of IONP, their agglomeration, and subsequent precipitation. The high oil recovery efficiency of the secondary imbibition by the shear-thinning Pickering emulsion (Fig.14b) is associated with a relatively high pressure drop across the porous medium, also characterized by strong fluctuations (Fig.14a). The fluctuations are due to the broad range of the viscosity for a shear-thinning fluid (Fig.15): (i) across a single pore [28], as the shear stress varies from zero (at the pore centre) to a maximum (at the pore walls); (ii) throughout the pore network [29,30] between the high and low flow rate regions.

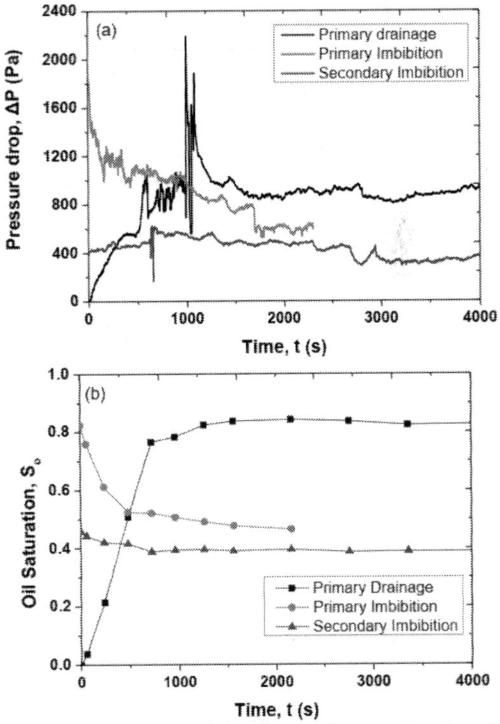

Fig.13. (a) Pressure drop, and (b) oil saturation, as a function of time. IONP suspension with C_{Fe}=0.75 g/L was injected at the secondary imbibition step.

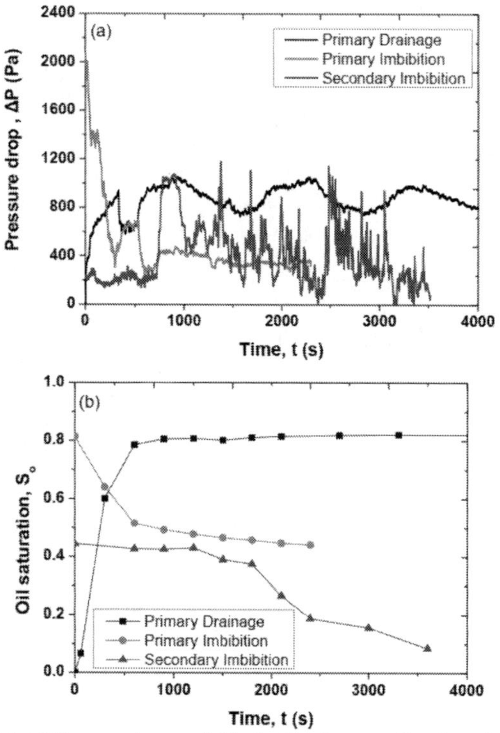

Fig.14. (a) Pressure drop, and (b) oil saturation, as a function of time. Pickering emulsion synthesized from IONP suspension with C_{Fe}=0.25 g/L and C_{NaCl}=0.25M was injected at the secondary imbibition step.

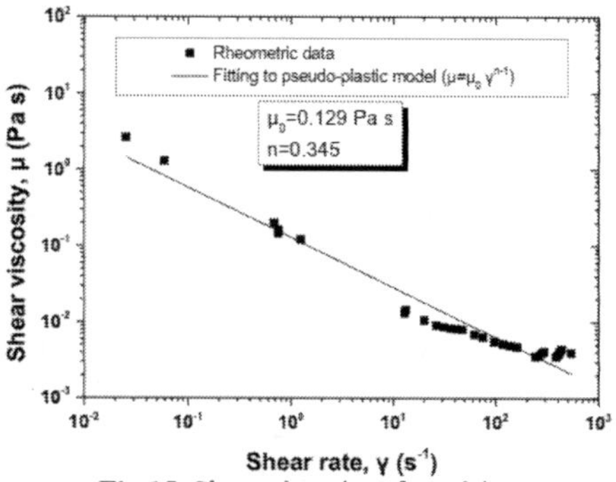

Fig.15. Shear viscosity of emulsion.

Table 4. Summary of results from EOR tests

Displacing fluid in secondary imbibition step	Drainage, S_o	Primary Imbibition, S_o	Secondary Imbibition, S_o	Oil recovery efficiency (%)
IONP C_{Fe}=0.25 g/L	0.815	0.453	0.36	20.5
IONP C_{Fe}=0.25 g/L,C_{NaCl}=0.25M	0.815	0.410	0.386	6.1
Pickering emulsion C_{Fe}=0.25g/L, C_{NaCl}=0.25M)	0.820	0.441	0.085	80.7
IONP C_{Fe}=0.5 g/L	0.818	0.459	0.405	11.7
IONP C_{Fe}=0.75 g/L	0.830	0.465	0.38692	16.7

4 Conclusions

The potential to increase the residual oil efficiency by injecting suspensions of iron oxide nanoparticles synthesized and stabilized by the polyphenols of plant extracts (parsley) is investigated with visualization tests on a glass-etched pore network. Over the investigated experimental conditions, imbibition with IONP suspensions is dominated by the capillary fingering flow pattern. The nanoparticles mobilize trapped oil by transferring it from upstream to downstream through a mechanism of successive steps of drainage (local increase of oil saturation) / imbibition (local decrease of oil saturation). It seems that the oil recovery efficiency of secondary imbibition tests is favored when using IONP suspensions without the presence of NaCl (Table 4). On the other hand, the oil recovery efficiency increases respectably when using Pickering emulsions stabilized by IONP.

Acknowledgements: The research project was supported by the Hellenic Foundation for Research and Innovation (H.F.R.I.) under the "1st Call for H.F.R.I. Research Projects to support Faculty members and Researchers and the procurement of high-cost research equipment" (Project Number: HFRI-FM17-361, acronym: EOR-PNP).

References

1. M.S. Kamal, A.A. Adewunmi, A.A. Sultan, M.F. Al-Hamad, U. Mehmood. J. Nanomat. **2473175**, 1-16 (2017)
2. H. ShamsiJazeyi, C.A. Miller, M.S. Wong, J.M. Tour, R. Verduzco. Appl. Polymer Sci. **131**, 40576, pp.1-13 (2014)
3. T. Sharma, G.S. Kumar, J.S. Sangwai **129**, 221-232 (2015)
4. T. Sharma, S.K. Govindarajan, J. Sangwai. Ind. Eng. Chem. Res. 54, 1576-1584 (2015)
5. N. Bizmark, M.A. Ioannidis, Langmuir **31**, 9282-9289 (2015)
6. N. Kumar, T. Gaur, A. Mandal. J. Ind. Eng. Chem. **54**, 304-315 (2017).
7. N. Saleh, T. Sarbu, K. Sirk, G.V. Lowry, K. Matyjaszewski, R.D. Tilton. Langmuir **21**, 9873-9878 (2005)
8. T. Saigal, H. Dong, K. Matyjaszewski, R.D. Tilton. Langmuir **26 (19)**, 15200-15209 (2010)
9. N. Bizmark, M.A. Ioannidis, Langmuir **33**, 10568-10576 (2017)
10. T. Saigal, A. Yoshikawa, D. Kloss, M. Kato, P. L. Golas, K. Matyjaszewski, R.D. Tilton, J. Coll. Interface Sci. **394**, 284-292 (2013)
11. X. Li, H. ShamsiJazeyi, S.L. Pesek, A. Agrawal, B. Hammouda, R. Verduzco. Soft Matter **10**, 2008-2015 (2014)
12. K.Y. Yoon, Z. Li, B.M. Neilson, W. Lee, C. Huh, S.L. Bryant, C.W. Bielawski, K.P. Johnston. Macromolecules **45**, 5157-5166 (2012)
13. N.J. Alvarez, S.L. Anna, T. Saigal, R.D. Tilton, L.M. Walker. Langmuir **28**, 8052-8063 (2012)

14. C. Tsakiroglou, K. Terzi, A. Sikinioti-Lock, K. Hajdu, C. Aggelopoulos. Sci. Total Env. **563–564**, 866–878 (2016)
15. K. Terzi, A. Sikinioti-Lock, A. Gkelios, D. Tzavara, A. Skouras, C. Aggelopoulos, P. Klepetsanis, S. Antimisiaris, C.D. Tsakiroglou. Coll. & Surf. A: Physicochem. & Eng. Aspects **506**, 711-722 (2016)
16. C.D. Tsakiroglou, A. Sikinioti-Lock, K. Terzi, M.A. Theodoropoulou, *Chem. Eng. Sci.* **192**, 391-413 (2018).
17. D. Arab, A. Kantzas, S.L. Bryant J. Pet. Sci. Eng. **163** 217–242 (2018)
18. A. Sebastian, A. Nangia M.N.V. Prasad. J. Clean. Prod. 174, 355-366 (2018)
19. J. Lin, X. Weng, R. Dharmarajan, Z. Chen. Chemosphere **169,** 413-417 (2017)
20. J. Fowsiya. G. Madhumitha. N.A. Al-Dhabi, M.V. Arasu, J. Photochem. & Photobiol. B: Biology **162**, 395–401 (2016)
21. M. Karavasilis, C.D. Tsakiroglou, Emerging Science Journal **3(6)**, 344-360 (2019)
22. M.A. Theodoropoulou, V. Sygouni, V. Karoutsos, C.D. Tsakiroglou, Int. J. Multiphase Flow **31,** 1155-1180 (2005)
23. M. Geremu, Y. B. Tola and A. Sualeh, Chem. Biol. Technol. Agric. **3:25**, pp.1-6 (2016)
24. V. Kamath, P. Chandra, G.P. Jeppu, Int. J. Phytorem. **22,** 1278-1294 (2020).
25. J.D. Berry, M.J. Neeson, R.R. Dagastine, D.Y.C. Chan, R.F. Tabor, J. Coll. Inter. Sci. **454** 226–237 (2015); http://opencolloids.com
26. C.D. Tsakiroglou, M. Theodoropoulou, V. Karoutsos, , AIChE J., **49,** 2472-2486 (2003).
27. A. Katende, F. Sagala, J. Molec. Liq. **278**, 627-649 (2019)
28. C.D. Tsakiroglou, J. Non-Newt. Fluid Mech. **105**, 79-110 (2002)
29. C.D. Tsakiroglou, J. Non-Newt. Fluid Mech. **117,** 1-23 (2004).
30. C.D. Tsakiroglou, M. Theodoropoulou, V. Karoutsos, D. Papanicolaou, V. Sygouni, J. Coll. Interface Sci. **267,** 217-232 (2003).

 https://doi.org/10.1051/e3sconf/202336701010

Digital Rocks Portal (Digital Porous Media): connecting data, simulation and community

Maša Prodanović[1,*], *Maria* Esteva[2], *James* McClure[3], *Bernard* C. Chang[1], *Javier* E. Santos[4], *Anuradha* Radhakrishnan[1], *Ankita* Singh[5] and *Hasan* Khan[6]

[1]Hildebrand Department of Petroleum and Geosystems Engineering, The University of Texas at Austin, Austin, TX, USA
[2]Texas Advanced Computing Center, The University of Texas at Austin, Austin, TX, USA
[3]Advanced Research Computing, Virginia Tech, VT, USA
[4]Center for Nonlinear Studies & Earth and Environmental Sciences, Los Alamos National Laboratory, Los Alamos, NM, USA
[5]University of New South Wales alumna, Sydney, Australia
[6]Petroleum Engineering Department, King Fahd University of Petroleum and Minerals, Dhahran, Saudi Arabia

Abstract. Digital Rocks Portal (DRP, https://www.digitalrocksportal.org) organizes and preserves imaged datasets and experimental measurements of porous materials in subsurface, and beyond, with the mission to connect them to simulation and analysis, as well as educate the research community. We have over 150 projects represented in more than 200 publications, and an active community that reuses the data, most recently in multiple machine learning applications for automating image analysis as well as the prediction of transport. Such automation is crucial for performing formation evaluation tasks in near-real time. We present benchmark datasets that have played a role in recent machine learning prediction successes in the field. We further discuss the vision for further research advances, educational materials, as well as growth and sustainability plan of this digital rock physics community resource. In particular, we are in the process of expanding into a broader repository of engineered porous materials, specifically those for energy storage and the portal will transition to Digital Porous Media (DPM) in near future.

1 Introduction

The field of digital rock physics combines imaged porous media (such as rocks and soil) to simulations and other research methods such as machine learning (ML) and data analysis to understand transport and deformation in subsurface. Applications include formation evaluation, management of groundwater resources, carbon sequestration, enhanced oil recovery, and contaminant transport [1]. Typical geological systems are composed of a broad spectrum of porous media with properties such as permeability varying by orders of magnitude within an individual system. The advances in high-resolution imaging techniques (x-ray tomography, scattered electron and optical microscopy) have provided a wealth of 2D and 3D datasets that reveal the microstructure of rocks, soil or model media on scales ranging from nanometers to centimeters, alas analysis and use of such images in simulation is computationally demanding [2]. Our practical questions nevertheless require upscaling to the field scale. Therefore, efficient connection of curated digital rocks data and scalable computing and machine learning (ML) applications is necessary for applications [3].

We here describe the organization of Digital Rocks Portal (DRP), a well-adopted portal for curation and visualization of large imaged datasets of geomaterials. DRP's mission is to organize and preserve imaged datasets and experimental measurements of porous materials in subsurface, and then connect them to research methods. We report on our efforts in educating the research community by organizing workshops and mini-courses in data preservation, analysis and visualization.

Deep learning methods and convolutional neural networks (CNNs) have many applications in image processing and flow prediction [4]. For this, reliable "ground truth" benchmark datasets are needed because (1) supervised deep learning methods require a large amount of validated data to train models; and (2) the capabilities of the trained classifiers must be assessed quantitatively. Applying learning algorithms to the large volume of data stored within DRP is currently rendering scientific advances. We successfully applied deep learning algorithms on predicting 3D velocity fields in porous media images [5,6]. The implementation was done in TensorFlow [7] using high performance resources (HPC) at Texas Advanced Computing Center (TACC). The work in [5] sampled a wide variety of images from DRP as well as adding a benchmark collection [8], has specifically enabled deep learning training with geologic porous media whose heterogeneity (including fractures/cracks) likely surpasses any engineered porous materials, and, should data from energy storage materials become available, the results should immediately transfer. This exemplifies a workflow that can enable future real time permeability estimation based on imaged datasets, as the computational time of a deep learning prediction is in seconds compared to hours on HPC.

* Corresponding author: masha@utexas.edu

The parallel work at this conference [9] is adapting the algorithm for electrical resistivity prediction.

Visualization algorithms of cross-sections, volumes and surfaces of 3D data (e.g. marching cubes algorithm) are well-known and available in a number of advanced visualizing platforms (e.g. ParaView, MayaVi, ImageJ, Dragonfly, to name a few that are open source). However, the complexity of the pore structure and processes captured within pose a challenge in visualizing the data, both from memory and computational perspective. That said, while this challenge is meant to reuse data from geosciences and subsurface engineering, all of the educational concepts are very much applicable to any scientific visualization endeavor: for instance, in biomedical field, heart or blood vessels in a human body, are often imaged using computed tomography (or simulated based on those images) and have particulate flow (blood flow, that is) within. In addition to the materials from the recent visualization contest reusing DRP datasets [10], we provide additional general input/output tools and Jupyter Notebooks examples visualizing a 3D velocity field as well as porous media surfaces in multi-fluid situation exemplified in this paper [11].

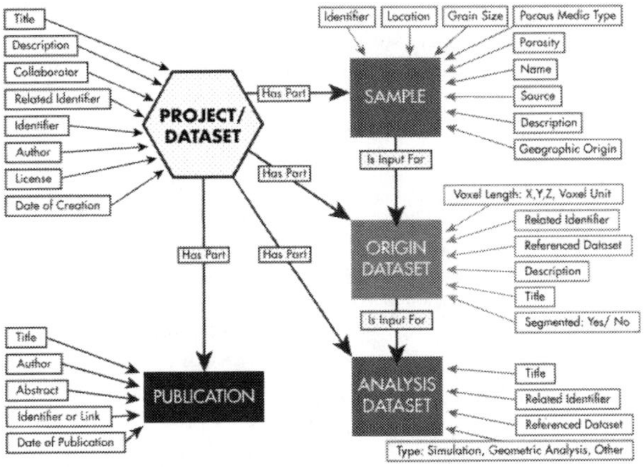

Fig.1. DRP Data model is foundation for the database linking web content and files on HPC storage.

Transitioning to renewable energy from the hydrocarbon-based sources that currently meet 65% of energy demands will require significant engineering effort on several fronts: (1) mitigating the effects of air pollution with solutions such as subsurface carbon dioxide sequestration; and (2) scaling up renewable energy based on the maturation of advanced energy devices, geothermal, and other solutions. Complex microstructure is a common element for many of these physical problems. We discuss our sustainability plan that includes the paid membership fees scheme for Digital Rocks Portal, thus ensuring its financial stability, (2) strengthening our visualization and analysis services and (3) generalizing and broadening the user base of DRP to Digital Porous Media (DPM) portal to energy storage materials. Geometric approaches developed to enhance digital rock images have excellent crossover potential to other complex material structures [12]. High performance methods to enhance experimental data collection workflows are also broadly generalizable [13].

2 Portal organization

Launched in September 2015, DRP [14] is an established data repository, trusted by the community. DRP is specialized for porous media as opposed to platforms such as Mendeley Data [15], Dryad [16] or Energy Data Exchange [17] that store data of any origin. The portal is specifically equipped with tools tailored to organize, visualize, analyze, and publish digital images of porous materials. Its data model shown in Fig.1. is the foundation for curating, describing and displaying the datasets. The data model enables imaged data curation workflows including adding provenance information and metadata used for rendering the images, and organizing the dataset in relation to its corresponding research activity such as observations, experiments, and simulations. Thus, the model is the basis for the datasets representations on the landing pages that displays provenance as material entities (samples) and processes (experimental or simulation) from which data derives. This results in a tree like representation that is visible on the landing page of any project, which was evaluated through user studies [18] and facilitates understanding of and access to complex and large datasets [19]. Example projects are shown in Fig.2. and Fig.3.

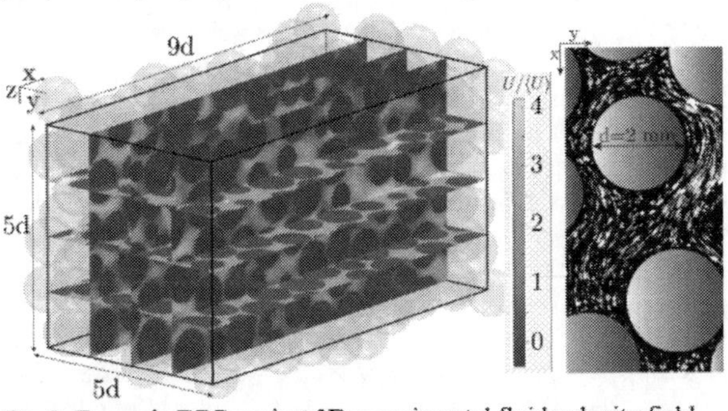

Fig.2. Example DRP project 3D experimental fluid velocity field within a sphere packing (model granular or soil structure) obtained particle velocimetry [20]. To our knowledge, this is the only open 3D velocimetry dataset, and has been the bases for the winning visualization entry in DRP visualization contest (see below).

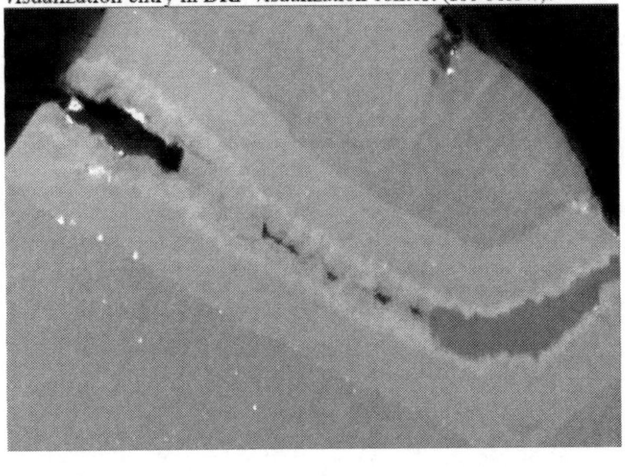

Fig.3. Example DRP project [21] with 3D x-ray microtomography images as well as XRF mineral maps of a partially mineralized vein contained from a core sample (depth of ~10,400 ft) in the Upper Wolfcamp formation. The sample was studied as part of the effort of the Department of Energy's Center for Nanoscale Controls on Geologic CO_2.

DRP does not assume a specific image format at this point. This is because subsurface porous media heterogeneity requires a variety of imaging modalities ranging from microscopy to x-ray microtomography to medical computed tomography with no commonly accepted data formats. As a result, DRP has built-in capabilities for translating and visualizing raw data after import, which distinguishes it from other imaged data repositories.

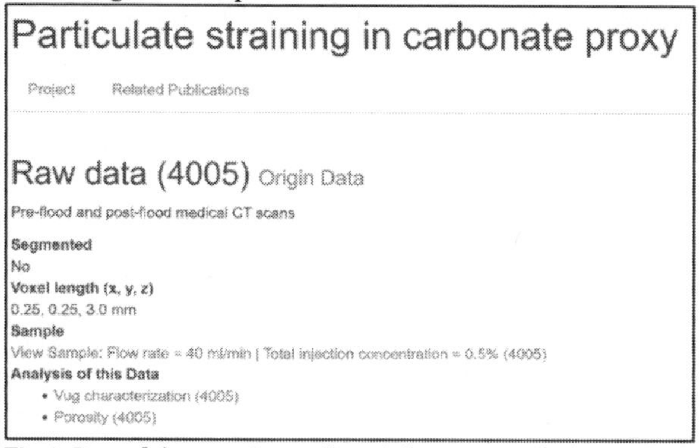

Fig.4. Part of the metadata collected for an example dataset [22] including voxel length that relates the image to sample physical size.

Data curation is assisted through the publication pipeline interface that carefully guides users through the process. Metadata collected during this process is essential for full documentation of the dataset. Images by themselves contain structural information collected at a certain time, but not more than that and this step is essential. Fig.4. and Fig.5. exemplify this information for an example project. Note that only part of the information collected is shown for brevity. Prior to publication and before assigning a digital object identifier (DOI), the datasets are verified for completeness and accuracy by a live curator, again unlike many data portals that allow upload of compressed archives. Current list of curators includes the majority of the authors of this publication and we are open to volunteers. DOIs are provided through The University of Texas Libraries subscription to DataCite Fabrica [23].

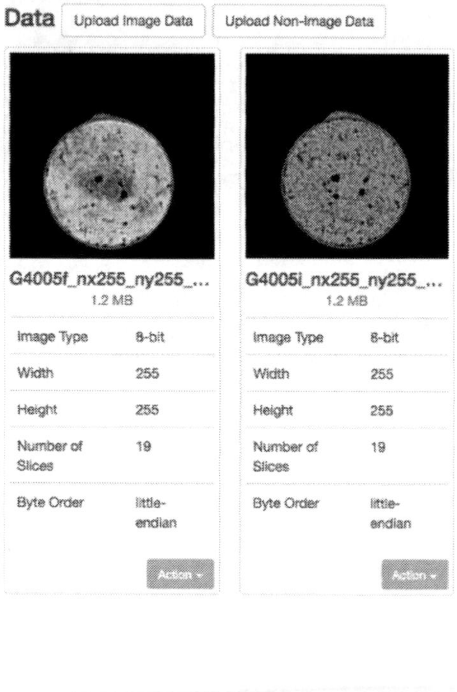

Fig.5. Second part showing successful "ingestion" of two datasets of the same project as shown in Fig.4. ([22]) that are part of of the same experiment (as proved by the image cross-sections. Action button allows download, shows a video of all slices (in 3D), and basic image statistics such as histogram. "Action" button also has a button that allows connection to HPC cluster on TACC that opens ParaView for visualization. This software stack is however currently being updated.

DRP is implemented within the reliable HPC environment deployed and maintained by Texas Advanced Computing Center, and data storage is supported by The University of Texas System Research Cyberinfrastructure (UTRC) in the storage resource Corral. This infrastructure assures data security and persistence, interoperability with different HPC systems for visualization and analysis (TACC Analysis Portal, [24] currently on Lonestar6 and Stampede2 clusters) and continuing operation of the software stack, and thus it contributes to the portal's sustainability.

3 Portal impact and user engagement

3.1. Metrics

As of June 2022, DRP has published 154 projects and 200 related publications [25], ranging in size from 10MB to 728GB. DRP accepts international contributions. Our current query country of the institutional affiliation of the first author of the current DRP projects we find 48% from the U.S., 18.2% from UK, 9.5% from Australia, and the reminder from China, Belgium, Brazil, Canada, South Korea, Japan, Russia and Switzerland.

Since its launch, we have been keen in promoting DRP and in understanding its impact. We implemented Search Engine Optimization (SEO) strategies such as using schema.org vocabularies and Google-friendly metadata in the repository landing page. Using Google Analytics, we monitor an increase in total visits over the years as well as consistent spikes in usage, which coincide with academic events that we undertake (workshops, presentations, publications). We also implemented a data download count (of individual datasets) to each of our dataset pages: it is available under link "Usage Information" embedded in each project landing webpage. Through Google Scholar Alerts we follow mentions and citations to the repository and to individual datasets.

3.2 Digital Object Identifiers (DOIs) and increasing visibility of data

Journal publications have had DOIs for a long time, but providing them for research data or code is a relatively new adoption thanks to efforts such as Findable, Accessible, Interoperability, and Reusable (FAIR) data efforts launched in 2016 [26]. We have found that authors have variable or incomplete referencing styles for datasets in papers, whether they are referencing their own data or reusing someone else's. Many, for instance, include only the name of the platform where the data is posted. Note that including the dataset DOI and a proper reference improves the findability, as **search engines presently do not find data as efficiently as they do papers (if at all).** One reason for this is that datasets come in variable formats and are not easily searchable for information the way .pdf or text formats are.

We thus next describe strategies that readers can use to improve other authors finding their data.

Increasingly, journals request that you publish your paper along with the data that supports your findings. For example, the AGU journals follow this FAIR data practice. Curating data for publication demands significant efforts on the part of the researchers that have to organize and describe it so that others can recreate their results or reuse the data for a different purpose. At the same time, data publications can bring significant rewards in terms of citations and exposure of your work. To help you boost the discoverability and consequently the reuse of your data there are some basic strategies that you can follow:

To enhance discoverability:

- Add all the information (metadata) that DRP asks for, including your ORCID [27] ID (that uniquely identifies you in case you change affiliation or someone else shares your name) if you have it, and who funded your project.
- In your data description use terms that you consider people will search for when looking for this type of data. Highlight the unique characteristics of your dataset and suggest how your data could be useful to others and for what purposes.
- All these metadata are indexed by search engines such as Google, improving the possibilities that your data is discovered by broader audiences.

To facilitate data reuse:

- Add documentation in the form of readme files or reports to clearly explain how you obtained and processed the data. This stimulates the understandability of your publication and consequent data reuse.
- You can include code snippets as "Non-image data" to your datasets, or better yet, publish the code on GitHub and add it as a related publication in DRP.
- If you have them prior to publishing the data, add papers and proceedings that discuss the data as related publications. After your dataset is published you can request DRP curators to add your new publications by sending an email to masha@utexas.edu.

3.3 Newsletters, project highlights and educational material

Engaging the research community is necessary if the data will become reused and thus actively contribute to science and engineering. This can be done through (classical) conference presentations, but also contests, newsletters and social media activities that promote data sharing, and reuse, data cyberinfrastructure, and overall data education through the published materials.

To that end, we launched newsletters in October 2020 and have since published them at the rate of two to three per academic semester. See [25] for links to all newsletters as well as links to a various news articles written about DRP. YouTube playlist on Digital Rocks Portal Visualization [28] contains various visualizations of the datasets, as well as a growing list of interviews we recorded to highlight projects on DRP.

Data reuse contests is another activity whose goal was to stimulate engagement and demonstrate the research potential of these datasets. organized a successful a virtual visualization mini-course, followed by a visualization challenge in early 2021 with three categories (video, static image and 3D printed porous microstructure). This is described in the next section.

3.4. Visualization mini course and contest

In October 2020, a porous media visualization competition was organized which consisted of a mini course (taught using Jupyter Notebooks), followed by a challenge with three categories with monetary awards to promote data reuse and create visualization templates for porous materials (known for their complexity). The task was to reuse any 3D dataset from the DRP to create a static image, video, or 3D printed visualization. Porous media are challenging to visualize due to complexity of pore/grain/fluid surfaces and interfaces, and this creates a resource for those who wish to learn advanced 3D visualization. Resources and materials were stored in DRP to help the participants. The events were sponsored by Southern Big Data Hub, Object Research Systems, Kitware and Dassault Systèmes. Tables 1 through 3 show the list of projects and the datasets that were reused from DRP.

Table 1. Winners of the static image category

Title & winner	Dataset reused	Visualization
Smart mineral mapper, **Julie Kim,** Princeton University	Eagle Ford shale: synchrotron-based element and mineral maps	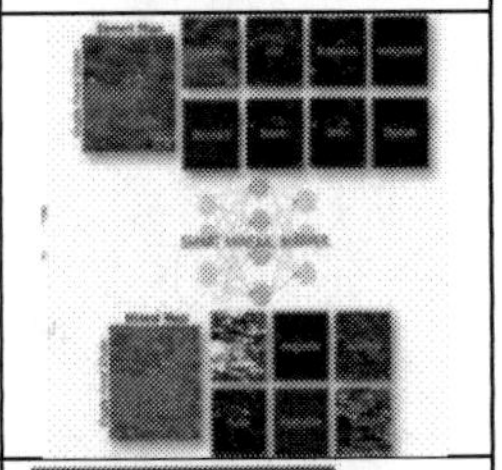
Contact Angle Evolution, **Laura Dalton,** West Virginia University and National Energy Technology Laboratory	Bentheimer and nugget residual saturation micro-computed tomography data	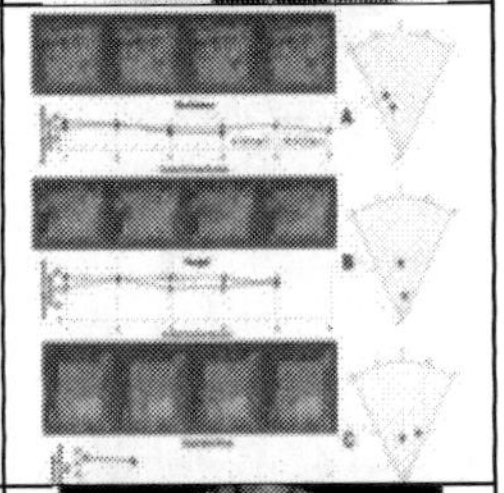
Identifying savonnieres carbonate ooliths, **Miral Tawfik,** Pennsylvania State University	Savonnières carbonate	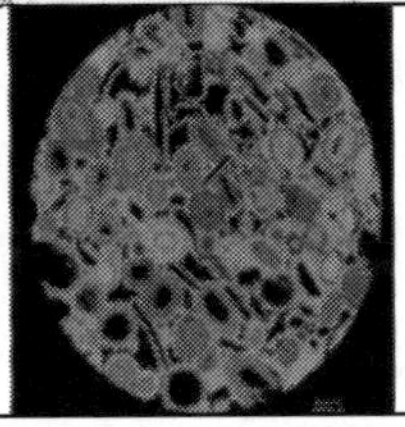

Table 2. Winners of the video category

Title & winner	Dataset reused	Visualization
Pore Aventura: 3D velocity field explorer, Dr. Mathieu Souzy, University of Twente	Experimental 3D velocity field in random sphere packing	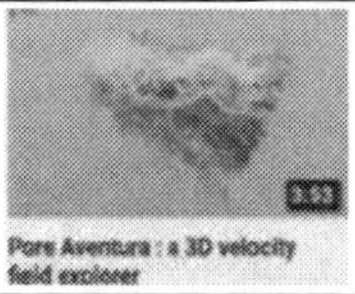
Flow visualization in porous media, **Bin Wang**, Louisiana State University	Experimental 3D velocity field in random sphere packing	
Fluid displacement in a fracture, **Honggeun Jo**, The University of Texas at Austin	Fractures with variable wettability and roughness	
Belgian fieldstone pores and solids, **David Rippner**, University of California, Davis	Belgian fieldstone	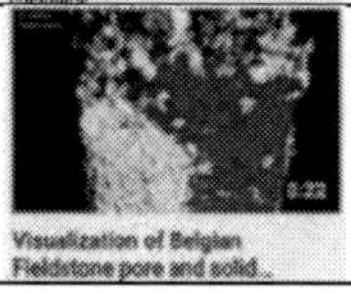

Table 3. Winners of the 3D print category

Title & winner	Dataset reused	Visualization
Sphere pack juxtaposed with the non-wetting fluid at 40% saturation, **Javier Santos,** The University of Texas at Austin	Finney packing of spheres	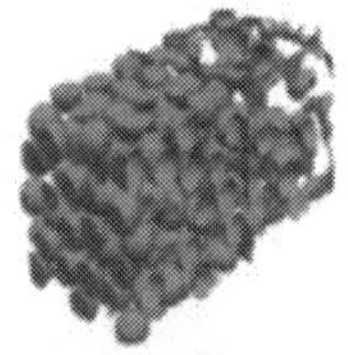
Visualization of sand/pore and a sphere packing surface, Yanbin Gong, University of Wyoming	Uniform quartz-Silver nanoparticle injection experiment and Finney packing of spheres	
Visualization of sandstone and carbonate pore space surface, Dr. Lingyun Kong, Energy and Environment Research Center	Doddington sandstone and Massangis Jaune carbonate	

4 Benchmark datasets

In order to push the state of the art for machine learning (ML) forward and provide comparisons between ML techniques, it would be very useful to have a large number of labeled data (the results of expensive full-physics simulations) to build models that work in the complexity of real-world pore structures. In other fields, large and diverse datasets [29,30] have sparked revolutions in ML by providing a common focal point that allows techniques to be compared against one another. Broadly speaking, this can be attributed to two factors. Firstly, unprecedented size and scale of data pushes the state of the art forward. Secondly, the public availability of these datasets makes easier the comparison of different modeling techniques, so that the relative advantages and disadvantages of different approaches can be more clearly assessed. As an example, in porous media, the choice of fluid solver and numerical implementation thereof is a confounding variable with unknown effects on ML model characteristics.

We are working for DRP to become the a resource where benchmark tests that represent challenging, complex porous media in the context of image analysis, 3D simulation, empirical functional forms, and machine learning can be found. This diversity should include properties such as porous media lithology, boundary conditions, geometric resolution, and physical processes simulated. A large-scale collection of pore-scale data is critical for developing advanced ML algorithms that generalize to unseen samples from many real-life sources. Additionally, it (1) fosters advancement in our understanding of microscopic physics, (2) enables creation of new upscaling relationships, (3) provides a point of comparison to benchmark other simulators of physical processes, and (4) guides ML, which is still an up-and-coming field.

To that end, we have recently added a catalog of 3D geometries [8,31], as well as the companion collection of simulations [32,33], and structural properties of samples. The dataset includes 1736 flow and electrical simulation results on 217 digital samples, which required more than 500 core years of computation time. This data can be used for many purposes, such as constructing empirical models, validating new simulation codes, and developing machine learning algorithms that closely match the extensive purely-physical simulation. The dataset consists of simulations based on 217 digital rock samples, each containing two different domain sizes 256^3 and 480^3. For each digital sample, there are five nanoconfinement simulations, one fluid flow simulation, ten geometric features, electrical conductivity data, topological descriptors, and .png files visualizing cross-sections of the 3D data. The companion SCA paper (Chang et al.) discusses how the simulations are used for training neural networks for prediction of electrical properties.

There are also benchmark datasets posted by other authors in the area of image analysis algorithms. Three sandstone and carbonate collections [34–36] have been used for improving resolution of micro-tomography images, especially in the presence of microporosity. Further, there is an extensive dataset [37] for comparison of both traditional and ML image denoising algorithms.

5 Visualization examples

We created two workflows for 3D visualization of surfaces and vector (velocity) fields and publish them in conjunction with this paper on GitHub [10,11]. The workflows download and use the datasets from the portal.

In the examples that follow we relied on Python-based PyVista visualization software and provide both regular Python code (that is somewhat more robust, as it allows PyVista to open a separate window with the plot) as well as Jupyter Notebook. Jupyter Notebooks are excellent teaching tool and embed the results into web browser. That makes them versatile, but an extension is needed to make it work. The figures below show a glimpse into what is available within the worksheets; they can be adjusted to any other data on DRP.

Notably, while these worksheets are currently set up to download the data and visualize it, **they also work on TACC Visualization and Analysis portal** [24], **in particular on Lonestar6 HPC environment that can access DRP data storage directly without any download, and can analyze large datasets beyond the capability of an individual workstation.**

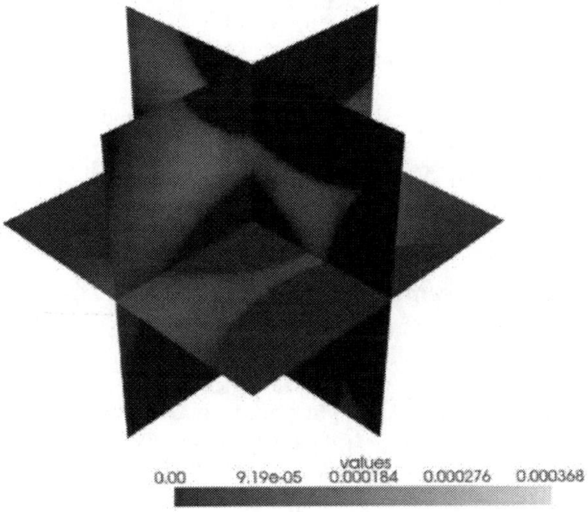

Fig. 6. Subset of the entire DRP velocimetry dataset from Fig.2 visualizing absolute values of velocities in PyVista. See [11]

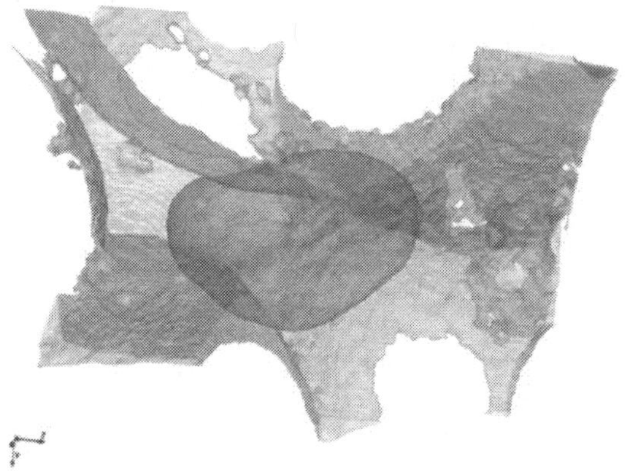

Fig. 7. Surface visualization in PyVista of an oil blob (in transparent green) within a Ketton limestone sample (transparent gray). The high resolution x-ray microtomography experimental data [38] was used in the example and the visualization code is available [11].

6 Related software

The data in Digital Rocks Portal can be characterized and visualized using various data analysis software, and previous section provides one example. Beyond that, open source software such as ImageJ, ParaView, Visit or DragonFly is available, as well as commercial options (e.g. MATLAB, PerGeos).

Transport properties of porous materials, however, require simulation with different type of solvers. If segmented (binarized) images are available, then images can be direct input into simulation that produces a scalar or vector field of interest. The following table includes some open source example libraries.

Table 4. Example open source simulation software using DRP data.

Software	Application / method	Platform
LBPM [39]	Velocity fields, multiphase flow, lattice Boltzmann method	Unix, Parallel, C/C++ and MPI
Palabos [40]	Velocity fields, multiphase flow, lattice Boltzmann method	Unix, Parallel, C/C++ and MPI
MPLBM-UT [41]	Pre-processing and post-processing tools for single- and multi- phase flow with Palabos	Windows/Unix, Python
LSMPQS 1.0 [42]	Capillarity dominated fluid displacement, level set method	Unix, Serial C & Fortran + OpenMPI
NIST [43] Laplace Solver	Generalized Laplace Equation Solver (e.g. utilized for el. conductivity)	Unix, Parallel
PoresPy [44]	Image analysis and pore network extraction, many algorithms	Unix/Windows, Python, Serial
OpenPNM [45,46]	Pore network modeling (extract network with PoresPy)	Unix/Windows, Python, Serial

7 Sustainability model

Sustainability is understood as a combination of technical and administrative strategies to ensure that users can rely on ongoing data infrastructure services that promote research innovation. While we support open science and specifically open data, neither is unfortunately free and a business model has to ultimately be developed for sustainability. The portal is on the way to scale up to a go-to location for benchmark datasets, and as such cannot be maintained in an ad-hoc manner and requires consistent financing to provide required services.

7.1. User fees

We formulated a business plan and begin adoption in 2021. Until that point users publishing data in DRP where not charged any fees as initial development (2015-2018) was supported by an NSF grant, and to this day rather limited support is available through Digital Porous Media Industry Affiliate Program at The University of Texas at Austin. In May and June 2021 DRP initiated a request for comments about establishing a funding model based on a) one-time publication fees or b) a DRP subscription plan to start in September of 2021. The community discussion was initiated via the DRP newsletter to all users (available on [25]) as well as through personal and email communications to the biggest DRP data providers. Documentation on DRP user agreement changes [47] has complete details of the proposed fee structure. Responses were publicly posted without personal identifiable information in DRP newsletter from July 13, 2021, linked to [25]. Five groups that publish regularly expressed support for the fee structure after the necessity was explained. The push back from the rest of the researchers came from the fact that they do not have open data fees built into their budgets. The fee structure is in Table 2 and the new user agreement is available on [14].

Table 5. The user fee structure implemented in September 2021.

Individual (one time per project)	**< 2 GB**	**2-500 GB**	**500 GB – 1 TB**
Open data publication fee	Free	$1,500	$ 3,000
Institutional membership (annual)	**< 500 GB**	**500 GB - 1TB**	**1 – 3TB**
Academic/non-profit	$ 5,000	$ 7,500	$ 10,000
Company/private	$ 7,500	$ 10,000	$ 12,500

7.2. Donations and sponsorships

Anyone can donate to or support Digital Rocks Portal: this is tax-deductible and easily done via credit card using UT Austin's Giving platform. Contributions of $100 or more are thanked with a promotional mug, mouse pad or a calendar. Table 6 below provides details on sponsorship levels.

Table 6. Donation and sponsorship levels.

Level	**Type**	**Details**
$20+	Any time giving	Tax deductible and easily done via credit card.
$500	Newsletter sponsor	You and/or your company will be featured in next newsletter.
$1,500	Project sponsor	The donation will: (1) make one of the Folk McBride high resolution thin sections examples available for research, (2) permanently place a thank-you note to you/your company on the project landing page, and (3) make you an owner of promotional items with that project imagery. Check this presentation for suggested datasets. Any other (existing or new) project can be sponsored under the same terms.
$3,000 or more	Student/fellow sponsor	Sponsor a student fellow (suggested $3000-5000) who will help upload data from partner institutions and research groups and create advanced visualizations to accompany them in the semester following the donation.

7.3. Expanding the user base

Last but not the least, many energy storage and materials engineering samples are porous and require image analysis, simulation and ML techniques that are comparable to digital rocks physics. Technology transfer and change of the name to Digital Porous Media is imminent.

Acknowledgements

Foundational development of this portal was obtained through the National Science Foundation, NSF EAR CAREER grant 1255622 "Advanced interface methods in heterogeneous porous materials: a multi-disciplinary and multi-scale framework" as well as NSF EarthCube IA 1541008 from 2015-18. Since the end of 2018, limited support is available through the Digital Porous Media (formerly Digital Rock Petrophysics) Industry Affiliate Program at the University of Texas at Austin. All are gratefully acknowledged.

Programming support has been provided by the Texas Advanced Computing Center (Keith Strmiska) and the administrative support by Center for Petroleum and Geosystems Engineering (now Center for Subsurface Energy and the Environment).

References

[1] M.J. Blunt, Multiphase flow in permeable media: a pore-scale perspective, Cambridge University Press, Cambridge, United Kingdom ; New York, NY, 2017.

[2] A. Mehmani, R. Verma, M. Prodanović, Pore-scale modeling of carbonates, Marine and Petroleum Geology. 114 (2020) 104141. https://doi.org/10.1016/j.marpetgeo.2019.104141.

[3] Y.D. Wang, M.J. Blunt, R.T. Armstrong, P. Mostaghimi, Deep learning in pore scale imaging and modeling, Earth-Science Reviews. 215 (2021) 103555. https://doi.org/10.1016/j.earscirev.2021.103555.

[4] A. Krizhevsky, I. Sutskever, G.E. Hinton, ImageNet Classification with Deep Convolutional Neural Networks, in: F. Pereira, C.J.C. Burges, L. Bottou, K.Q. Weinberger (Eds.), Advances in Neural Information Processing Systems 25,

Curran Associates, Inc., 2012: pp. 1097–1105. http://papers.nips.cc/paper/4824-imagenet-classification-with-deep-convolutional-neural-networks.pdf (accessed August 15, 2019).

[5] J.E. Santos, Y. Yin, H. Jo, W. Pan, Q. Kang, H.S. Viswanathan, M. Prodanović, M.J. Pyrcz, N. Lubbers, Computationally Efficient Multiscale Neural Networks Applied to Fluid Flow in Complex 3D Porous Media, Transp Porous Med. (2021). https://doi.org/10.1007/s11242-021-01617-y.

[6] J.E. Santos, D. Xu, H. Jo, C.J. Landry, M. Prodanović, M.J. Pyrcz, PoreFlow-Net: a 3D convolutional neural network to predict fluid flow through porous media, Advances in Water Resources. (2020) 103539. https://doi.org/10.1016/j.advwatres.2020.103539.

[7] TensorFlow, TensorFlow. (n.d.). https://www.tensorflow.org/ (accessed July 3, 2022).

[8] J. Santos, Y. Yin, M. Prodanović, H. Khan, N. Lubbers, 3D Collection of Binary Images, (2021). https://doi.org/10.17612/NXGK-K611.

[9] B. Chang, J.E. Santos, R.A. Victor, M. Prodanović, ElRock-Net: Assessing the Utility of Machine Learning to Initialize 3D Electric Potential Simulations, in: Society of Core Analysists Conference Proceedings, The Society of Core Analysts (SCA), Austin, Texas, USA, 2022: pp. SCA2022-T049.

[10] M. Prodanovic, J.E. McClure, DRP Visualization Minicourse, 2020. https://github.com/dr-masha/drp_visualization_mini_course.

[11] B. Chang, F. Farias, M. Prodanovic, Digital Porous Media tools (dpm_tools), n.d. https://github.com/BC-Chang/dpm_tools.

[12] Y. Niu, Y. Da Wang, P. Mostaghimi, J.E. McClure, J. Yin, R.T. Armstrong, Geometrical-Based Generative Adversarial Network to Enhance Digital Rock Image Quality, Phys. Rev. Applied. 15 (2021) 064033. https://doi.org/10.1103/PhysRevApplied.15.064033.

[13] J.E. McClure, J. Yin, R.T. Armstrong, K.C. Maheshwari, S. Wilkinson, L. Vlcek, Y. Da Wang, M.A. Berrill, M. Rivers, Toward Real-Time Analysis of Synchrotron Micro-Tomography Data: Accelerating Experimental Workflows with AI and HPC, in: J. Nichols, B. Verastegui, A. 'Barney' Maccabe, O. Hernandez, S. Parete-Koon, T. Ahearn (Eds.), Driving Scientific and Engineering Discoveries Through the Convergence of HPC, Big Data and AI, Springer International Publishing, Cham, 2020: pp. 226–239. https://doi.org/10.1007/978-3-030-63393-6_15.

[14] M. Prodanovic, M. Esteva, M. Hanlon, G. Nanda, P. Agarwal, Digital Rocks Portal, (2015). https://doi.org/10.17612/P7CC7K.

[15] Mendeley Data, (n.d.). https://data.mendeley.com/ (accessed September 1, 2019).

[16] Dryad, (n.d.). http://datadryad.org/ (accessed July 3, 2022).

[17] Energy Data Exchange (EDX), (n.d.). https://edx.netl.doe.gov/ (accessed July 3, 2022).

[18] S. Sweat, A. Ranganath, M. Esteva, M. Prodanović, User Guided Design: Building Confidence in Engineering Data Publication, in: University of Texas at Austin, Denver, CO, 2016. https://doi.org/10.15781/T27940X4W.

[19] S. Fuchs, D. Crandall, M. Gill, J. Moore, B. Kutchko, Foamed Cement API RP 10B-4, (2018). https://doi.org/10.17612/p7gd4q.

[20] M. Souzy, H. Lhuissier, Y. Méheust, T. Le Borgne, B. Metzger, Experimental 3D Velocity Field in Random Sphere Packing, (2020). https://doi.org/10.17612/HDP8-0149.

[21] C. Peters, S. Hajireziae, J. Kim, D. Crandall, Mineralized fracture in a mudrock shale, (2021). https://doi.org/10.17612/P4WH-W011.

[22] H. Khan, M. Prodanovic, D. Dicarlo, Particulate straining in carbonate proxy, Digital Rocks Portal, 2018. https://doi.org/10.17612/P7D96T.

[23] DataCite Fabrica, DataCite Fabrica. (n.d.). https://doi.datacite.org/ (accessed May 13, 2022).

[24] TACC Visualization Portal, (n.d.). https://vis.tacc.utexas.edu/.

[25] Digital Rocks Portal About section, (n.d.). https://www.digitalrocksportal.org/about/.

[26] M.D. Wilkinson, M. Dumontier, Ij.J. Aalbersberg, G. Appleton, M. Axton, A. Baak, N. Blomberg, J.-W. Boiten, L.B. da Silva Santos, P.E. Bourne, J. Bouwman, A.J. Brookes, T. Clark, M. Crosas, I. Dillo, O. Dumon, S. Edmunds, C.T. Evelo, R. Finkers, A. Gonzalez-Beltran, A.J.G. Gray, P. Groth, C. Goble, J.S. Grethe, J. Heringa, P.A.C. 't Hoen, R. Hooft, T. Kuhn, R. Kok, J. Kok, S.J. Lusher, M.E. Martone, A. Mons, A.L. Packer, B. Persson, P. Rocca-Serra, M. Roos, R. van Schaik, S.-A. Sansone, E. Schultes, T. Sengstag, T. Slater, G. Strawn, M.A. Swertz, M. Thompson, J. van der Lei, E. van Mulligen, J. Velterop, A. Waagmeester, P. Wittenburg, K. Wolstencroft, J. Zhao, B. Mons, The FAIR Guiding Principles for scientific data management and stewardship, Scientific Data. 3 (2016) 160018. https://doi.org/10.1038/sdata.2016.18.

[27] ORCID, (n.d.). https://orcid.org/ (accessed May 14, 2022).

[28] Digital Rocks Portal: Visualization - YouTube, (n.d.). https://www.youtube.com/playlist?list=PL7QcRCNp33XA7BBvpNbwsDZjLB-mezkUu (accessed May 16, 2022).

[29] ImageNet, (n.d.). https://image-net.org/index (accessed October 20, 2021).

[30] Quantum-Machine.org: Home, (n.d.). http://quantum-machine.org/ (accessed May 16, 2022).

[31] J.E. Santos, M.J. Pyrcz, M. Prodanović, 3D Dataset of binary images: A collection of synthetically created digital rock images of complex media, Data in Brief. 40 (2022) 107797. https://doi.org/10.1016/j.dib.2022.107797.

[32] J. Santos, Chang, Bernard, Gigliotti, Alex, Kang, Qinjun, Lubbers, Nicholas, Viswanathan, Hari, Prodanovic, Masa, 3D Dataset of Simulations, (2021). https://doi.org/10.17612/93PD-Y471.

[33] J.E. Santos, Chang, Bernard, Gigliotti, Alex, Y. Yin, W. Song, Prodanovic , Masa, Q. Kang, Lubbers, Nicholas, H.S. Viswanathan, A Dataset of 3D Structural and Simulated Transport Properties of Complex Porous Media, Nature Scientific Data. Under Review (n.d.).

[34] Y.D. Wang, R. Armstrong, P. Mostaghimi, A Diverse Super Resolution Dataset of Digital Rocks (DeepRock-SR): Sandstone, Carbonate, and Coal, (2019). https://doi.org/10.17612/S3M9-E024.

[35] Y. Niu, Armstrong, Ryan, Mostaghimi, Peyman, Unpaired super-resolution on micro-CT sandstone by using cycle-consistent generative adversarial network, (2020). https://doi.org/10.17612/VZTT-YX38.

[36] N. Alqahtani, P. Mostaghimi, R. Armstrong, A Multi-Resolution Complex Carbonates Micro-CT Dataset (MRCCM), (2021). https://doi.org/10.17612/3T36-Q704.

[37] M. Tawfik, Z. Karpyn, S.X. Huang, scCO2-Brine-Glass Dataset for Comparing Image Denoising Algorithms, (2022). https://doi.org/10.17612/A1QA-2A25.

[38] K. Singh, M.J. Blunt, High resolution X-ray micro-tomography datasets for in-situ effective contact angle analysis in carbonate rocks, (2018). https://doi.org/10.17612/P7D95F.

[39] J.E. McClure, Z. Li, M. Berrill, T. Ramstad, The LBPM software package for simulating multiphase flow on digital images of porous rocks, Comput Geosci. 25 (2021) 871–895. https://doi.org/10.1007/s10596-020-10028-9.

[40] Palabos CFD and Complex Physics, (n.d.). http://www.palabos.org/ (accessed January 16, 2014).

[41] J.E. Santos, A. Gigliotti, A. Bihani, C. Landry, M.A. Hesse, M.J. Pyrcz, M. Prodanović, MPLBM-UT: Multiphase LBM library for permeable media analysis, SoftwareX. 18 (2022) 101097. https://doi.org/10.1016/j.softx.2022.101097.
[42] R. Verma, M. Prodanovic, Level Set Method based Progressive Quasi-Static (LSMPQS) software 1.0, 2018. https://github.com/rahulverma88/lsmpqs-1.0.
[43] National Institute of Standards and Technology, NIST. (n.d.). https://www.nist.gov/ (accessed May 18, 2022).
[44] J.T. Gostick, Versatile and efficient pore network extraction method using marker-based watershed segmentation, Phys. Rev. E. 96 (2017) 023307. https://doi.org/10.1103/PhysRevE.96.023307.
[45] OpenPNM | Home | An Open Source Pore Network Modeling Package, (n.d.). http://openpnm.org/ (accessed November 2, 2015).
[46] J. Gostick, M. Aghighi, J. Hinebaugh, T. Tranter, M.A. Hoeh, H. Day, B. Spellacy, M.H. Sharqawy, A. Bazylak, A. Burns, W. Lehnert, A. Putz, OpenPNM: A Pore Network Modeling Package, Computing in Science & Engineering. 18 (2016) 60–74. https://doi.org/10.1109/MCSE.2016.49.
[47] Digital Rocks Portal User Agreement Documentation, (n.d.). https://www.digitalrocksportal.org/user-agreement/.

Pore-scale Analysis of CO_2-brine Displacement in Berea Sandstone and Its Implications to CO_2 Injectivity

Guangyuan Sun*, *Zhuang* Sun, *Andrew* Fager and *Bernd* Crouse

Dassault Systèmes, San Francisco, USA

Abstract. For storage in deep saline formations, where CO_2 is injected into the pore spaces of rocks previously occupied by saline groundwater (brine), relative permeability is a key input parameter for predictive models. CO_2 injectivity is considered to reach the maximum value at the CO_2 endpoint relative permeability when brine saturation becomes irreducible. The objective of this study is to investigate the effect of viscosity ratio, interfacial tension and wettability on relative permeability during CO_2-brine drainage. A multiphase lattice Boltzmann model (LBM) is employed to numerically measure pore-scale dynamics in CO_2-brine flow in the sample of Berea sandstone. CO_2/brine with interfacial tension from 30 to 45 mN/m and viscosity ratio from 0.05 to 0.17 (the range of values expected for typical storage reservoirs conditions) are carried out to systematically assess the influence on the relative permeability curves. Although CO_2 storage in sandstone saline aquifers is predominantly water wet, there are contradictory results as to the magnitude of the contact angle and its variation with fluid conditions. Therefore, the range of wetting conditions is studied to gain a better insight into the effect of wettability on supercritical CO_2 displacement. In this study, it is observed that interfacial tension variations play a trivial impact while both of viscosity ratio and wettability are likely to have a significant effect on relative permeability curves under representative condition of storage reservoirs. We also perform a near-wellbore scale geomechanics analysis to investigate the impact of relative permeability on CO_2 injectivity. The result shows that water-wet condition facilitates the CO_2 injection when there is no fracture induced.

1 Introduction

It is well accepted that global warming is due to the increase in atmospheric concentrations of greenhouse gases, such as carbon dioxide (CO_2). To achieve mid- and long-term targets in reducing CO_2 emissions and their intensity, various mitigation approaches are explored, foremost among them being CO_2 capture, utilization and storage (CCUS), which will play a significant role in the first half of this century if reduction targets are to be met [1-2]. Current methods to reduce atmospheric emissions of anthropogenic greenhouse gases often include the capture of CO_2 and subsequent injection into deep subsurface formations [3-5]. Such injection scenarios involve multiple fluids and lead to multiphase flow problems. Common approaches involve injection of CO_2 into deep formations, such that the CO_2 will be under sufficiently high pressure and temperature to remain in a supercritical state. For temperatures T and pressures P above the critical point (T_c = 31.1°C and P_c = 7.38 MPa), the injected CO_2 will have significantly higher density than that of gaseous CO_2, although still less than that of the resident formation water, usually brine. The viscosity of the CO_2 is also significantly less than that of the brine. Therefore, migration of the CO_2 away from the injection well involves both gravity override and unfavorable mobility ratio.

The potential for CO_2 storage in saline aquifers largely depends on the storage capacity and the CO_2 injectivity of the site. The storage capacity of aquifers is governed by aquifer volume, porosity, and, most importantly, volumetric as well as microscopic displacement efficiencies. Injectivity, on the other hand, is governed by permeability, relative permeability, fracture characteristics of the rock, and rock compressibility [6].

The primary independent variables on multiphase flow, other than the rock structure, are pressure, temperature and salinity. The fluid properties such as viscosity, density, contact angle and interfacial tension can be described as secondary variables, depending on the first ones. Relative permeability, irreducible water and residual CO_2 saturations may be considered as tertiary variables, as these are in turn controlled by the fluid properties [7].

The viscosity ratio is defined as the ratio of the viscosity of the non-wetting phase to the wetting phase. The effect of viscosity ratio on relative permeability is not clear, as the observations reported in the literature are contradictory. Relative permeability is shown to be both independent of [8] or to increase with viscosity ratio [9-11] in experimental oil/water systems. However, Lefebvre du Prey et al. [12] showed that for a high viscosity ratio, the relative permeability of the lower viscosity fluid significantly decreases during drainage. Downie and Crane [13] suggest that a subsequent decrease in viscosity ratio does not return the relative permeability to its original value, thus concluding that viscosity ratio is not the only factor to change relative permeability. Bachu and Bennion [7] found that an increase of interfacial tension or a decrease in the viscosity ratio brings about similar changes of relative permeability of brine, irreducible water and residual CO_2 saturations that is,

* Corresponding author: Guangyuan.SUN@3ds.com

increasing relative permeability and decreasing residual saturations.

Interfacial tensions for the CO_2-brine system typically range from 25 to 55 mN/m at the conditions for subsurface storage [14]. The data to investigate the effect of interfacial tension on relative permeability for the CO_2-brine system is limited. Bennion and Bachu [7] measured relative permeability curves of drainage and imbibition with interfacial tensions in the range 19.8 to 56.2 mN/m in a water-wet sandstone. Unlike the oil-brine system, a significant increase in permeability is observed for decreasing interfacial tension. The imbibition curves become more linear with decreasing interfacial tension, suggesting there is an increase in total relative mobility and the flow at low interfacial tension behaves miscible. Both endpoint permeability of CO_2 and residual CO_2 saturation increase as interfacial tension decreases. However, they also suggested that the effects of viscosity ratio and interfacial tension cannot be separated, as the viscosity ratio simultaneously increases (0.02, 0.05 and 0.1) as interfacial tension decreases (56, 34, 20 mN/m). The discrepancy about the influence of interfacial tension could potentially result from the low viscosity ratio of the CO_2-brine system ($<<1$) as opposed to the high viscosity ratio of the oil-water system ($>>1$) [15]. The oil-water system is much more strongly controlled by viscous forces, which have the effect of decreasing the influence of interfacial tension on distribution of the wetting and non-wetting fluids.

For CO_2 storage in sandstone saline aquifers, the reservoirs under consideration are expected to be predominantly water-wet [16-18] but the results are contradictory as to the magnitude of the contact angle and its variation with fluid conditions. Espinoza and Santamarina [16] observed that quartz and calcite in CO_2-brine system is strongly water-wet with no impact of changing pressure and a weak dependence on salinity up to 3.5M NaCl. Chiquet et al. [19] have also observed the system to be water-wet on quartz with little impact of both pressure and salinity from atmospheric up to 10 MPa pressure and for 0.01-1M NaCl solutions. However, a transition was observed from water- to intermediate-wet with increasing CO_2 pressure and NaCl and the system appeared to be mixed-wet across the entire range of conditions during imbibition. In microfluidic experiments with silica micromodels, Kim et al. [17] found contact angles ranging from water- to mixed-wet for the supercritical CO_2-brine system, as the salinity increases from 0.01 to 5M NaCl.

In this work, we present the results of a numerical study that carefully characterizes relative permeability of CO_2-brine in Berea sandstone using the steady-state method. We study the effect of key fluid properties (viscosity ratio, interfacial tension and wettability) on drainage displacement of brine by CO_2 injection at different flow conditions of reservoir. We also perform a near-wellbore scale geomechanics analysis to investigate the impact of relative permeability on CO_2 injectivity.

2 Measurements and procedure

In this section, we present detailed information regarding the rock sample, flow conditions, and simulation setup and procedure used in this study.

2.1 Micro-CT imaging and processing

A Berea sandstone sample was selected in the study. Micro-CT scans were carried out on a cylindrical core sample of 5 mm inch diameter and 10 mm inch length to acquire 3D images of the pore structure with resolution of 2.02μm/voxel. As the attenuation of X-rays within the sample are associated to the material density, the micro-CT scans provide grey-scale images with color values related to the sample material (Figure 1 left). The resulting 3D image was denoised using typical image processing techniques including contrast enhancement and low-pass filtering (Figure 1 middle). Threshold segmentation was applied resulting in a binary image (Figure 1 right) suitable to be used as input to the numerical simulations. The resolved porosity is 15.7% and an absolute permeability of 235 mD was obtained from single-phase flow simulation. An imaging domain of $500\times500\times500$ voxels is chosen for the multi-phase flow simulations.

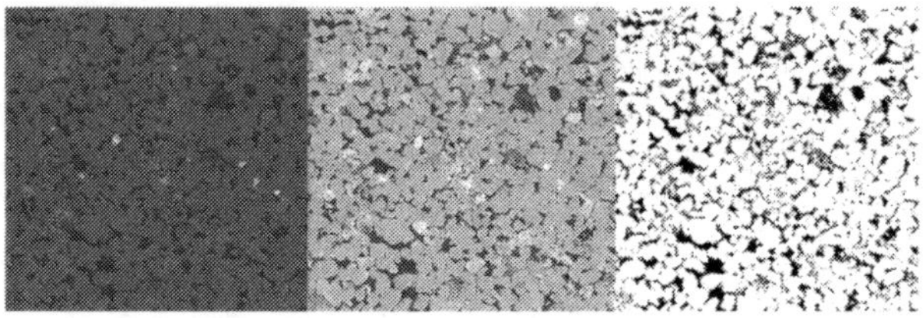

Fig. 1. Berea Sandstone micro-CT cross section: original (left), denoised (middle), and segmented (right) image slices.

2.2 Physical properties of system

To prepare for core flow condition, a primary drainage computation is performed to establish wettability distribution in pore space. It is assumed that during the drainage process, the smallest pores remain water-filled, while the non-wetting phase invades the remainder of the pore space. The pore surface can only become more repulsive to the water due to contact with non-wetting phase and the rock will achieve a "mixed wettability" state. More detail on these widely accepted mechanisms of aging or wettability alteration can be found in [20]. In the present numerical procedure, by assuming that 5% smallest pore volume still in contact with water after primary drainage, these pore surfaces are assigned a contact angle of 10°. The surfaces contacted by non-wetting phase are assigned a 30° and 170° contact angle for the water-wet and mixed-wet scenarios, respectively. The overall contact angle distributions are shown in Figure 2.

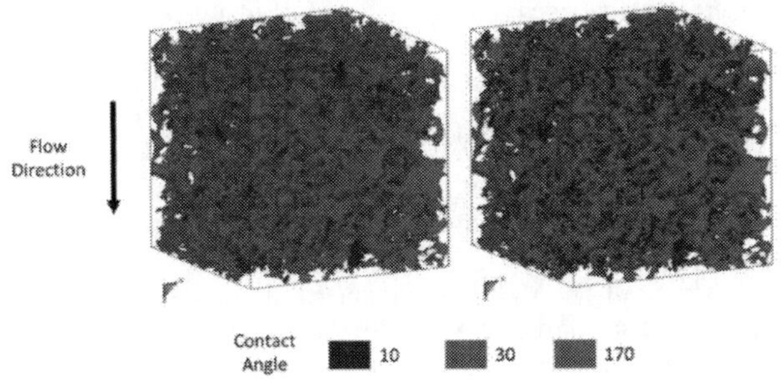

Fig. 2. Surface wettability in water- (left) and mixed-wet (right).

When it comes to fluid properties such as interfacial tension, viscosity and contact angle, it is straightforward to describe how these change with pressure, temperature and salinity. However, the design of a study to identify the individual contribution of each fluid property on the relative permeability remains a challenging exercise. As a result, the selections of fluid properties are at the same salinity (=1.98 mol/kg) and at the same time one or more of other parameters remain constant. It is common to use a dimensionless parameter such as capillary number (N_c) to combine these effects of interfacial tension and viscosity on relative permeability, which describes the ratio of viscous (governing flow) to capillary forces (governing trapping) acting in the pore space and is given by:

$$N_c = \frac{q\mu}{\sigma} \quad (1)$$

where q is the Darcy velocity, μ is the viscosity of the invading fluid and σ is the interfacial tension between CO_2 and brine. All the simulations reported in this study were performed under capillary dominated displacement regime with $N_c \sim 10^{-5}$ [21-22]. In section 3, the impact of the viscosity ratio, interfacial tension and wettability on relative permeability are discussed in more detail.

2.3 Multiphase numerical simulation

All flow simulations were performed using DigitalROCK™, which uses a numerical solver based on the Shan-Chen multi-phase lattice Boltzmann model [23]. This solver has been validated on a variety of fundamental benchmarks and real reservoir rock test cases [23-28]. It also includes the ability to deal with high viscosity ratio of the immiscible fluid phases [29].

A steady-state displacement method is applied to measure relative permeability in this work. A displacement simulation starts at initial 100% brine saturation. Once the simulation proceeds until total flow rate is converged, the permeability is measured. Then the mixture of CO_2 and brine with increased CO_2 fraction is injected until the certain CO_2 saturation level is reached. The injection stops and waits until the flow rate is converged to achieve the equilibrium. At this point the permeability of CO_2 and brine is measured again at the current saturation. The same process will be repeated until the residual condition is reached when the brine phase barely moves. Periodic boundary conditions and a driving force are applied in the flow direction. Additional details for this method can be found in [22, 25].

3 Pore-scale numerical simulation results

3.1 Interfacial tension effect

The interfacial tension of the CO_2-brine system at conditions covering the range relevant to CO_2 storage has been recently measured with the most comprehensive dataset provided by Li et al. [14]. As a thermophysical property, the interfacial tension is a function of pressure, temperature, and brine salinity. At conditions in subsurface storage, the interfacial tension ranges from 55 mN/m at high temperature, low pressure and high salinity to 20 mN/m at low temperature, high pressure and low salinity. Therefore, when it comes to evaluating the impact of interfacial tension on multiphase flow in the CO_2-brine system, the multi-dimensional nature of the relevant parameter space (density, viscosity and interfacial tension) are changing over the conditions of pressure, temperature and brine salinity. By holding one or more of these parameters (viscosity ratio and density ratio) constant, three values of interfacial tension chosen are applied within the range from 30 to 45 mN/m (Table 1) to examine the effect of interfacial tension on relative permeability (Figure 3).

Table 1. Conditions of relative permeability curves to assess the effect of interfacial tension [14].

Subsurface condition		Thermophysical fluid property		
Pressure [MPa]	Temperature [°C]	Interfacial tension [mN/m]	Viscosity ratio [-]	Density ratio [-]
7	62	45	0.04	0.14
8	35	37	0.04	0.63
20	72	30	0.08	0.63

The results illustrate that the dependence of the relative permeability displacement chracteristicstics (end-point saturation and shape) on interfacial tension under the certain range of conditions in storage reservoirs. Endpoint relative permeability to CO_2 during drainage decreases with increasing interfacial tension, and residual brine saturation increases with increasing interfacial tension. This confirms that for lower interfaical tension it should be easier for the non-wetting phase (CO_2) to pass through pore throats, leading to higher relative permeability. However, this suggests that variations in interfacial tension alone (for the range of conditions expected in storage reservoirs) are still little and will not lead to significant variation in the relative permeability characteristics of a given rock.

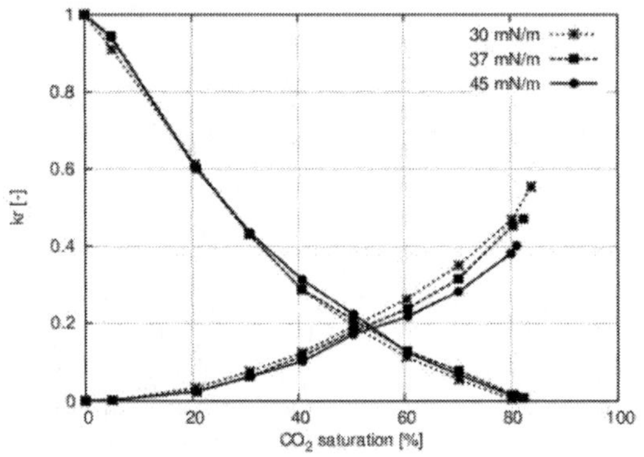

Fig. 3. Relative permeability curves measured for three different interfacial tensions.

3.2 Viscosity ratio effect

In this study, the viscosity ratio is defined as the ratio of the viscosity of CO_2 (non-wetting phase) to brine (wetting phase). CO_2 typically has lower viscosity than brine, and viscosity ratios from 0.02 to 0.2 are expected at geologically relevant conditions [30]. If injection of CO_2 into deep geological formations is considered, then the possible range of properties of the CO_2 can be estimated by considering typical conditions under which injection would occur. Because of the significant effect that the geothermal regime in a sedimentary basin has on CO_2-phase density and viscosity [31], we identify different basins in terms of magnitude of geothermal gradient and average land surface temperature. The average depth of the formation targeted for CO_2 storage range from about 1 km for shallow cold sites to 3 km for deep warm sites. Therefore, the properties of CO_2 and brine may vary, depending on pressure and temperature conditions. As representative condition of typical storage formations, surface temperature is 10 °C with geothermal gradient 25 °C/km for cold sites and 20 °C with geothermal gradient 45 °C/km for warm sites, respectively [30]. In both cases, a hydrostatic pressure gradient of 10.5 MPa/km is assumed that is typical for saline formation waters. Table 2 shows the conditions of pressure and temperature applied to study the effect of viscosity ratio. Under the conditions of deep cold formation, CO_2 injection is expected to be sufficiently deep to maintain the CO_2 in a supercritical state.

Table 2. Viscosity values for water and CO_2 at temperature (T) and pressure (P) conditions for representative deep old and shallow warm storage formation [30].

	Deep cold formation P: 31.5 MPa; T: 85 °C		Shallow warm formation P: 10.5 MPa; T: 65 °C	
	Viscosity [μPa·s]	Density [kg/m³]	Viscosity [μPa·s]	Density [kg/m³]
water	378	995	491	998
CO_2	63	740	24	292
Ratio	0.17	0.74	0.05	0.29

Relative permeability curves of CO_2-brine with different viscosity ratios are shown in Figure 4. As the viscosity contrast between the two fluids increases, the less viscous fluid (CO_2) becomes more mobilized through the pore space and therefore has a higher permeability. This illustrates the advantage of deep aquifers that provides the injection conditions for CO_2 with higher mobility and thus more storage of CO_2.

Although it highlights that the individual effect of interfacial tension and viscosity ratio cannot be separated, interfacial tension does not play a significant role within certain storage conditions as suggested in this study. As seen previously, the interfacial tension for CO_2-brine systems decreases with increasing pressure and increases with increasing temperature and salinity. Water viscosity increases with pressure and salinity, and decreases with increasing temperature, while CO_2 viscosity also increases with pressure and decreases with increasing temperature. Considering that the variation of CO_2 viscosity is stronger than that of brine viscosity, the viscosity ratio of CO_2 to brine varies in the opposite manner as interfacial tension.

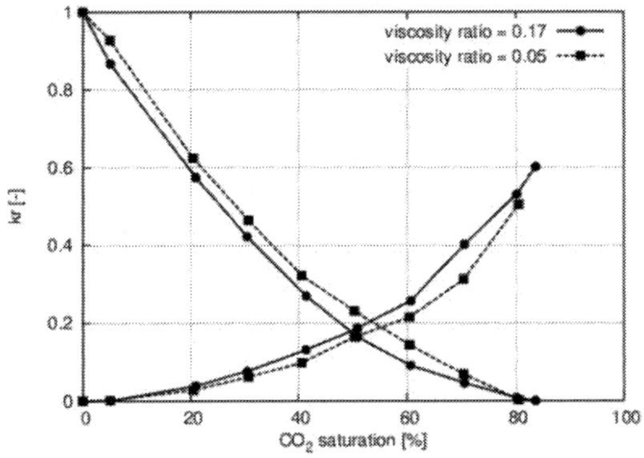

Fig. 4. Relative permeability curves measured for deep cold reservoir and shallow warm reservoir.

3.3 Wettability effect

Observations for CO_2-brine systems generally indicate a water-wet system for both sandstone and carbonate mineralogies, but there are contradictory results as to the magnitude of the contact angle and its variation with fluid conditions. In this study, two different wetting conditions (water-wet and mixed-wet) are chosen as defined in Section 2. The dependence of relative permeability on wetting condition are shown in Figure 5.

 https://doi.org/10.1051/e3sconf/202336701011

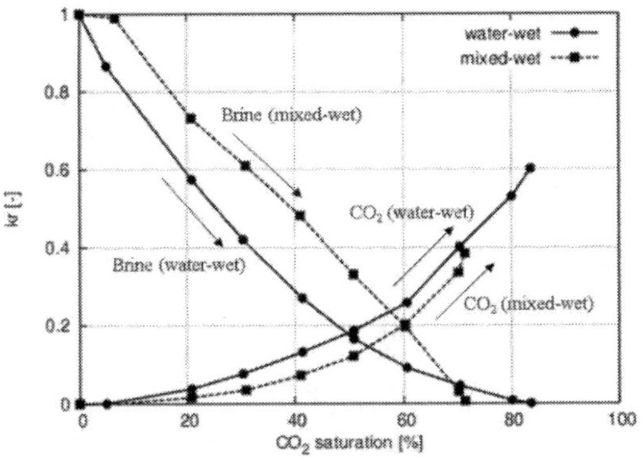

Fig. 5. Relative permeability curves measured for water-wet and mixed-wet condition.

During the displacement of water-wet system (drainage in this study), the rock is preferentially in contact with the water. As the nonwetting phase, CO_2 is generally located in the centers of the larger pores that forms continuous channels, while brine is found in the small pores and as a thin film on the rock surfaces. Eventually the remaining water is mainly found in three forms: (1) filling the smaller pores, (2) as a continuous film over the pore surface and (3) as trapped blobs surrounded by CO_2. In a mixed-wet system, the location of the two phases is reversed from the water-wet case. The injected CO_2 tends to imbibe into small and medium size pores, moving water into the large pores where it is easily displaced. The disconnected residual water remains in two forms: (1) small blobs in the centres of the larges pores, and (2) larger patches of water that are completely surrounding by CO_2. As above observed, the residual fluid distribution is shown in Figure 6.

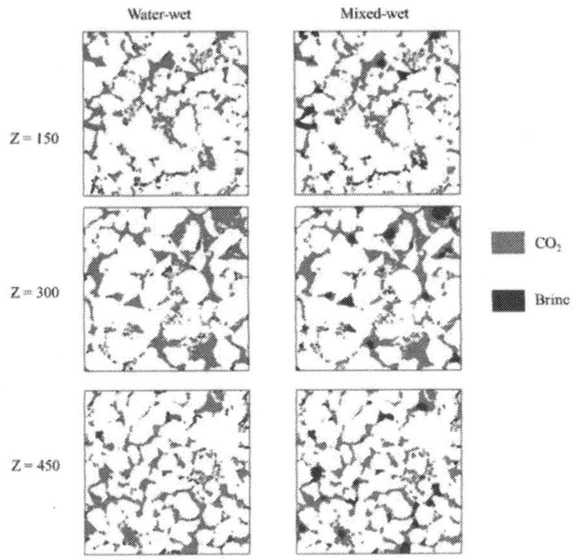

Fig. 6. Fluid distribution (red: CO_2; blue: brine) at residual stage along different depth in flow direction (Z) (left: water-wet; right: mixed-wet).

The differences in the relative permeability measured in water-wet and mixed-wet systems are caused by the differences in the fluid distributions. Consider a water-wet system, the brine does not significantly block the flow of CO_2, and the CO_2 permeability is relatively higher. This leads to a more efficient displacement than the one in a mixed-wet system and consequently lower residual water saturation. In contrast, the positions of the two fluids are reversed in mixed-wet system, CO_2 permeability is lower and brine permeability is higher because CO_2, as the wetting phase, tends to be located in the small pores and as a film on the surface, where it has little effect on the water flow. Therefore, the results of different wettability indicate that in CO_2-brine system, water-wet formation is able to store more CO_2 and more efficient injection is observed.

4 Near-wellbore analysis of CO_2 injectivity

We have shown the impact of wetting condition on the relative permeability curve in Section 3. In this section, we further perform a near-wellbore geomechanics analysis to investigate how the wettability impacts the CO_2 injectivity. The injectivity measures the ability of an injection well to receive injected fluid, and is a key parameter of CO_2 storage. It is defined as $\text{II} = q_i/\Delta P$, where II is the injectivity index, q_i is the injection rate, and ΔP is the pressure drop (bottom hole flowing pressure – reservoir pressure). CO_2 injectivity is considered to reach the maximum value at the CO_2 endpoint relative permeability when brine saturation becomes irreducible.

We set up the geomechanics model based on the finite element solver Abaqus™. The porous formation is assumed isotropic and poroelastic following the Biot's theory [32]. Fluid flow in the pore space follows the Darcy's law. Governing equations of deformation and pore fluid flow are based on well-established theories and can be found in many textbooks and literatures [33-35]. Therefore, they are not repeated in this study.

A large CO_2 injection rate may induce fractures near the wellbore (e.g. In Salah CO_2 storage project) [36]. We model the fracture propagation and fluid flow in the fractures based on the cohesive zone method (CZM). CZM characterizes the fracturing process as the damage evolution between two initially bonded surfaces with zero thickness, and has been widely used in fracture mechanics. Detailed description of the CZM can be found in the Abaqus Documentation and many other literatures [37].

Figure 7 shows a schematic of the geomechanics model. We adopt a plane strain model (zero strain in the direction normal to the plane, a typical assumption for near-wellbore fracturing process) with a maximum principal stress σ_{max} of 20 MPa and a minimum principal stress σ_{min} of 10 MPa. We assign no normal displacement conditions to all the boundaries. The domain size is 50 × 50 m and the injection well is located in the center of the domain.

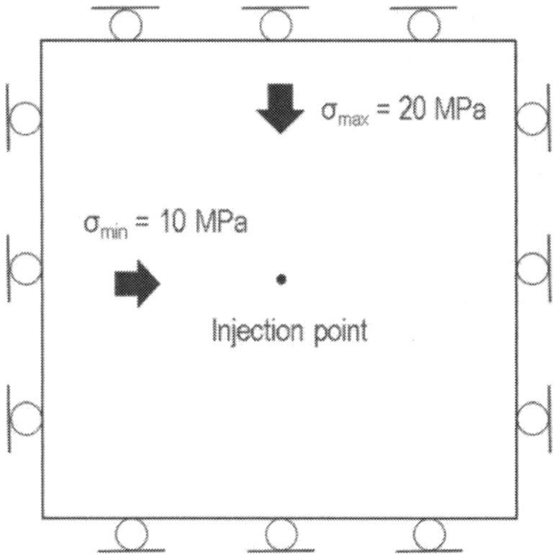

Fig. 7. Schematic of the geomechanics model. Domain size: 50 × 50 m.

In this section, we focus on the impact of wettability on the CO_2 injectivity. We consider two scenarios: water-wet and mixed-wet as presented in Section 3. Table 3 summarizes the geomechanics model parameters. The wetting condition will impact the relative permeability curve. All other parameters remain identical for these two case scenarios.

Table 3. Parameters of the geomechanics model

Parameters	Water-wet	Mixed-wet
Permeability (mD)	235	
Porosity	15.7%	
Residual water saturation (S_{wr})	16.4%	28.6%
Effective permeability of CO_2 at S_{wr} (mD)	141.1	90.4
Young's modulus (GPa)	15	
Poisson's ratio	0.25	

We perform a baseline simulation with a CO_2 injection rate of 0.1 m^3/s, water-wet condition. Figure 8 shows geomechanics simulation results including the pore pressure distribution at 100 s, the horizontal stress distribution at 100 s, and the evolution of injection pressure. CO_2 injection induces a fracture propagating perpendicular to the minimum principal stress direction. The injection pressure finally becomes stable at ~ 20 MPa.

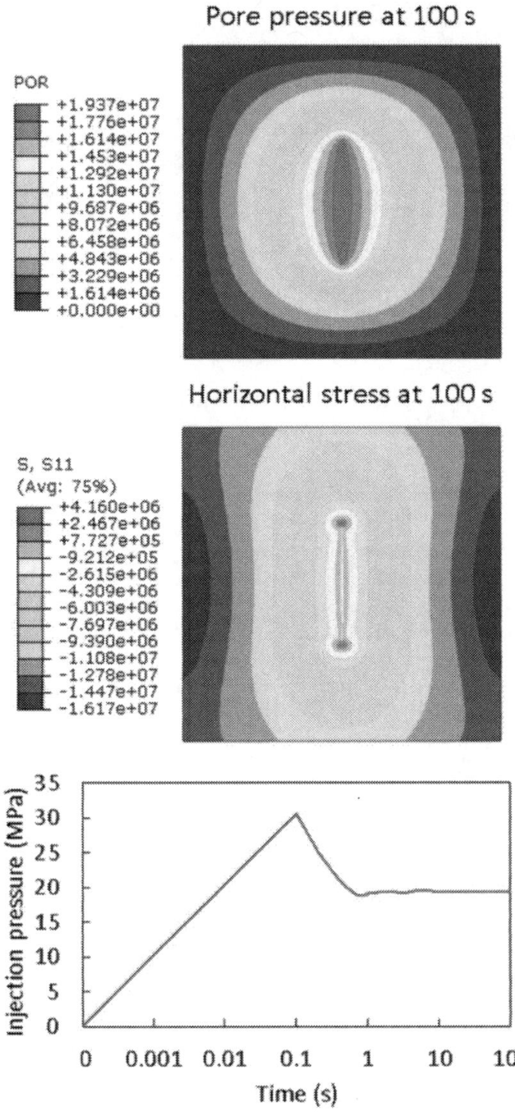

Fig. 8. Top: pore pressure distribution at the end of the numerical simulation (100 s). Middle: horizontal stress distribution at the end of the numerical simulation (100 s). Bottom: change of injection pressure over time. Time is plotted in a log scale to facilitate the visualization of pressure change. The injection pressure stabilizes at ~ 20 MPa. CO_2 injection rate is 0.1 m^3/s and it is a water-wet condition.

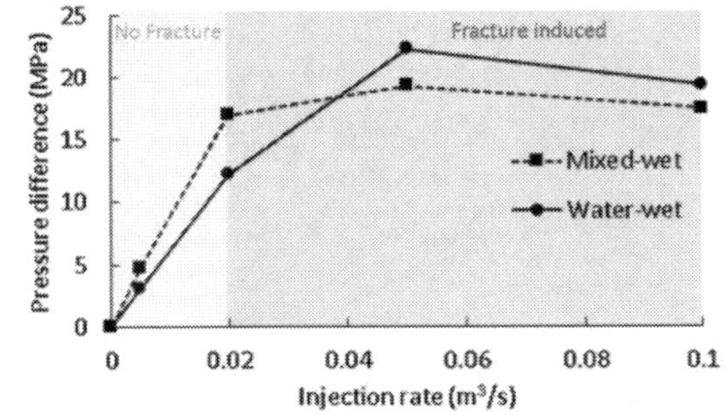

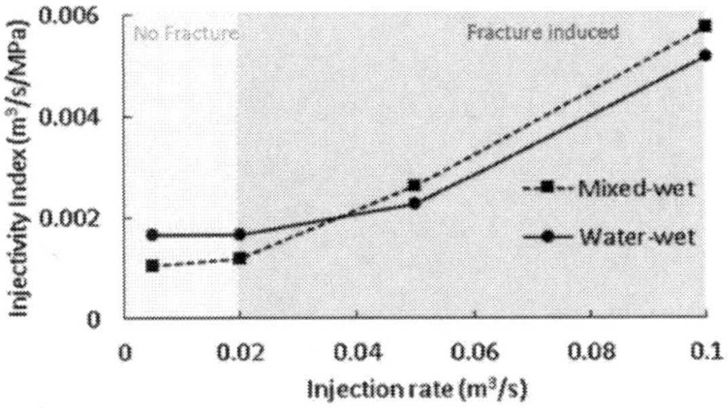

Fig. 9. Effect of wettability on CO_2 injectivity. The bottom figure is calculated from the top figure, where the injectivity index is defined as the injection rate over the pressure difference.

Figure 9 shows the geomechanics simulation results by varying the wetting condition and the injection rate. Each point corresponds to one geomechanics simulation. All other parameters remain identical. When the injection rate is relatively low (<= 0.02 m^3/s), there is no fracture and the fluid flow follows the Darcy's law. On the contrary, when the injection rate is high, a fracture is induced and the injectivity is significantly enhanced (manifested as a change in slope in the pressure difference - injection rate curve). Nevertheless, the induced fracture on the other hand may compromise the CO_2 containment, therefore, it is typically not favorable during a CO_2 injection.

We simulate two wetting conditions: water-wet and mixed-wet as discussed in Section 3. Water-wet condition facilitates the CO_2 fluid flow by offering the large pores and continuous flowing channels, which exhibits as a higher relative permeability of CO_2 on the relative permeability curve. Therefore, water-wet condition tends to have a higher injectivity compared to the mixed-wet condition when the injection rate is relatively low (no induced fracture). On the other hand, mixed-wet condition of a low CO_2 relative permeability facilitates the fracturing process. Therefore, mixed-wet condition tends to have a higher injectivity compared to the water-wet condition when the injection rate is high (with induced fracture), which indicates that the induced fracture starts to dominate the injectivity behavior. In summary, the wetting condition has an opposite impact on the CO_2 injectivity before and after the fracture was induced. Given that CO_2 injection generally tends to avoid fracturing process near the wellbore, a water-wet condition would be preferred from an injectivity perspective.

5 Conclusion

Using a DigitalROCK simulation approach, numerical simulation was carried out in Berea sandstone sample to study the impact of fluid properties (interfacial tension, viscosity ratio, and wettability) on steady-state relative permeability. As all these fluid properties depend on temperature, pressure and salinity, isolating the independent impact of them on relative permeability needs to hold one or more of other parameters constant. A weak dependency of displacement characteristics (end point relative permeability and shape of relative permeability and residual brine saturation) on interfacial tension was observed within typical range of storage formation condition. In contrast, viscosity ratio between CO_2 and brine shows clear dependency. Endpoint relative permeability to CO_2 decreases with increasing interfacial tension and decreasing viscosity ratio, while residual brine saturation increases with increasing interfacial tension and decreasing viscosity ratio. As it is common to consider storage formation as water-wet and the variation of contact angle observed in previous experiments, different wettability (water-wet and mixed-wet) are explored to investigate its impact on relative permeability. It turns out that the wettability has strong influences on relative permeability and there will be more CO_2 injected with higher CO_2 mobility in water-wet condition. A near-wellbore geomechanics model further demonstrates that a water-wet condition is preferred from an injectivity perspective.

References

1. IEA (International Energy Agency) Prospects for CO2 Capture and Storage. IEA/OECD, Paris, 249 (2004)
2. IPCC (Intergovernmental Panel on Climate Change) IPCC special report on carbon dioxide capture and storage. In: Metz B, Davidson O, de Coninck HC, Loos M, Mayer LA, Cambridge University Press, Cambridge, 442 (2005)
3. S. Bachu, J. J. Adams, Energy Conversion and Management **44(20)**, 3151-3175 (2002)
4. S. Holloway, Annual Review of Energy and the Environment **26**, 145–166 (2001)
5. S. M. Klara, R. D. Srivastava, and H. G. McIlvried, Energy Conversion and Management **44**, 2699-2712 (2003)
6. Y. Cinar, A. Riaz, & H. A. Tchelepi, SPE Journal, **14(04)**, 588-594 (2009)
7. S. Bachu, B. Bennion, Environmental Geology, **54(8)**, 1707–1722 (2008)
8. M.C. Leverett, Transactions of the AIME, **132(1)**, 149-171 (1939)
9. S. T. Yuster, Proceedings of the 3rd World Petroleum Congress, **2**, 437-445 (1951)
10. A. S. Odeh, Trans., AIME, **216**, 346-353 (1959)
11. R. Ehrlich, Transport in Porous Media, **11(3)**, 201-218 (1993)
12. E. J. Lefebvre du Prey, Society of Petroleum Engineers Journal, **13(01)**, 39-47 (1973)

13. J. Downie, F. E. Crane, Society of Petroleum Engineers Journal, **1(02)**, 59-60 (1961)
14. X. Li, E.S. Boek, G.C. Maitland, J. M. Trusler, Journal of Chemical & Engineering Data, **57(5)**, 1369-1375 (2012)
15. B. Bennion, S. Bachu, SPE/DOE Symposium on Improved Oil Recovery, 1-10, January (2006)
16. D.N. Espinoza, J.C. Santamarina, Water Resources Research, **46**, 7 (2010)
17. Y. Kim, J. Wan, T.J. Kneafsey, T.K. Tokunaga Environmental Science and Technology, **46 (7)**, 4228-4235 (2012)
18. J. W. Jung, J. Wan, Energy & Fuels, **26**, 6053-6059 (2012)
19. P. Chiquet, J.L. Daridon, D. Broseta, S. Thibeau, Energy Conversion and Management, **48**, 736-744 (2006)
20. J. S. Buckley, Y. Liu, S. Monsterleet, Society of Petroleum Engineers Journal, **3(01)**, 54-61 (1998)
21. R. Xu, B. Crouse, D.M. Freed, et al. SCA **2018-066** presented at at the International Symposium of the Society of Core Analysts held in Trondheim, Norway (2018)
22. A. Fager, G. Sun, R. Xu, et al. Journal of Petroleum Science and Engineering, **208**, 109435 (2022)
23. X. Shan, H. Chen, Physical Review E, **47(3)**, 181 (1993).
24. B. Crouse, D. M. Freed, N. et al. SCA **2016-058** presented at the International Symposium of the Society of Core Analysts held in Snow Mass, Colorado, USA (2016)
25. G.R. Jerauld, J. Fredrich, N. Lane, Q. Sheng, B. Crouse, D. M Freed, R. Xu, In SPE Abu Dhabi International Petroleum Exhibition & Conference. Society of Petroleum Engineers. November (2017)
26. H. Otomo, H. Fan, Y. Li, M. Dressler, I. Staroselsky, R. Zhang, H. Chen, Journal of Computational Science, **17**, 334-339 (2016)
27. H. Chen, C. Teixeira, K. Molvig, International Journal of Modern Physics C, **9(08)**, 1281-1292 (1998)
28. X. Shan, X.F. Yuan, H. Chen, Journal of Fluid Mechanics, **550**, 413-441 (2006)
29. H. Otomo, B. Crouse, M. Dressler, D. M. Freed, I. Staroselsky, R. Zhang, Chen, H, Computers & Fluids, **172**, 674-682 (2018)
30. J. M. Nordbotten, M. A. Celia, and S. Bachu, Transport in Porous Media, **58(3)**, 339-360 (2005)
31. S. Bachu, **Environmen**tal Geology, **44**, 277-289 (2003)
32. M.A. Biot, The Journal of the acoustical Society of america, **28(2)**, 179-191 (1956)
33. Jaeger, John Conrad, Neville GW Cook, Robert Zimmerman. *Fundamentals of rock mechanics* (John Wiley & Sons, 2009)
34. Zienkiewicz, Olgierd C., et al. *Computational geomechanics*, Vol. 613 (Chichester: Wiley, 1999)
35. E. Detournay, A.H.D. Cheng, *Fundamentals of poroelasticity*. In Analysis and design methods, 113-171 (Pergamon. 1993)
36. P. S. Ringrose, A. S. Mathieson, I. W. Wright, F. Selama, O. Hansen, R. Bissell, N. Saoula, and J. Midgley, Energy Procedia **37**, 6226-6236 (2013)
37. Abaqus. 2021. Abaqus 2021 online documentation.

An integrated petrophysical analysis based on NMR, organic geochemistry and mineralogy. The Vaca Muerta source rock-unconventional play at different thermal maturities

Diana Masiero[1*], *Marcos* Comerio[1], *Esteban* Domené[1], *Gabriela* Vila[1], *Bernarda* Epele[1], *Mariano* Cipollone[1], *Mariela* Silka[2], *Carlos* Camacho[3], *Lourdes* Vera López[2] and *Silvina* Chiappero[2]

[1]YPF-Tecnología S.A. (Y-TEC-CONICET), Buenos Aires, Argentina
[2]YPF S.A., Buenos Aires, Argentina
[3]EQUINOR ASA, Houston, TX, United States of America

Abstract. The Vaca Muerta Formation (Upper Jurassic–Lower Cretaceous) of the Neuquén Basin, Argentina is a world-class source rock and renowned as an unconventional reservoir for both oil and gas. The present study examines rotary sidewall cores representative of the oil and gas windows to analyze correlations between nuclear magnetic resonance (NMR T_2 and T_1-T_2 maps at 2 MHz), Gas Filled Porosity (GFP), Rock-Eval programmed pyrolysis and quantitative X-ray diffraction (XRD) mineralogy. Shale samples are characterized by high present-day total organic carbon contents ranging from 3 to 7.20% (mean: 4.30%) for Well A at the oil window and, 0.72 to 11.77% (mean: 5.86%) for Well B at the gas window. The analyzed set also covers a wide compositional spectrum from carbonate-rich (> 50% of calcite and minor dolomite), mixed carbonate-siliciclastic (30 to 50% of calcite), to siliciclastic-rich (< 30% of calcite) samples. Notably, the clay fraction is dominated by interstratified illite-smectite minerals with less than 10% of expandable layers. Total porosity, calculated from Gas Filled Porosity plus NMR T_2 Cumulative Distribution, is in the range of 10 to 20 porosity units. In this work we also present a new NMR sequence to detect solid-like organic matter with NMR at 2 MHz in an integrated workflow to characterize petrophysical properties of the Vaca Muerta Formation.

1 Introduction

Petrophysical studies are envisioned to characterize both conventional and unconventional reservoirs based on the evaluation of different properties including lithology, wettability, fluid saturation and composition, as well as porosity and permeability [1, 2]. Contrary to conventional reservoirs, a complete petrophysical analysis of unconventional plays, in particular shale rocks characterized by fine-grained components, presence of organic matter, and heterogenous lithologies is still challenging and new case studies are needed to obtain different workflow strategies.

The Vaca Muerta Formation of the Neuquén Basin Argentina, is a marine source rock with excellent characteristics from the perspective of source rock-unconventional plays: (1) high present-day TOC content (1 – 8% with peaks of 12%); (2) original hydrogen index (HIo) estimated to be approximately 680 mg HC/g TOC; (3) moderate depths of 3.150 m; (4) overpressure distribution in the order of 47–61 MPa, (5) lack of expandable clays in mature areas; and (6) different landing zones for the production of oil, wet gas and dry gas [3, 4, 5, 6, 7, 8].

In addition to the organic geochemical patterns of the Vaca Muerta Formation, petrophysical properties are key factors in hydrocarbon production. Several works correlate laboratory measurements and downhole log information to identify main reservoir properties [9, 10, 11]. However, in the Vaca Muerta Formation reservoir properties and fluid types seem to vary significantly both laterally and vertically suggesting that each interval of interest must be addressed and analyzed separately [11]. Nuclear magnetic resonance (NMR) is a widely used technique to characterize the porosity framework and fluid properties of shale plays due to its non-invasive nature and the possibility of performing in-situ measurements. Recently, the application of low field NMR laboratory measurements to detect both fluid (water and hydrocarbons) and solid-like organic matter components in Vaca Muerta samples was demonstrated, making low field NMR a powerful technique to improve the petrophysical characterization of this unit [12].

The present study examines two wells located in areas where according to regional maturity maps the unit ranges in maturity from the oil window to the gas window [6]. The

* Corresponding author: diana.m.masiero@ypftecnologia.com

 https://doi.org/10.1051/e3sconf/202336701012

main objective of this work is to obtain a workflow that allows a full and integrated petrophysical characterization of the Vaca Muerta Formation. This is achieved by combining different techniques and performing a detailed comparison and integration of the obtained results. The experimental techniques explored in this work include low-field NMR T_1-T_2 maps, organic geochemistry, gas filled porosity, microCT imaging and mineralogical analysis.

2 Data and methods

The present study shows petrophysical, mineralogical and organic geochemistry results integrated with 3D micro-scale reconstructions from rotary sidewall and core samples from two wells: Well A and Well B (Figure 1). The study includes shale samples (13 from Well A, 15 from Well B), but lithologies such as altered tuffs (1 sample, Well A) were also taken into consideration.

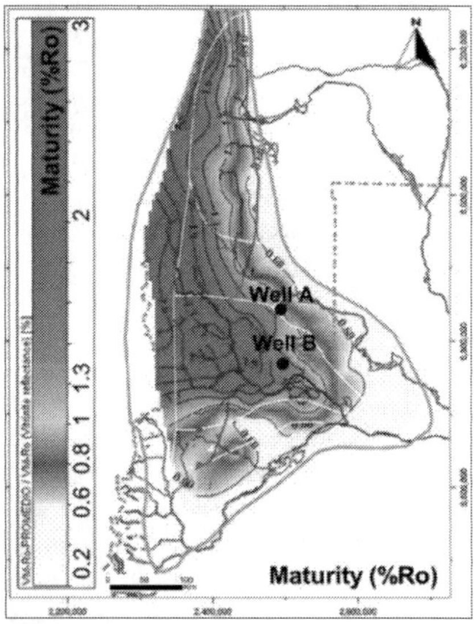

Fig. 1. Vaca Muerta Formation regional maturity map based on Brisson [6], and location of Well A and Well B analyzed in this work.

2.1 Workflow

In Fig. 2 the proposed workflow for characterization of rock samples from the Vaca Muerta formation is presented. This workflow can be applied to any non-conventional rock sample. Special emphasis is made on treating samples "as received" and performing non-destructive measurements to preserve as much sample as possible.

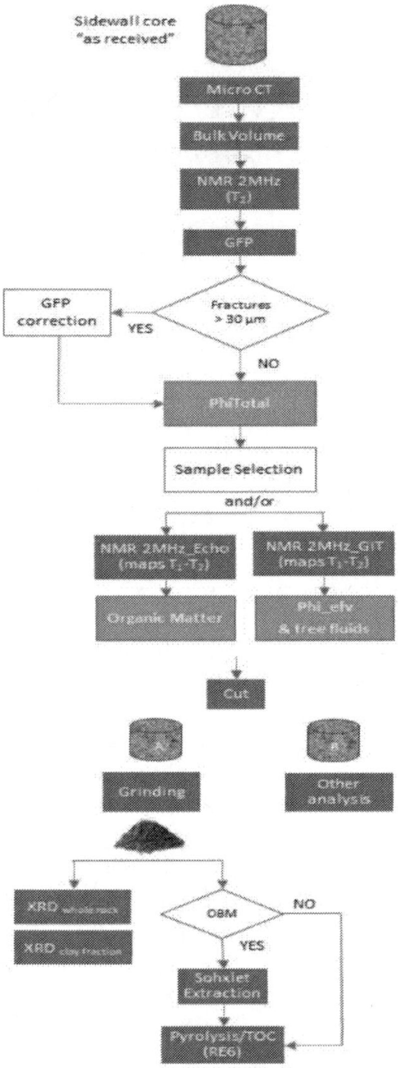

Fig. 2. The integrated workflow used for the analysis of sidewall core samples from Well A and Well B is presented.

2.2 Nuclear Magnetic Resonance (NMR)

NMR measurements were performed at 2.27 MHz with an Oxford Geospec2 analyzer. Samples were measured bulk as received; the average mass of the samples was 40–50 g, approximately. T_2 measurements were carried out with a standard CPMG sequence with echo time 100 µs. This experiment allows to quantify the fluid-filled pore space; total porosity is obtained by adding this value to the GFP porosity (see below).

T_1-T_2 maps were acquired with the SR-FID-ECHO-CPMG sequence introduced by Silletta et al. [12], a scheme that allows to probe the decay of both solid-like and liquid components of shale samples in a single measurement. In this setup, T_1 was encoded with a saturation recovery (SR) sequence with delay time from 21 µs to 390 ms in 50 logarithmic steps. In the T2 dimension, a CPMG pulse sequence was applied (echo time 200 ms, 500 echoes) sampling one data point at the echo tops. Furthermore, 18 points were sampled during the FID (dwell time 8 ms) before the first 180° pulse of the CPMG; to increase the FID acquisition time a spin echo of t = 200 ms was introduced before the CPMG echo train. The numerical inversion was performed applying the algorithm presented in [13] with exponential kernels in both dimensions. The zones in the T_1-T_2 maps were interpreted as in [12].

2.3 Gas Filled Porosity (GFP)

The empty pore space inside the rock sample is quantified using results from two separate experiments: a Hg pycnometer and GFP. The first, the Hg pycnometer, allows the characterization of the bulk volume. It uses a container filled with Hg whose density is characterized beforehand. Then the sample is fully submersed in the Hg and the change in weight is registered by a scale. The Archimedes method is used to estimate the outer volume of the sample.

The second experiment is known as GFP and is performed using a He Porosimeter from Vinci Technologies. The equipment itself measures grain volume. It has a sample chamber to house the sample and a reference chamber which is loaded with He at 200 psi; both chambers are connected, and the final stabilization pressure is detected. Therefore, the empty space in the sample chamber can be obtained, and by knowing the total volume of the chamber, the grain volume is deduced. It is important to point out that the time it takes the He to occupy the empty pore space will depend directly on the permeability of samples. Thus, lower permeability samples will take longer times than higher permeability ones. A long stabilization time (around 10 minutes) with a strict pressure tolerance (± 0.03 psi) for unconventional rocks is considered. We also repeat the experiment at least three times to make sure that measurements are repeatable.

By combining the bulk volume measurement of the rock with the grain volume, the empty pore space is quantified. This value refers to the part of the porosity that is empty (filled with air) and accessible from the outer surface of the rock.

2.4 X-ray diffraction (XRD)

For mineralogical analysis both bulk rock and clay fraction analysis were done. For bulk rock analysis samples were ground to a fine particle size using a Retsch MM400 mixer mill. X-ray diffraction patterns were collected on random powder specimens using a D8 Advance Bruker X-ray diffractometer (Ni-filtered CuKα, 40 kV, 40mA). A 5–70 deg 2θ range was acquired, with 0.02 deg steps and 0.5 second counting time. Diffraction data were analyzed using DIFFRAC.EVA software for identification of mineralogical phases. Weight percentages of observed phases were calculated using quantitative Rietveld analysis with DIFFRAC.TOPAS.

Determination of clay minerals was performed on the fraction less than 2 µm. Three oriented aggregates were prepared (air-dried, ethylene glycol solvated and calcinated) and their XRD patterns collected in the range of 3–30 deg 2θ with 0.02 deg steps and 0.5 second counting time. Identification of clay mineral phases was based on diagnostic peaks according to Moore and Reynolds [14] and their quantification using NEWMOD program as described in Moore and Reynolds [14].

2.5 Rock Eval programmed pyrolysis

The characterization of organic matter (OM) was performed using a Rock-Eval 6 [15]. Powdered whole rock samples were placed in an oven and first heated to 300°C under an inert atmosphere and then gradually pyrolyzed up to 650°C. After the pyrolysis was completed, samples are gradually heated up to 850°C in the presence of air. The S1 (free hydrocarbons) was not considered since samples were solvent extracted due to the use of oil-based mud during drilling operations. The S2 (hydrocarbons generated by kerogen cracking) and S3 (releasable oxygen) measures allowed to calculate hydrogen and oxygen indices (HI, OI) and estimate kerogen types (from I to IV), using the modified Van Krevelen diagram [16]. The thermal maturity was based on the determination of Tmax [17]. This parameter increases with the maturation degree of the organic matter and was used to obtain a thermal maturity estimation of the analyzed samples.

2.6 X-ray micro–Computed Tomography (microCT)

As a nondestructive technique, X-ray microcomputed tomography provides the required insight into opaque objects eliminating the need of sectioning. Specimens were imaged in a Bruker SkyScan 1173 High Energy desktop microCT at 130 kV and 61 µA with a 0.25 mm brass filter. The cubic voxel size was fixed at 35.72 µm and the samples were rotated around 360° at angular increments of 0.80°. The microCT radiograph projections were reconstructed using NRecon version 1.6.9.8 to obtain an axial slice image dataset. The 8-bit (256 grey values) image dataset was imported into DataViewer version 1.5.0 for visualization. The images obtained provide detailed visualization of the internal fabric and recognition of heterogeneities including fractures [18]. In grayscale images black correspond to air, dark grays to low X-ray absorbing minerals like quartz or feldspars and bright grays to higher X-ray absorbing phases such as calcite or, in white, pyrite. Segmentation and analysis of microCT images allow for the quantification of porosity associated with open fractures (> 30 µm thick) as observed for Well A (Figure 4). With image processing percentage of open fractures, in dark gray, was calculated with respect to the total volume. As pores in the internal structure have sizes lower than the resolution of the images, they are not considered in the segmentation. Thus, GFP was corrected only to include pores in the internal structure of the rock.

3 Results and Discussion

3.1 Well A

Shale samples are characterized by high present-day TOC contents ranging from 3–7.20% (mean: 4.30%), with maximum values registered in siliciclastic-rich (< 30% of calcite) samples (Table 1). The S2 peak which coincides with the remaining potential of hydrocarbon generation ranges from 2.84–12.75 mgHC/grock (mean: 5.75); whereas the HI shows values between 103–177 mgHC/gTOC (mean: 130). Rock Eval® Tmax values (458–462°C, mean: 460) point to analyzed samples being mature in the late oil window (Fig. 3). As expected, anomalous values of TOC (< 1%) and S2 (< 1 mg/g) were determined in altered tuff levels.

According to whole rock XRD analysis, siliciclastic-rich shales are mainly composed of quartz (22–41%, mean: 31), plagioclase (13–19%, mean: 16), and clay minerals (12–21%, mean: 16) with minor proportions of K-feldspar, pyrite and apatite. On the contrary, carbonate-rich samples show a predominance of calcite (50–60%, mean 53). Mixed carbonate-siliciclastic (30–50% of calcite) shales exhibit variable proportion of quartz (18–25%, mean: 23), plagioclase (9–12%, mean: 11), and clay minerals (9–15%, mean: 11). The composition of the clay fraction does not display vertical differences with illite-smectite (expandable layers < 10%) as the dominant phase.

In some siliciclastic-rich samples, microCT images reveal original depositional features with very thin parallel lamination and organized in lenticular fabrics. Massive fabrics are observed in carbonate-siliciclastic and carbonate-rich samples (Fig. 4). There is no evidence of internal disposition, possibly because the main components are smaller than the resolution achieved. MicroCT was an important tool to resolve the occurrence of fracture porosity, considering they derive from the extraction of samples from the well. In such cases, GFP values were corrected by subtracting the porosity associated with fractures. Due to experimental limitations on the microCT resolution, only fractures greater than 30 µm thick were considered.

The combination of GFP and T_2 measurements yield for shale samples values in the order of 12 – 19 porosity units (pu), with maximum porosities registered in siliciclastic shales (Table 1).

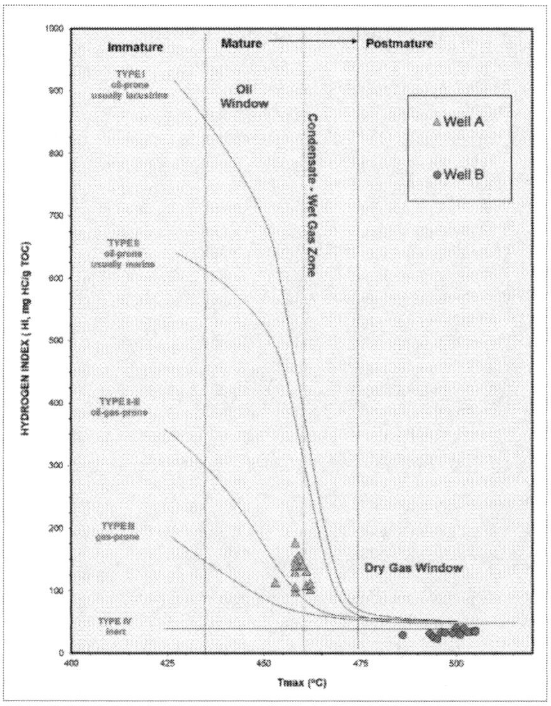

Fig. 3. HI vs. Tmax diagram for the analyzed samples from Well A and Well B (type lines modified from Espitalié et al. [16]).

Table 1. Summary with results for Well A. Minimum, maximum and mean values are presented.

	Siliciclastic shales (n: 5)	Carbonate-siliciclastic shales (n: 3)	Carbonate shales (n: 5)
TOC [%]	3 – 7 (5)	3 – 5 (4)	2 – 4 (3)
Carbonates [%]	15 – 29 (25)	31 – 48 (45)	52 – 61 (57)
Qz + Feld [%]	40 – 62 (49)	31 – 39 (35)	27 – 34 (30)
Clays [%]	12 – 21 (16)	9– 15 (11)	7 – 9 (8)
GFP_AR [pu]	3 – 6 (4)	4 – 6 (5)	3 – 6 (4)
PhiRMN_AR [pu]	10 – 13 (12)	10– 13 (11)	8 – 11 (9)
PhiTotal_AR [pu]	14 – 19 (16)	15 – 17 (16)	12 – 16 (14)

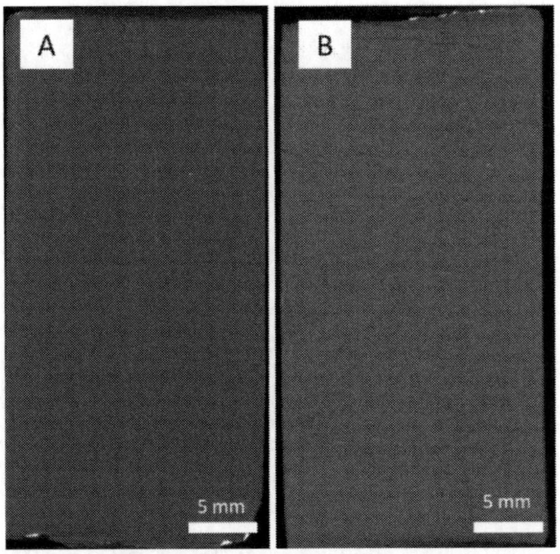

Fig. 4. Examples of microCT images from Well A samples. A and B) Massive aspect observed in carbonate-siliciclastic and carbonate-rich samples.

Further characterization of the components inside the rock was achieved by NMR measurements. Fluid contribution to porosity was obtained from T_2 NMR measurements (Table 1). T_1-T_2 maps give further information on the hydrogen rich components present in the rock. A typical map for Well A can be seen in Fig. 5. The following zone identification is adopted from [12]: zone 1 corresponds to hydroxyl groups present in the inorganic components of shales (clay or other minerals); zone 2 is associated with solid organic matter, kerogen, bitumen and light hydrocarbons inside the kerogen network; zone 3 is attributed to clays; zone 4 is corresponds to moveable hydrocarbons.

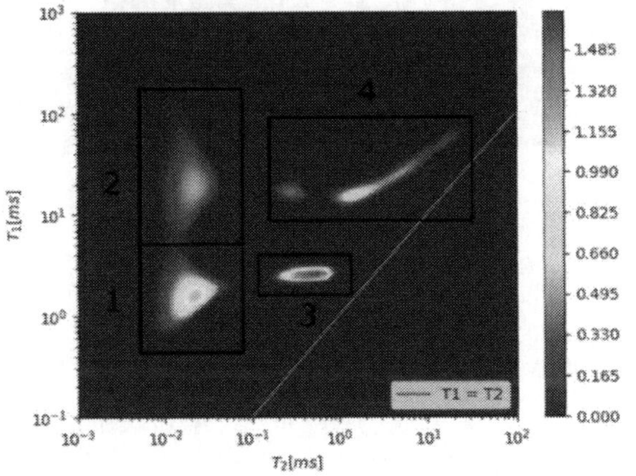

Fig. 5. NMR T_1-T_2 map for a representative sample of Well A. Four zones are clearly marked by black rectangles.

Quantification of NMR signal per zone is performed for each of the 13 samples from Well A. NMR results are integrated with XRD and Rock-Eval measurements. Zone 3 is correlated with clay content from XRD (Fig. 6); zone 2 is correlated with TOC values from Rock Eval pyrolysis (Fig. 7). Both integrations show a very good correlation. This indicates that the three experimental techniques are comparable. Also, since both XRD and Rock-Eval only consider less than 2–3 grams of mass, while NMR is performed on whole rock (approximately 40 grams for Well A samples), the good correlation points to representativity of subsample extraction and homogeneity of the sample.

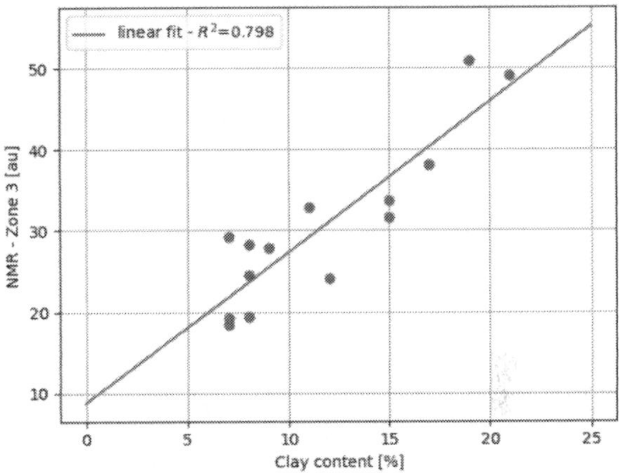

Fig. 6. NMR signal from zone 3 is plotted against clay content in the rock from XRD. A linear fit is performed (red line). R^2 stands for the coefficient of determination of the least-squares.

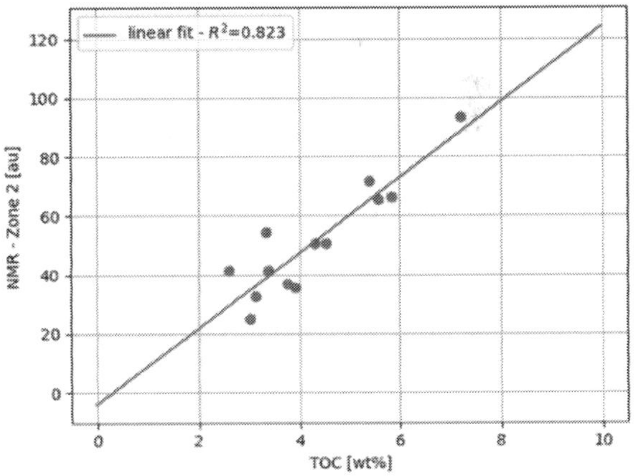

Fig. 7. NMR signal from zone 2 is plotted against TOC value from Rock-Eval pyrolysis. A linear fit is performed (red line). R^2 stands for the coefficient of determination of the least-squares.

3.2 Well B

Shale samples from Well B also show high present day TOC contents (0.72–11.77%, mean: 5.86), with S2 peak ranging from 0.21–4.14 mgHC/grock (mean: 1.93) and HI of 23–40 mgHC/gTOC (mean: 32). Rock Eval® Tmax values (486–506°C, mean: 499) are higher than those presented for Well A, indicating that in the position of Well B the unit reached the wet gas window (Fig. 3).

Based on the adopted compositional scheme, most analyzed samples are siliciclastic-rich (< 30% of calcite) with the highest TOC contents (Table 2). They are composed of quartz (15–37%, mean: 26), plagioclase (9–34%, mean: 20), and clay minerals (18–43%, mean: 28) with minor proportions of K-feldspar, pyrite and apatite. Two samples of the analyzed set correspond to the mixed carbonate-siliciclastic (30–50% of calcite), whereas only one represents the carbonate-rich (> 50% of calcite) shales. However, the latter shows important micro-fractures filled with calcite which impacted in the calculated whole rock composition. Calcite mm-thick micro-fractures were observed under microCT acquisitions as well as interbedded laminae showing compositional and grain size changes (Fig. 8).

According to GFP and T_2 measurements shale samples show porosity values of 10 – 17 pu (Table 2).

Table 2. Summary with results for Well B. Minimum, maximum and mean values are presented.

	Siliciclastic shales (n: 12)	Carbonate-siliciclastic shales (n: 2)	Carbonate shales (n: 1)
TOC [%]	1 – 12 (6)	5– 6	6
Carbonates [%]	12 – 29 (19)	43 – 45	60
Qz + Feld [%]	36 – 53 (46)	41 – 43	27
Clays [%]	18 – 43 (28)	8– 11	10
GFP_AR [pu]	2– 10 (7)	7 – 8	6
PhiRMN_AR [pu]	6 – 10 (7)	5	4,5
PhiTotal_AR [pu]	11 – 17 (14)	12 – 13	10,5

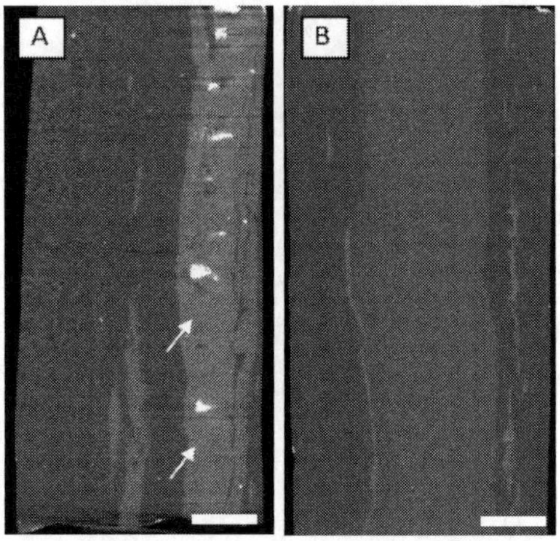

Fig. 8. Examples of microCT images from Well B samples. A) Organic-rich, siliciclastic shale with calcite microfractures (white arrows). B) Siliciclastic shale with internal heterogenies as result of compositional and grain-size changes. Scale bar = 5 mm.

A similar NMR workflow was performed on samples from Well B, where T_2 measurements account for the porosity associated with fluids in the rock, while NMR T_1-T_2 maps allow for component typification. In Fig. 9 a representative T_1-T_2 map for a sample in Well B is shown. Four different zones are marked with black rectangles.

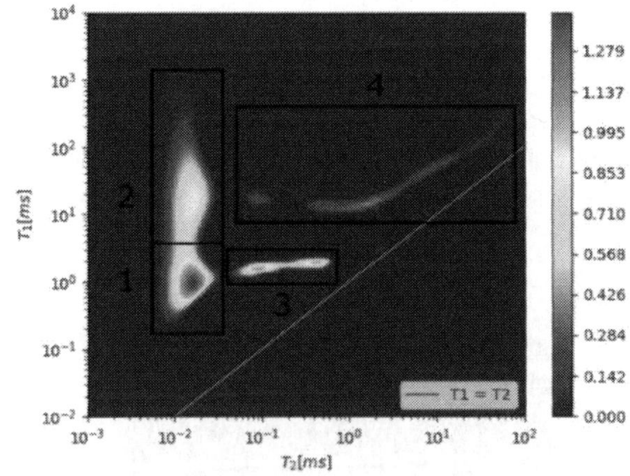

Fig. 9. NMR T_1-T_2 map for a representative sample of Well B. Four zones are clearly marked by black rectangles.

The NMR signal corresponding to each zone is calculated and the results are compared to XRD and Rock-Eval pyrolysis. Zone 3 is compared to clay content (Fig. 10) while zone 2 is compared to TOC (Fig. 11). As in the case of Well A, NMR signal from zone 3 strongly correlates to clay content obtained from XRD analysis. In the case of zone 2 and TOC values, the correlation is not as strong, and the data points show a larger dispersion compared with Well A. We consider that different variables might have controlled the observed uncertainty: (1) Wells represent different maturity window, oil window for Well A and gas window for Well B; (2) the analyzed set for both Wells does not strictly represent the same organic-rich intervals of the Vaca Muerta Formation;

and (3) the internal heterogeneities documented through microCT images. Another possible reason for this discrepancy in correlation could come from the difference in measured mass in Rock-Eval (less than 2–3 grams) and NMR (whole rock, more than 40 grams). Accordingly, further work is necessary to find the cause of the dispersion in values of NMR zone 2 for Well B samples.

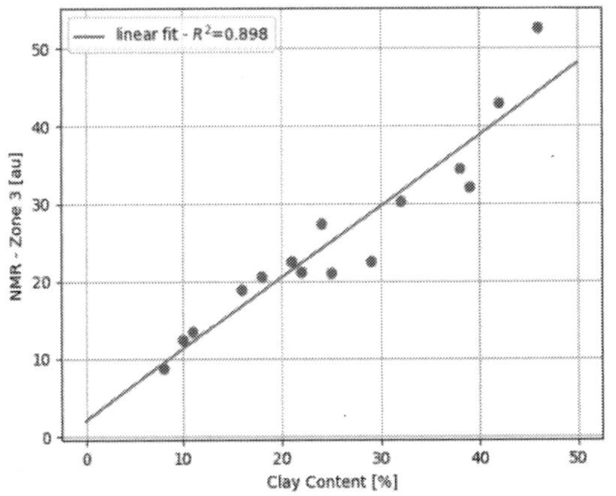

Fig. 10. NMR signal from zone 3 is plotted against clay content in the rock from XRD. A linear fit is performed (red line). R^2 stands for the coefficient of determination of the least-squares.

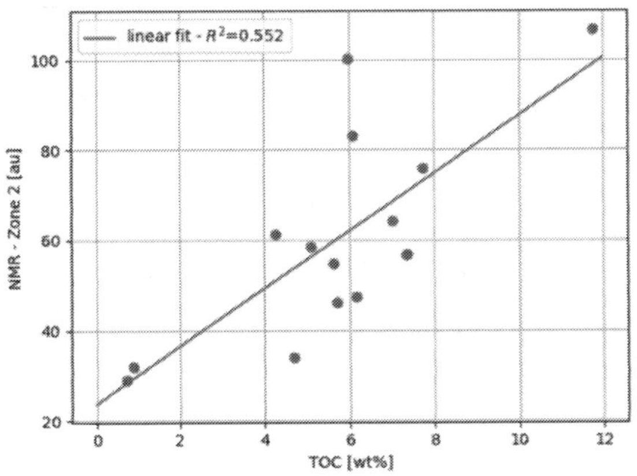

Fig. 11. NMR signal from zone 2 is plotted against TOC value from Rock-Eval pyrolysis. A linear fit is performed (red line). R^2 stands for the coefficient of determination of the least-squares.

4 Conclusions

The present study analyzed samples of the Vaca Muerta Fm based on RMN T_1-T_2 maps at 2 MHz combined with GFP, XRD, Rock-Eval pyrolysis and microCT data with the intent of defining a multi-scale and multidisciplinary petrophysical workflow that can be used for the evaluation of sidewall and core samples at different thermal maturities.

The unit is an heterogenous system independent of the scale of work. GFP and T_2 measurements have shown total porosity values ranging from 10 pu to 19 pu, with maximum porosities recorded in siliciclastic shales (< 30% of calcite). Rock-Eval pyrolysis confirm that a part of the analyzed siliciclastic shales has the largest organic carbon contents.

MicroCT images reveal different types of heterogeneities including open fractures as those recorded in Well A. In such cases, indirect measurements of parameters (e.g., porosity by GFP) do not account for the origin of pores and, for example, do not determine whether the sample is fractured or not. Imaging techniques, such as microCT, together with Digital Rock Analysis (DRA), allow the study of heterogeneities in the rock, as those described in this work, and are of great importance to discriminate open or closed fractures and achieve more reliable porosity measurements.

The presence of heterogeneities as well as other variables, such as the difference in measured mass, might impact in the correlation between NMR and direct rock analysis (e.g., TOC% via pyrolysis Rock Eval). In the future, new measurement strategies can be refined to reliably discern the causes of data dispersion.

This study is published with the permission of YPF, EQUINOR and Y-TEC. We are especially grateful to C. Smal, M. Sánchez, G. Bobrovsky and J. P. Alvarez (Y-TEC) and V. Consoli, D. Ceccon, M. Foster, C. Bernhardt, M. Fasola, I. Brisson and G. Sagasti (YPF) for allowing us to publish the present data.

Special thanks for the technical staff at Y-TEC (J. Juárez, D. Robledo, G. Otegui, J. Acosta, K. Irvicelli, I. Loyza, R. Panno) for the preparation and acquisition of core samples and data analyzed in this study.

References

1. R.L. Kleinberg, W.E. Kenyon, P.P. Mitra, J. Magn. Reson Ser A, **108**, 206-214 (1994)
2. G.R. Coates, L. Xiao, M. Prammer. Nmr Logging, 251, (1999)
3. M. Uliana, L. Legarreta, Journal of Petroleum Geology, **16**, 397–420, (1993)
4. G. Sagasti, M. Foster, D. Hryb, A. Ortiz, V. Lazzari, *Unconventional Resources Technology Conference*, 797-816, (2014)
5. R.F. Dominguez, M. Di Benedetto, *Unconventional Resources Technology Conference*, 5114-5128, (2019)
6. I.E Brisson, M.E. Fasola, H.J. Villar, *Integrated geology of unconventionals: The case of the Vaca Muerta play, Argentina,* AAPG Memoir **121**, 297-328, (2020)
7. J.B. Spacapan, M. Comerio, I. Brisson, E. Rocha, M. Cipollone, J.C. Hidalgo, Basin Research, **33**, 3183-3211 (2021)
8. J.B. Spacapan, M. Comerio, R. Ruiz, E. Rocha, Journal of Petroleum Geology **45**, 219-248, (2022)

9. A. Askenazi, P. Biscayart, M. Cáneva, S. Montenegro, S., M. Moreno, M, Society of Petroleum Engeneers (SPE), (2013)
10. S. Cuervo, E. Lombardo, D. Vallejo, L. Crousse, C. Hernandez, L. Mosse, Unconventional Resources Technology Conference (URTeC), 778-796, (2016)
11. A.C. Ortiz, L. Crousse, C. Bernhardt, D. Vallejo, L. Mosse, Integrated geology of unconventionals: The case of the Vaca Muerta play, Argentina, AAPG Memoir **121**, 329-350, (2020)
12. E.V. Silletta, G.S. Vila, E.A. Domené, M.I. Velasco, Fuel, **312**, 122863, (2022)
13. P. Teal, C. Eccles C., Inverse Probl., **31**, 045010, (2015)
14. D.M. Moore, R.C. Reynolds, *X-ray diffraction and the identification and analysis of clay minerals,* (1989)
15. E. Lafargue, F. Marquis, D. Pillot, Revue de l'institut français du pétrole, **53**, 421-437, (1998)
16. J., Espitalié, G. Deroo, F. Marquis, Revue de l'institut Français du Pétrole, **41**, 73-89, (1986)
17. K.E. Peters, AAPG Bulletin, **70**, 318–329, (1986)
18. M. Comerio, D.E. Fernández, N. Rendtorff, M. Cipollone, P.E. Zalba, P.J. Pazos, AAPG Bulletin, **104**, 1679-1705, (2020)

Some Useful Guidelines for Whole Core CT-Scanning for Petrophysical Applications

Shameem Siddiqui[1,2*]

[1]Halliburton Consulting, 3000 N. Sam Houston Pkwy, E, Houston, TX 770, USA
[2]ResVisions LLC, 16902 Cook Landing Drive, Richmond, TX 77407

Abstract. From the time CT-scanning was introduced in the oil and gas industry, its use could be divided into SCAL and routine type applications. The first involves flowing fluids through the cores, whereas the second involves scanning, mainly the whole cores, in their native (preserved or unpreserved) state. For whole core CT-scanning, there has been a constant shift from qualitative to quantitative analysis. The size of whole cores is ideal for scanning them with a 'converted' medical CT-scanner but obtaining meaningful data requires more than just artifact-free images. In this paper, useful guidelines are offered for extracting meaningful quantitative data from whole core CT-scanning. Today's fast, multi-slice medical CT-scanners generally provide sufficient image resolution for most of the whole core related applications. However, they usually require some modifications to the equipment and the calibration procedures. Special solid phantoms matching the size and density should be used for calibrating the scanner at different X-ray energies. A set of calibration 'standards' is necessary for converting the CT numbers into density and Z_{eff}(Effective Atomic numbers), which should be scanned under the similar 'environment' as the whole cores. Using pilot scans is highly recommended for every core tube to reduce uncertainties. For dual energy scanning, it is important to take scans at the exact same locations, selecting the right energy pair and corresponding calibration tables. For image processing, important guidelines include aligning the first and last slices for selecting the largest region-of-interest, quality controlling each slice and assigning depths to each slice before converting from CT numbers to bulk density and Z_{eff}. Additional information is provided for scanning with industrial CT-scanners which have both advantages and disadvantages. Guidelines such as the ones mentioned above allow the data from whole core CT imaging useful for a number of applications. Such data were successfully used for bulk density and porosity determination, heterogeneity quantification, lithology determination, dual-energy based mineralogy detection, density-based micro-imaging, core-log correlations and depth matching, fracture characterization, formation damage evaluation, and many others. Several examples are included in this paper. Adherence to strict guidelines helps extracting artifact-free, meaningful data on whole cores that can improve our understanding of rocks. This paper demonstrates at least 10 different applications of CT-scanners along with the proper guidelines that can be useful for engineers and geoscientists.

1 Introduction

CT-scanning has been used for core characterization and flow visualization in reservoir rocks for over 35 years. In the flow visualization category, CT-scanning has been successfully used for porosity and pore volume compressibility determination, two- and three-phase flow visualization and quantification, and for evaluating core treatments such as acidizing, foam injection, EOR fluid injection, etc. CT-based core plug screening has been one of the most popular early uses of the technology. Although the other core characterization applications came a little later, CT-based characterization includes evaluating core quality, identifying and quantifying bulk lithology, measuring density and porosity, integrating CT-derived density and porosity with log-derived data, depth matching, quantifying heterogeneity detection of fractures, visualization and evaluation of mud or treatment fluid invasion, etc. While most of these petrophysical applications apply equally to whole cores and plugs, the focus will be on whole core CT-scanning only.

Cores collected from reservoirs during the exploration and development phases can provide very important petrophysical, paleontological, sedimentological, petro graphic, and diagenetic information and therefore, utmost care should be taken to avoid damaging the cores. Some of the best practices in core handling involves onsite sectioning of the cores, while inside their aluminum or fiberglass sleeves, into 3 ft (about 1 m) long sections, marking the depths and identifying tops (typically with two parallel lines, red and black), putting caps on both ends and then putting these sleeves in reservoir compatible brine inside a PVC core box in order to preserve wettability (Fig. 1). The core box, without brine, also serves to protect the core integrity during transportation. After carefully transporting the core boxes to the core handling facility, CT-scanning should be one of the first operations executed on core samples because core integrity is compromised once the core is extracted from its sleeve (Fig. 2) for core description, slabbing and plugging. This way the three-dimensional, mostly density-related data remains available for any future qualitative and quantitative use, even after the core becomes subjected to extensive slabbing and plugging.

Fig. 1. Photograph of PVC core boxes at the loading dock of a CT Lab.

* Corresponding author: siddiqui.shameem@gmail.com

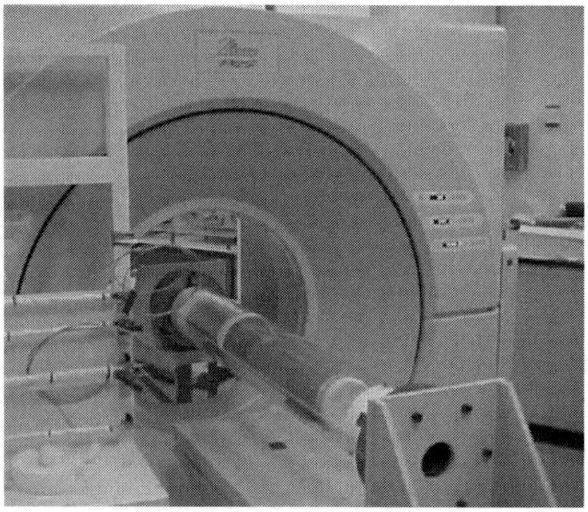

Fig. 2. Fiberglass core sleeve with rubber caps at the two ends placed concentrically in the Plexiglas housing of a converted medical CT-scanner before scanning.

Both medical and industrial CT-scanners can be used for whole core CT-scanning, with each one having its own advantages and disadvantages. A comparison between the two types of scanners for various petrophysical applications is presented in Ref. [1]. Industrial CT-scanners, which include the micro and nano CT-scanners, offer a wide range of X-ray voltages and intensities, and provide excellent spatial resolution, sometimes to the pore level. They are more preferred for evaluating small plugs and cuttings rather than whole cores, mainly because the scanning speed is very slow and the samples must be rotated in a vertical position during scanning, which may be detrimental to the sample integrity, especially for unconsolidated or naturally fractured cores.

Medical CT-scanners, with their large weight capacity, reasonable spatial resolution (typically 0.3 to 0.5 mm pixel length) and fast speed (especially with multi-slice technology), are ideal for scanning whole cores for various petrophysical applications. Although a properly calibrated Medical CT-scanner is capable of generating high quality images for most reservoir rocks, the data are not always sufficient for quantitative analysis that may be required for advanced petrophysical applications. In the next section, a discussion is made on the necessary modifications that would make a medical CT-scanner useful for scanning various types of reservoir rocks and for getting meaningful quantitative data for integrated reservoir studies.

2 CT-Scanner Modifications Needed for Petrophysical Applications

The selection of the medical scanner for petrophysical applications is of primary importance. Some of the points that should be considered during the selection process include maximum voltage, whether or not it is dual-scan capable, number of slices (for multi-slice CT-scanners), capability to do pilot scans and their lengths, etc. Medical CT-scanners are designed for human subjects and exposure of soft tissues, where maximum X-ray voltage, scanning speed are all very important considerations. Due to safety and intellectual property issues, many manufacturers are generally not willing to share advanced calibration and data transfer information with the non-medical clients and it is therefore, essential to obtain necessary permissions before purchasing a new scanner from a particular vendor. Sometimes it is more convenient to go through a third-party vendor with good connections with the CT-scanner manufacturers and good familiarity of petrophysical CT-scan requirements.

In general, modifications are needed in three major areas to convert a medical CT-scanner into one that can be used for petrophysical applications. The first one is the modification of the patient bed into a sturdier metallic table. The strong cantilever arm used to support the patent bed is not designed to align a 3 ft long core box, and since core alignment is crucial for petrophysical measurements, it is better to replace the soft patent bed with a steel table fitted with a pair of vises to ensure the proper centering of the core box or core tube. Fig. 2 shows how a vise mounted on the modified patent table (right) and another vise mounted on an added rear table are used to align a Plexiglas tube, which in turn, holds and centers the fiberglass core sleeve with the help of foamed plastic rings.

The second required modification is that of the internal calibration tables. Having the access to the calibration tables and permissions to modify them in order to scan the rocks is very important. Normally the manufacturers calibrate the scanner with a "water phantom" containing a radiopaque salt (typically sodium iodide) in water, which is sufficient for scanning human subjects. Reservoir rocks with typical grain density of 2.65 (quartz) to 2.86 (dolomite) g/cm^3 require calibration with solid phantoms of similar grain density, such as fused quartz, Macor, etc; it ensures artifact-free images at the different anticipated energy levels (X-ray voltage/intensity combinations). An experienced user or a designated service technician should be able to generate these calibration tables so that they can be used by the operator readily for different materials (e.g., carbonates or clastic rocks), different sizes of scanned objects (e.g., 6" diameter core box, 4.5" diameter core sleeve, 1.5" diameter core plugs, etc.), as well as for different energy levels. Every time the X-ray tube is replaced, new calibration tables must be created again.

The third modification relates to transferring data from the CT-scanner. The image processing software that are usually packaged with the CT-scanners are only good for general viewing and quality control (QC), as far as petrophysical applications are concerned. So, transferring data to workstations loaded with specialized image processing software is very important. In the past, different manufacturers used different data formats and some of them would deliberately make it difficult to track the data (due to patient data security concerns) or to transfer the data as straight files.

Today the process has mostly been streamlined and most manufacturers conform to the DICOM format, which is also readable by most of the image processing programs designed for petrophysical analysis of CT data. Many CT-scanners allow direct retrieval of data on USB drives rather than going through the network, which simplifies matters further.

3 Use of Calibration Standards

For whole core CT-scanning, usually the actual cores remain out of sight unless it is decided to scan bare cores, extruded from their aluminum or fiberglass sleeves. They are either devoid of any liquid or have some initial liquid saturation. Therefore, a set of calibration 'standards' is necessary to extract meaningful data such as density, porosity, etc. from the CT images. The only quantitative data that a CT-scan slice provides is the CT number (CTN or Hounsfield Units, HU) for each voxel (volume element, 3D pixel). CT number is a function of both Z_{eff} (Photoelectric Effect, predominant at low energies such as 80 kV) and electron density (Compton Scattering, predominant at high energies such as 140 kV). The medical CT-scanners are generally calibrated to read a value of -1000 for air and 0 for water. Wellington and Vinegar [2] suggested using 100 kV as the threshold for these two effects.

Most medical CT-scanners are limited at 150 kV as the highest voltage in order to prevent severe damage to human soft tissues and therefore, it is a common practice in petrophysical CT-scanning with a medical CT-scanner to use either 140 or 150 kV (for single energy scanning), which allows good X-ray penetration for larger whole cores such as the 4" diameter ones in their sleeves and core boxes.

In order to convert CT numbers into bulk density for the scanned cores, Wellington and Vinegar [2] proposed to use a calibration curve similar to the one in Fig. 3, using clean sandstone samples, used as 'standards' and with known measured bulk densities, from different quarries or reservoirs. This works relatively well for extruded dry cores. However, for preserved cores inside fiberglass/aluminum sleeves and/or core boxes, it is strongly recommended to scan the standards in the same 'environment' as the actual cores to avoid having to make adjustments.

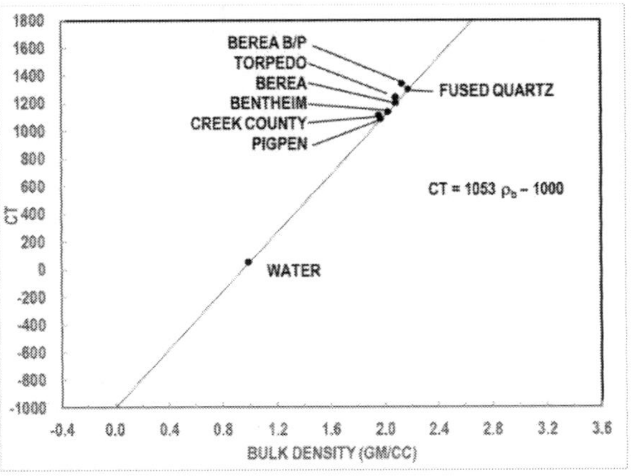

Fig. 3. Calibration curve of bulk density vs. CT numbers for sandstones. After Wellington and Vinegar [2].

For developed reservoirs, it is always possible to obtain bulk densities of a few cleaned (using Soxhlet extraction) representative core plugs and use them as 'standards' in addition to the solid 'standards' such as fused quartz, Macor, etc. prior to a whole core CT-scan job. Fig. 4 shows some of the 'standards' that are scanned inside the blank core sleeves prior to a whole core CT-scanning project. Scanning of these 'standards' does not take too much additional time and it should be done regularly, as these data are extremely important for conversion of CT data into bulk density.

For this case, with only one representative cleaned sandstone 'standard', the CT-derived bulk density, $(\rho_b)_{CT}$ (also called pseudo bulk density, PBD), is calculated using a straight-line relationship between CTN and ρ_b, for air and the representative cleaned sandstone plug, respectively. Both should be within a circular region of interest (ROI) inside the rock matrix for the core contained in the sleeve and core box, as shown in Eq. 1. Here A is the slope of the ρ_b vs. CTN curve and B is the intercept.

$$(\rho_b)_{CT} = A*CTN + B \qquad (1)$$

Fig. 4. Calibration 'standards' - quartz, Macor, Berea and a representative cleaned sandstone (top). Blank PVC core box and fiberglass core sleeve (bottom).

It is strongly recommended to scan multiple calibration standards each time - both solid standards and porous rock standards, although the latter group works best for converting data for reservoir rocks. In the calibration curve of ρ_b vs. CTN (inverse of the graph shown in Fig. 3), it is recommended to use 0.001225 g/cm^3 as the density of air rather than 0, as it represents air density at room temperature and at 1 atm, at sea level. Also, there is no need to draw data for water as it may skew the data in one direction or the other. It is not recommended to mix data for carbonate and sandstone reservoir rocks in the same calibration plot. If multiple calibration standards (say, representing different facies of a clastic reservoir) are to be scanned, only data for those standards should be plotted along with the air data.

The solid 'standards' mainly help observing the drift in the X-ray tube performance, but they are usually not so good for converting CTN into $(\rho_b)_{CT}$ since they do not have any porosity. In the early days of CT-scanner use for routine and SCAL (special core analysis) evaluation of reservoir rocks, it was common to wrap small rods (about ½ inch in diameter) of different materials or tubes filled with different liquids as 'standards' around the core holder or core sleeves. Better results can be obtained by scanning solid plugs of about 1½ inch diameter placed tightly next to one another, as shown in Fig. 5, mainly because one can use a larger region of interest

(ROI) and also, the standards can be easily placed in the same environment that the actual sample is in.

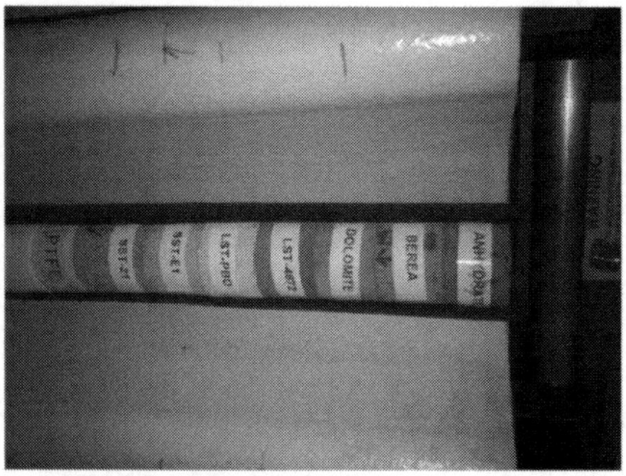

Fig. 5. 1½ inch diameter calibration 'standards' placed next to one another inside a rubber sleeve for bare whole cores and plugs.

4 Matching CT Density Data with Logs

One of the challenges of whole core CT-scanning is matching CT based density, $(\rho_b)_{CT}$ data with bulk density log (RHOB) data. Three points should be considered for that – the first one is selection of proper calibration standards to convert the data, which is described in the previous section; the second one is fluid effect, which will be discussed below; and the third is the depth shifting issue, which will also be discussed later.

When logging is done in the wellbore, it is generally filled with mud and the adjacent formation is fully saturated with native fluids, with some degrees of invasion by the filtrate. Since the radius of investigation of density log is only about 6 inches, it mostly measures bulk density of the invaded zone, filled with a combination of fluids. When a core is sectioned at the wellsite, the difference between reservoir pressure and atmospheric pressure (except with specific pressure coring) causes a good part of the reservoir fluid contained in the pores to be expelled and replaced by air, which is harder to remove even when the sectioned core is placed with or without sleeve, capped or uncapped, inside the brine-filled core box. Leaks during the transportation and storage prior to CT-scanning may also introduce more air into the cores being scanned. So, chances are high for the scanned core to have its pore space filled with air rather than liquids. Because the conversion of CTN into ρ_b (Eq. 1) involves dry calibration standards, the CT-derived bulk density, $(\rho_b)_{CT}$ is essentially a bulk density with air-filled pores. A correction must be applied to the $(\rho_b)_{CT}$ data in order to match the log-derived bulk density, $(\rho_b)_{log}$, which, applies to liquid-filled bulk density and this is shown in Eq. 2.

$$(\rho_b)_{log} = (\rho_b)_{CT} + \phi_{CT}*(\rho_{water} - \rho_{air}) \qquad (2)$$

Here, ρ_{water} and ρ_{air} are the densities of water and air (in g/cm³), respectively, and ϕ_{CT} is the porosity of the CT slice. The CT-derived porosity, ϕ_{CT}, can be calculated from the bulk density and matrix density, ρ_{matrix}, through the relation shown in Eq. 3.

$$\phi_{CT} = (\rho_{matrix} - (\rho_b)_{log}) / (\rho_{matrix} - \rho_{water}) \qquad (3)$$

This so-called liquid correction involves an iterative process because at the initial stages of calculation, the $(\rho_b)_{log}$ is not known and therefore, must be estimated with an initial value for porosity. For matrix density the values shown in Table 1 are generally used.

Table 1. Matrix Density of Primary Minerals

	Primary Mineral Name	Matrix Density, g/cm³
1	Calcite	2.71
2	Dolomite	2.86
3	Anhydrite	2.95
4	Quartz	2.65
5	Clay/Shale	2.65

For the liquid correction process, it is also recommended to use tube average or facies average porosity data rather than individual slice porosity data, otherwise the calculations become very tedious, especially if large number of CT slices are involved (i.e., for thick reservoirs). The steps for this iterative process, using EXCEL, are given below:

1. Create a table with the following columns - the depth of the CT-slice, $(\rho_b)_{log}$ from Eq. 2, ϕ_{CT} from Eq. 3. The calculation of $(\rho_b)_{log}$ should also include A, B and the respective CTN values used in Eq. 1.
2. Create the following additional columns – average depth of the core tube, average $(\rho_b)_{log}$ for core tube and average ϕ_{CT} for core tube.
3. Create an additional column for average porosity containing the same value for a core tube or for certain facies. Create a last column in which the numerical data are pasted from the previous column. Note that the porosity value for liquid correction in Eq. 2 points to the very last column only.
4. For the first pass, fill up the entire last column with one initial value, say 0.15, for the entire CT-scanned interval. Because of the way the equations are set up, the second column will have an initial value of $(\rho_b)_{log}$, which will then be used to calculate a ϕ_{CT} for each of the slices. The second-to-last column will then get updated with a new value for each core tube (or facies, covered by several core tubes). Manually copying the entire column date and pasting the numerical value into the last column will start the second pass, and so on.
5. At the end, the data in both the second-to-last and the last columns will be the same and at that time the second column will contain the liquid-corrected $(\rho_b)_{log}$ data and the third column, the corresponding ϕ_{CT} data. The depths in the first column, with the CT-derived density and porosity data can then be used for comparing against logs.

If some additional slice quality control is also done to ignore the CT artifacts and missing/bad data, this workflow can produce excellent CT versus log correlations. It should be noted that typical vertical resolution of a logging tool is about 6 inches whereas multi-slice CT-scanners can take several slices per inch, making a continuous map of CT numbers within each slice. Unless there are operational limitations for taking single scans, it is recommended to scan the whole cores at fixed intervals such as 0.5-inch, 1-inch, 2-inch, etc. This reduces the number of slices and thus, the image processing time; and the resolution is sufficient for slice-by-slice calculations for core to log comparison.

The last piece of work that needs to be done is the depth correction. It is quite common to have some mismatch between driller's depths (that all cores use) and logging depth. This causes a problem for core-log integration, core sample selection for SCAL and routine core analyses, well treatment design, etc. In typical core operations, the Core Gamma tool is used to align core depths with log depths. It works on matching the spikes seen in shale streaks in clastic rocks in both logs and cores. It generally works better if the cores are already extruded from the sleeves. For carbonate cores, which mainly contain calcite and dolomite, gamma ray variations are not very noticeable but the large density difference between calcite and dolomite makes CT a better tool for depth matching for carbonates compared to Core Gamma [2, 3]. Overall, the depth accuracy improves a lot if the depth of the top of the core in each tube is accurately known through pilot scans (also called scout scans). For scanners without the pilot scan feature, tilting the core box to move the core sleeve inside the core box to one side, prior to the CT-scans, and scanning from approximately the same position for all core boxes may result in reduction of location errors.

For dual energy CT-scanning, one of the important factors to consider is the positioning of the samples at the exact same location. Some of the most recent CT-scanners can take a high-energy scan and a low-energy scan in one single pass and for those scanners positioning is not an issue. However, in the case of older CT-scanners requiring an elaborate process for changing voltage settings for dual energy CT-scanning, the scanning must start from the same starting point and to be repeated at each slice location. Minor shifts can make the dual-energy CT-scanning analysis very difficult. It should be remembered that in order to ensure artifact-free data, appropriate calibration tables should be used for each energy and sometimes, it is more than just pushing the 140 kV or 80 kV button on the control panel.

5 CT Image Artifacts

Just like its medical counterpart, petrophysical CT-scanning also has to deal with several CT image artifacts during data processing. The most common ones are streaks (aliasing), ring, and beam hardening [4, 5]. Streaks appear as dark lines that radiate away from sharp corners. For the cylindrical cores this is usually not a problem as the cross section is circular. However, if the core inside the sleeve or core box is fractured naturally or during the coring process, it is possible to get sharp edges, and streaks can be observed. For the extruded, dry cores, if they are already slabbed before scanning, streaks may also be encountered from the sharp edges. Also, if core plugs are scanned in sagittal position, which presents a rectangular cross-section to the scanner, X-shaped streaks may be observed [Fig. 6]. There is hardly any post-processing correction for the streaks and during data analysis slices containing streaks should not be considered.

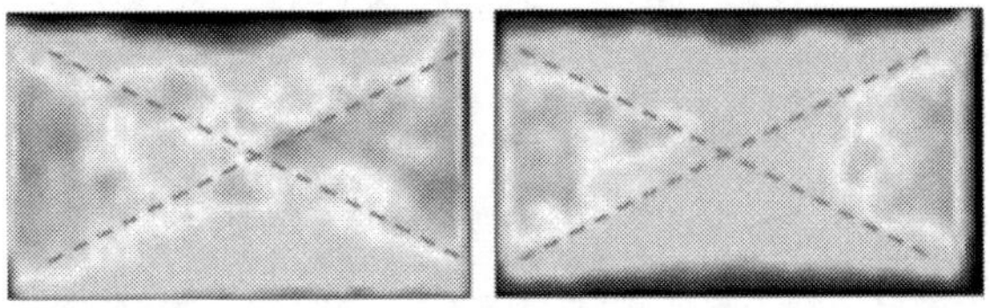

Fig. 6. Examples of X-shaped streaks in core plugs (sagittal case).

The ring artifacts, which appear as a series of off-centered concentric rings in some CT slices, are rare for medical CT-scanners, but are more common for industrial CT-scanners. It is caused by the imperfect detector elements at a specific position in the detector array. The large number of detectors used by the medical scanners as well as good centering of core samples within the CT gantry makes them rare.

The last and the most common artifact seen during whole core CT-scanning is the beam hardening artifact. The filtering of a polychromatic X-ray beam by the scanned object causes beam-hardening (cupping) artifacts. Because of beam-hardening, a perfectly homogeneous object appears to have a dense periphery. If one draws a profile on the CT slice, going from one end of the object to the other end through the center, the profile looks like a cup.

It is preferable to eliminate all the artifacts before the image is acquired through improved remedial actions and improved calibrations. In general, the medical CT-scanner calibration process with relatively large circular phantoms (having a slightly larger size than the largest object scanned) appropriate for the density range of the objects to be scanned helps minimize or eliminate this artifact.

Beam hardening is more common in bare core images rather than images of cores within core boxes and core sleeves because the polychromatic X-ray beams get pre-hardened before penetrating the rock materials. Insufficient penetration due to lower energy for large objects can also have the appearance of beam hardening on the images. In Fig. 6, the calibration was right for the vuggy carbonate core, pictured on the left, which shows no beam hardening. However, the image on the right shows strong beam hardening due to the calibration phantom not being appropriate for the material (anhydrite) in the core slice.

Beam hardening directly affects whole core CT-scanning results and forces one to use a smaller ROI, away from the edges, thus reducing the sample population for the voxels. Some image processing software include the beam hardening correction which may be used if there are no other alternatives. Since beam hardening may not be present in all the slices, the quantitative data from the beam hardening-corrected images may have some consistency issues when compared against non-corrected images.

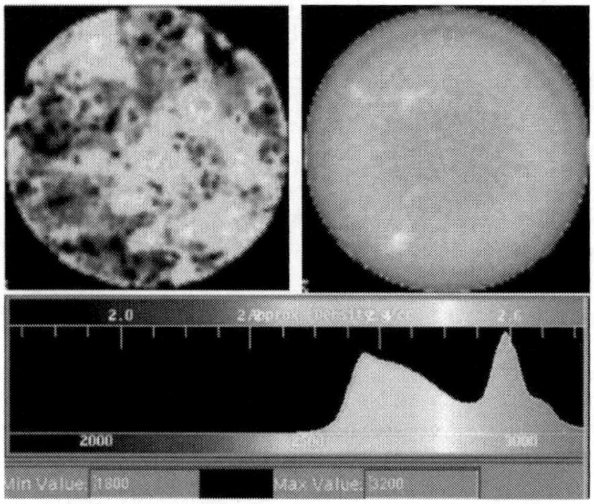

Fig. 7. Example of slices without (left) and with beam hardening (right). Histogram with overall range for all the slices scanned in the batch is shown at the bottom.

6 Image Processing

There are several image processing software packages, both commercial and non-commercial, for processing the CT-scan images. They are also compatible with multiple image data formats that are commonly used by the CT-scanners. Most of this software is designed for medical imaging of human subjects and may therefore not be able to cater to the need of petrophysical CT-scanning. It is not within the scope of this paper to review various software available and describe their advantages. However, some of the essential or useful features for petrophysical CT-scanning software are listed below:

1. Ability to import and read multiple image data formats generated by various manufacturers of CT-scanners. It should also be able to select multiple images at will from a group of image files.
2. Ability to digitally align the slices as it is almost impossible to have perfect alignment of slices for a core tube before scanning.
3. Ability to do ROI statistics with multiple ROI options such as circular, elliptical, polygonal, square, etc. The statistical data should include mean, minimum, maximum, standard deviation, and histograms. It should also allow drawing profiles on any slice and extract the CTN on the profile.
4. Ability to use an array calculator with functions such as add, subtract, multiply, divide, absolute, log, exponential etc., and also to apply various filters. This feature allows reporting of whole core data in terms of density or porosity units; and SCAL data in terms of saturations; rather than CTN.
5. Ability to use multiple color schemes such as grayscale, heat, rainbow, etc.; and some continuous and discrete color schemes; and the ability to create new color scales, if needed.
6. Ability to create orthogonal slabs (similar to slabbed cores), both horizontal and vertical.
7. Ability to do segmentation and thresholding.
8. Ability to generate 3D volumes (volume rendering such as iso-surface plots) from CT-slices with ability to pan, rotate, move etc.

Some of the desired advanced features specific to routine and also, SCAL type petrophysical CT-scanning should include the following:

9. Beam hardening correction.
10. Dual-energy CT calculation (Bulk density and Z_{eff}) using analysis like reference [6].
11. Individual fluid phase calculations using advanced image subtraction, with radiopaque solution (dopant) mixed in oil, water and gas phases.

7 List of Some of the Applications of Whole Core CT-scanning

CT-scanning has been used for evaluating reservoir rocks for over 35 years and many good reviews of various routine and SCAL type of applications are available in the literature [4, 7, 8 and 9]. While more and more applications are coming up, CT is currently being used for at least the following applications for the whole cores:

1. Bulk density determination: details are discussed in this paper.
2. Porosity determination from bulk density: details are discussed in this paper.
3. Quick preview of the core tubes for sample selection: Fig 8 from Ref. [11] shows density variations in CT slices taken at 2-inch intervals for 20 core tubes covering about 60 ft of a reservoir. This provides a quick preview of the condition of the preserved cores and helps in determining what samples are intact and can be used for plugging.
4. Depth matching: It is discussed in Refs. [2, 3] and Fig. 9 is taken from Ref. [10]. The images on the left are the slab images (horizontal and vertical) for five core tubes. The log data (RHOB) are shown with the red lines, the CT-derived density data (PBD) are shown with black lines, for the tube average data and with green dots, for the slices.
5. Formation damage due to mud: Fig. 10 is taken from Ref. [11] and it shows a CT slice of a 4 in diameter whole core from a sandstone reservoir. The barite-laden mud is shown to penetrate deep inside the core. The yellow outer ring is the aluminum sleeve, the bright red inner ring is the barite mud at the periphery and the dark blue material is the sandstone core. The black channels filled with mud are the fractures, possibly induced during coring. Ref. [12] has more examples of mud invasion.
6. Heterogeneity quantification and SCAL sample selection: Fig. 11, which is taken from Ref. [13], shows two types of heterogeneity that can be defined by analyzing CT-scan slices. The nine slice images at the left (inter-slice heterogeneity case) show the presence of a distinct anhydrite nodule in the last three slices. The nine slice images at the right (intraslice heterogeneity case) show large CTN variations within each slice, with relatively small

channels available for flow. Both of these plugs were determined unsuitable for SCAL tests.

7. Lithology variation: Fig. 12, which is taken from Ref. [10], is a composite image showing 19 CT-Scan slices on the left, taken 2-inch apart from each other, and representing a 3 ft section of preserved sponge core from a sandstone reservoir. The bright color in the middle is due to lithology change from sand to shale.
8. Dual-energy based mineral characterization: Fig. 13, which is taken from Ref. [6,], shows the application of dual-energy CT-scanning for mineral characterization in a carbonate reservoir. By scanning these cores at the same slice locations at 80 kV and 140 kV, respectively, it was possible to detect the streak containing dolomites. Dual energy CT is also discussed in Ref. 14.
9. CT density-based FMI type logs: Fig. 14, taken from Refs. [15, 10] shows how CT-slice image data can be used to generate density-based FMI-type images. The example image, shown in the rightmost panel, was created by combining continuous CT-scan data for a 34 ft section of core through a process of collecting the surface slice data and unrolling them. The cores were depth-shifted prior to placing next to the FMI logs.
10. Fracture Characterization: Fig. 15, which is taken from Ref. [10] shows a 360-degree core photograph next to a CT-generated unrolled image of a fractured core that was also plugged. The CT image displays the fracture with dark blue lines, and it also shows additional fractures not visible on the surface.

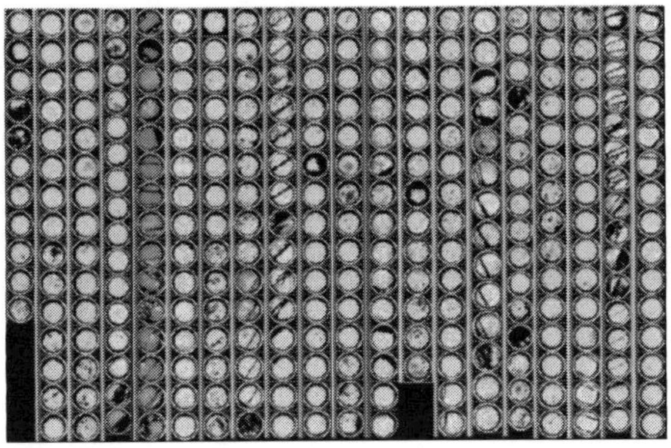

Fig. 8. Example of quick preview of the condition of the preserved cores.

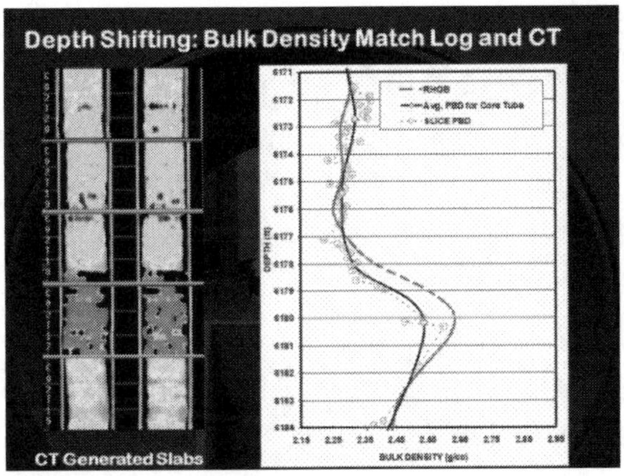

Fig. 9. Example showing how CT generated density data are used for depth shifting in a 15-ft section of a carbonate core.

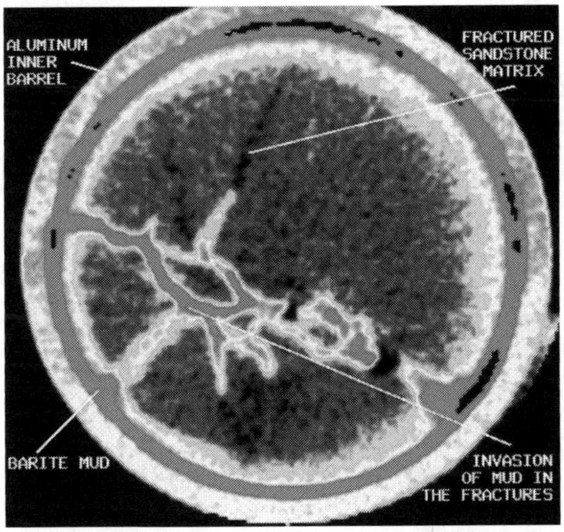

Fig. 10. Example of mud invasion in a core during the coring process.

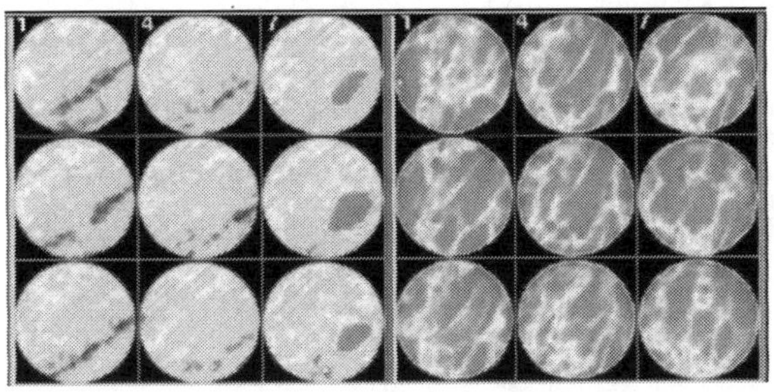

Fig. 11. CT-Scan slices for two core plugs showing case of inter-slice heterogeneity (left images), and intra-slice heterogeneity (right images).

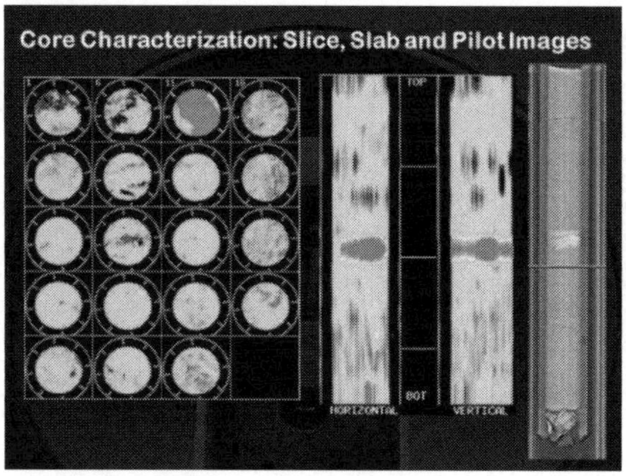

Fig. 12. Example of lithology variation within a core tube.

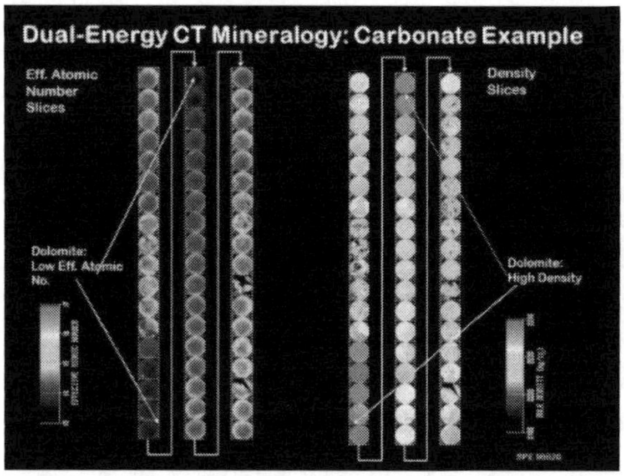

Fig. 13. Example of dual-energy based mineralogy determination using CT-Scan slices for three core tubes taken from a carbonate reservoir containing both calcite and dolomite.

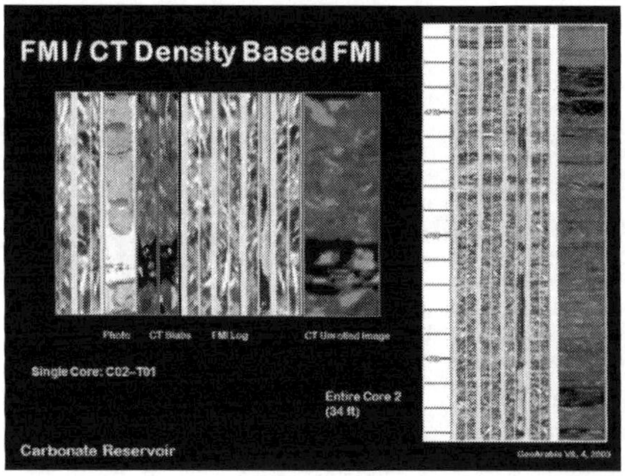

Fig. 14. Example of CT-generated density-based FMI (formation micro imager log) type images (rightmost panel).

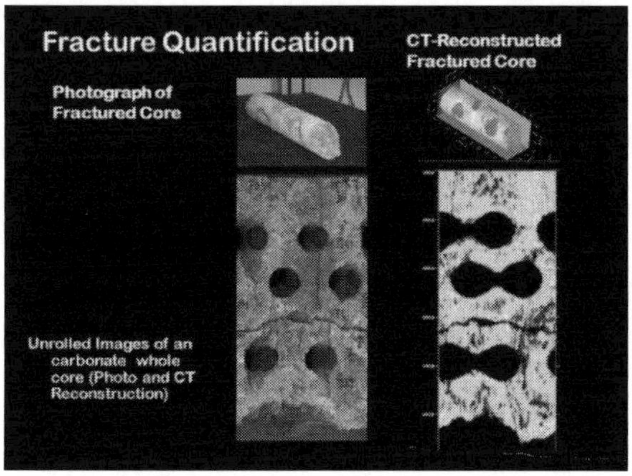

Fig. 15. Example of fracture characterization in whole cores.

8 Conclusions

In this paper an attempt has been made to present some of the interesting applications of computerized tomography (CT) for evaluating whole cores. Starting with what changes are needed to convert a purely medical CT-scanner into one that can generate artifact-free images of reservoir rock samples, this paper covers discussions on calibration standards and how they can be used to generate bulk density and porosity data. Step-by-step instructions on how to use an iterative approach to correct for liquid effects for matching the CT-derived bulk density and porosity data with the logs, are also provided.

This paper covers some common CT artifacts that may be encountered while using a medical CT-scanner for whole core CT-scanning. Necessary features for the petrophysical use of a CT image analysis software are also discussed. Finally, a list of ten specific applications applicable to whole core CT-scanning, with examples and references, is presented with the hope that more and more new applications of this very useful non-destructive imaging technology will be developed by the future petroleum engineers and geoscientists in order to broaden our understanding of the reservoir rocks.

9 Acknowledgments

The author expresses his deep appreciation for his mentors Abraham Grader, Paul Hicks and Phillip Halleck for their help and support. The author also thanks his former colleagues at the Saudi Aramco R&D Center in Dhahran, KSA, who made it possible to explore new ideas. The author appreciates the support of Bob Kehl of Kehlco, Inc. for his *VoxelCalc* petrophysical image processing software, which was used for generating some of the images.

Nomenclature

Z_{eff}	Effective Atomic Number
CTN	CT Number
RHOB	Log-derived bulk density, g/cm^3
$(\rho_b)_{log}$	Log-derived bulk density, g/cm^3
A, B	Slope and intercept of ρ_b vs. CTN graph
$(\rho_b)_{CT}$	CT-derived bulk density, g/cm^3
$(\phi)_{CT}$	CT-derived porosity, fraction
ρ_{matrix}	Matrix (grain) density, g/cm^3
ρ_{air}	Air density, g/cm^3

References

1. M.R.H. Sarker and S. Siddiqui. "Advances in Micro-CT Based Evaluation of Reservoir Rocks." Paper presented at the SPE Saudi Arabia Section Technical Symposium, Al-Khobar, Saudi Arabia, May 2009. doi: https://doi.org/10.2118/126039-MS
2. S.L. Wellington and H.J. Vinegar. "X-Ray Computerized Tomography." JPT, August 1987, p. 885.
3. S. Siddiqui. "Method for depth-matching using computerized tomography." US Patent number: 6876721, 2003.
4. S. Akin, and A. R. Kovscek. "Computed tomography in petroleum engineering research." In Applications of X-ray computed tomography in the geosciences. F. Mees; R. Swennen; M. Van Geet; P. Jacobs (Eds.), Jan. 2003. doi: https://doi.org/10.1144/GSL.SP.2003.215.01.01.
5. R.P. Kehl and S Siddiqui. "Two Less-Used Applications of Petrophysical CT-Scanning." in Advances in Computed Tomography for Geomaterials: GeoX 2010, pp.180-188.doi: https://doi.org/10.1002/9781118557723.ch22.
6. S. Siddiqui, S. and A.A. Khamees. "Dual-Energy CT-Scanning Applications in Rock Characterization."" Paper presented at the SPE Annual Technical Conference and Exhibition, Houston, Texas, September 2004. doi: https://doi.org/10.2118/90520-MS.
7. E.M. Withjack, C. Devier, and G. Michael. "The Role of X-Ray Computed Tomography in Core Analysis." Paper presented at the SPE Western Regional/AAPG Pacific Section Joint Meeting, Long Beach, California, May 2003, doi: https://doi.org/10.2118/83467-MS.
8. E.M. Withjack. "Computed Tomography for Rock-Property Determination and Fluid-Flow Visualization." SPE Formation Evaluation, December, 1988, p. 696.
9. A. Kantzas, "Investigation of Physical Properties of Porous Rocks and Fluid Flow Phenomena in Porous Media Using Computer Assisted Tomography." In Situ, Vol. 14, No. 1, 1990, p. 77.
10. S. Siddiqui, "Understanding Reservoir Rocks through Computerized Tomography (CT) Imaging." SPE Distinguished Lecturer Presentation, Society of Petroleum Engineers, 2013-14.
11. S. Siddiqui, "Application of Computerized Tomography in Core Analysis at Saudi Aramco." Saudi Aramco Journal of Technology, Winter 2001, pp. 2-14.
12. H.J. Vinegar and S.L. Wellington. "Method of Determining Drilling Fluid Invasion." US Patent No. 4540882A, 1983.
13. S. Siddiqui, T.M. Okasha, J.J. Funk, and A.M. Al-Harbi. "Improvements in the selection criteria for the representative special core analysis samples." SPEREE 9, **06** (2006), pp. 647-653.
14. J.G.C. Coenen, and J.G. Maas. "Material Classification by Dual-Energy Computerized X-ray Tomography." Proceedings of the International Symposium on Computerized Tomography for Industrial Applications, Berlin, Germany, 1994, pp. 120–127.
15. G.W.G. Hughes, S. Siddiqui, and R.K. Sadler. "Shu'aiba rudist taphonomy using computerised tomography and image logs, Shaybah field, Saudi Arabia." GeoArabia 8, **4**, pp. 585-596.

Modification of the SDR equation for permeability prediction

Andreas Weller[1,*], and *Zeyu* Zhang[1]

[1]Technische Universität Clausthal, Institut für Geophysik, 38678 Clausthal-Zellerfeld, Germany

Abstract. We evaluate the potential of permeability prediction of the original and modified versions of the Schlumberger-Doll Research (SDR) equation that is applied to data of nuclear magnetic resonance (NMR) relaxometry. Different definitions of characteristic relaxation time are considered. In a further modification, pore radius replaces the characteristic relaxation time in the original SDR equation. Only a good estimate of the surface relaxivity enables a reliable transformation from the relaxation time distribution (RTD) into the pore radius distribution (PRD). A recently published approach considers the specific surface area per unit pore volume (S_{por}) and the weighted harmonic mean (T_{2hm}) of the RTD for the determination of an individual value of surface relaxivity for each sample. The evaluation is performed for three sample sets originating from different sandstone formations. The good predictive quality of the original SDR equation is confirmed. However, the prefactors have to be adjusted by calibrating with core data. We demonstrate that the use of average pore radii instead of average relaxation times enables a better permeability prediction. The weighted harmonic mean of the PRD, which proves to be a reliable proxy for the effective hydraulic radius, provides a high quality permeability prediction.

1 Introduction

Permeability prediction for reservoir rocks is still a challenge in geophysical exploration. Porosity and pore radius are the most relevant parameters used in models of permeability prediction. A variety of petrophysical experiments or logging tools provide reliable porosity values. The effective hydraulic radius r_{eff} controls the fluid flow through porous rocks with a certain pore radius distribution (PRD).

Nuclear magnetic resonance (NMR) relaxometry provides an estimate of porosity and a relaxation time distribution (RTD). The maximum or mean values of the RTD are regarded as proxies for the effective hydraulic radius. The original Schlumberger-Doll Research (SDR) equation relates the weighted geometric mean of the transverse relaxation time and porosity to permeability. In the common form of SDR equation, the relaxation time is raised to the 2nd power and porosity to the 4th power.

Using NMR data of samples originating from three sandstone formations, we investigate the relationships between different characteristic relaxation times and r_{eff}. In a further step, we evaluate the predictive quality of the original and modified SDR equations.

The transformation of characteristic relaxation times into characteristic pore radii requires the knowledge of the surface relaxivity [1]. Using the procedure of [2], we determine an individual value of surface relaxivity for each sample. We investigate whether the different characteristic pore radii can be used as suitable proxies for r_{eff}. In a further modification of the SDR equation, we replace the characteristic relaxation time by a characteristic pore radius. We demonstrate that this substantial modification enables an improvement of permeability prediction.

2 Theory

Numerous theoretical approaches for permeability estimation of porous rocks have been presented. They link parameters of pore space geometry to permeability. A simple model, which considers the fluid flow through capillary bundles with uniform pore radius, relates permeability to the pore radius r and the resistivity formation factor F (e.g., [3, 4]):

$$k = r^2/(8F). \tag{1}$$

Using the first Archie equation [5], which describes the relationship between F and porosity ϕ:

$$F = \frac{1}{\phi^m}, \tag{2}$$

where m is the cementation exponent, we get a permeability prediction model that considers the two parameters pore radius r and porosity ϕ:

$$k = \frac{1}{8}\phi^m r^2. \tag{3}$$

The porosity exponent m proves to be variable. For straight and parallel capillaries, an exponent $m = 1$ should be selected, whereas $m = 2$ is a good choice for consolidated sandstones. Larger exponents m are found especially for vuggy reservoir rocks (e.g., [6, 7]). Considering the physical dimensions of k (in m^2) and r (in m), the exponent of r seems to be fixed at two. A more general form of equation 3 reads

$$k = a\phi^b r^2, \tag{4}$$

with two adjustable parameters: the prefactor a and the porosity exponent b. Equation 4 suggests that the quality of

* Corresponding author: andreas.weller@tu-clausthal.de

permeability prediction depends on reliable values of porosity and pore radius. There are several procedures applicable in the laboratory or in well logging that provide good estimates of porosity. Considering the wide variation of pore radii in porous rocks, the determination of the relevant characteristic pore radius is challenging.

The application of equation 4 requires a good proxy for the pore radius r. Regarding equation 1, the effective hydraulic radius

$$r_{eff} = \sqrt{8Fk} \quad (5)$$

can be determined for rock samples with known values of F and k. In this study, we use the effective hydraulic radius concept [8] to evaluate the predictive quality of proxies of the pore radius in equation 4.

The NMR method records the decay of magnetization after an excitation impulse has changed the magnetic orientation of the hydrogen nuclei. The measured transverse decay curve is transformed in a discrete distribution of relaxation times $b_i(T_{2i})$, where the index i is the sequential number of the predefined relaxation time T_2, and b_i is the resulting normalized amplitude. The normalization considers that the sum of all individual b_i equals one [2].

The Schlumberger-Doll Research (SDR) equation for permeability prediction from NMR relaxometry reads (e.g., [9]):

$$k^* = a_1\, \phi^4\, \overline{T_2}^2, \quad (6)$$

where k^* is the predicted permeability and $\overline{T_2}$ a characteristic relaxation time that is determined from the transverse relaxation time distribution. The empirical coefficient a_1 is adjusted for lithology by calibrating with core data. Equation 6 corresponds to the general form of equation 4 with the porosity exponent $b = 4$. The characteristic relaxation time $\overline{T_2}$ is used as a proxy for the pore radius. Different characteristic values are determined from the relaxation time distribution $b_i(T_{2i})$. The original form of the SDR equation applies the weighted geometric mean as a characteristic relaxation time:

$$T_{2lm} = \exp(\textstyle\sum_{i=1}^{n} b_i \ln T_{2i}). \quad (7)$$

Since the geometric mean uses a weighting of the logarithms of T_2, several authors use the name log-mean (e.g., [9]). We adopt this term and use the standard subscript lm for this weighted mean value. Other characteristic values are the weighted harmonic mean T_{2hm} that is calculated by

$$T_{2hm} = \frac{1}{\sum_{i=1}^{n} \frac{b_i}{T_{2i}}}, \quad (8)$$

or the weighted arithmetic mean

$$T_{2am} = \textstyle\sum_i^n b_i T_{2i}. \quad (9)$$

T_{2peak} is another characteristic relaxation time $\overline{T_2}$ that corresponds to the maximum of the relaxation time distribution.

We evaluate the relationships between different characteristic relaxation times $\overline{T_2}$ and r_{eff} for selected sample sets in order to identify the most suitable characteristic relaxation time. The prefactors a_i in equation 6 are adjusted for the individual sample sets.

It has been shown in a variety of studies (e.g., [9, 10]) that a linear relationship between a characteristic pore radius $\bar{r}$ and characteristic relaxation time $\overline{T_2}$ can be assumed:

$$\bar{r} = 2\rho\overline{T_2}, \quad (10)$$

where ρ is the transverse surface relaxivity. We use the approach of [2] to determine an individual value of surface relaxivity for each sample:

$$\rho = \left[S_{por} \times \left(\frac{2\,T_{2min}}{\lambda_2}\right)^{(D_t - D_f)} \times T_{2hm}\right]^{\left(\frac{-1}{1+D_t-D_f}\right)}, \quad (11)$$

where D_f is the fractal dimension and $D_t = 2$ the topological dimension of the pore surface area. The specific surface area per unit pore volume S_{por} is determined from the nitrogen adsorption method with the resolution $\lambda_2 = 0.4$ nm.

Using equation 10 with different characteristic relaxation times and the individual values of surface relaxivity of each sample, we get the weighted characteristic pore radii r_{lm}, r_{hm}, and r_{am}. T_{2peak} is converted into r_{dom} in the same way. We investigate whether the resulting characteristic pore radii are suitable proxies for r_{eff}.

3 Samples and methods

The evaluation of the original and modified versions of the SDR equation is based on data from three different sandstone formations:

1) The first set of samples originates from the Shahejie Formation in the Dongpu Depression, Wenliu Oilfield, being located in the Northeast of Henan Province, China, between latitudes N 35° and N 36° and longitudes E 114°30′ and E115°30′. The geological age of this formation is Eocene Epoch, Paleogene Period (approx. 35-56 Ma old). The samples have been collected in six boreholes in the depth interval between 3332 m and 3738 m. This set of 24 samples was used in studies related to permeability prediction, which were reported in [2] and [11].

2) The second set of samples was collected from the outcrops of Araba Formation in Wadi Saal, east central Sinai, Egypt, between latitude N 29° and N 30° and longitude E 34° and E 35°. This formation is of Carboniferous age or older (Lower Paleozoic). Considering the weak clay content, most of the investigated samples can be classified as clean sandstones. Petrographically, the 14 sandstone samples of the studied Araba Formation can be classified into three facies associations: (i) quartz arenite, (ii) sublithic quartz arenite, and (iii) calcareous quartz arenite. These facies are characterized by the presence of some pyrite and iron oxides as cement [12].

3) The third set, which consists of 20 samples, originates from the Cretaceous Nubia Formation, which reaches an approximate thickness of 40 m. Samples were taken in Wadi Ras El-Esh in the Red Sea area of Eastern Desert, Egypt, at longitude N 27° 27 ′58 ″ and latitude E 33° 30′ 51″. These sandstone beds are yellowish, highly dissected, cross-bedded, and rarely fossiliferous. The sandstone samples were differentiated into five microfacies

https://doi.org/10.1051/e3sconf/202336701014

associations: quartz arenite, lithic quartz arenite, calcareous quartz arenite, quartz wacke, and gypsum quartz arenite. This formation includes both clean and clay-cemented sandstones.

The sample sets of the Araba and Nubia Formations have been integrated in a previous study on permeability prediction that is published in [13].

Cylindrical plugs with a diameter of about 25 mm and a length of about 30 mm have been prepared from all samples. This plug size is suitable for the required petrophysical experiments. Gas permeability k is determined in a steady state flow experiment following the guidelines of [14]. The resistivity formation factor F was measured in an electrical experiment with the plugs saturated with a high salinity brine (~ 15 S/m). The specific surface area per unit mass was determined by nitrogen adsorption method.

The NMR experiments were performed on the MARAN 7 instrument that is operating at a Lamor frequency of 7 MHz. The samples were saturated with a low salinity brine (~ 0.5 g/l NaCl) before the measurements. The signal decay curve, which was measured in the transverse mode, was transformed into a relaxation time distribution (RTD) using a least-squares algorithm that minimizes the deviation between measured and calculated curves. The weighted mean values T_{lm}, T_{hm}, and T_{am} of the RTD consider the interval between $T_{2min} = \tau_{echo} = 0.6$ ms and $T_{2max} = 3$ s.

4 Results

The relationships between characteristic relaxation times and r_{eff} are compared for the three sandstone formations in double logarithmic plots as shown in Figures 1 and 2 for T_{lm} and T_{hm}, respectively. A power law fitting was performed for each sample set. The resulting equations and the coefficient of determination (R^2) are displayed in the legend.

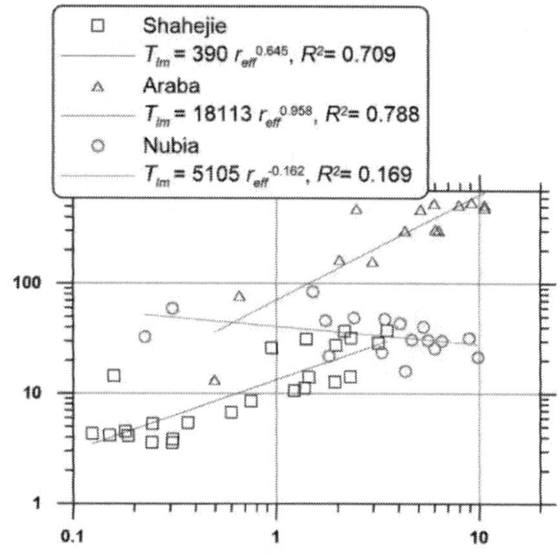

Fig. 1. Relationships between r_{eff} and T_{lm} for three sandstone formations.

The relationships between the logarithm of r_{eff} and the logarithm of the characteristic pore radii r_{lm} and r_{hm} are shown in Figure 3 and 4, respectively. The fitting equation assumes a linear relationship between the characteristic pore radius and r_{eff}.

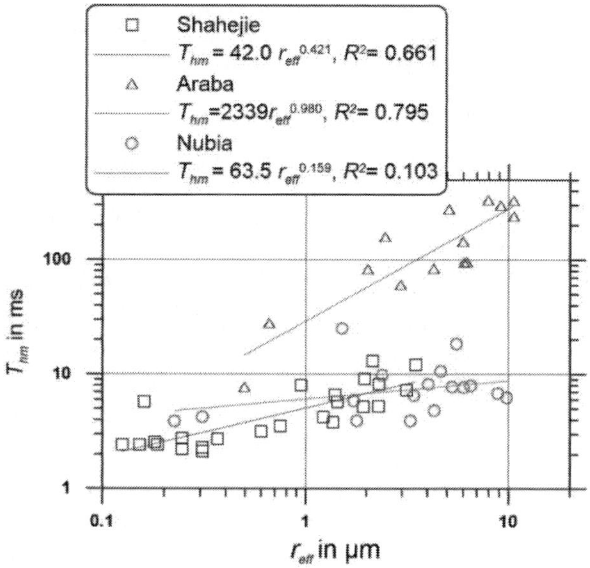

Fig. 2. Relationships between r_{eff} with T_{hm} for three sandstone formations.

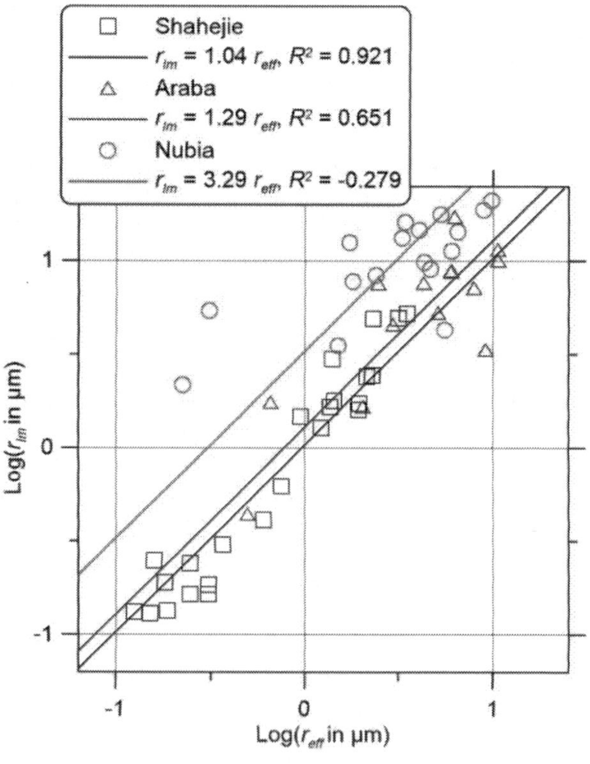

Fig. 3. Relationship between r_{eff} and characteristic pore radii r_{lm} for three sandstone formations.

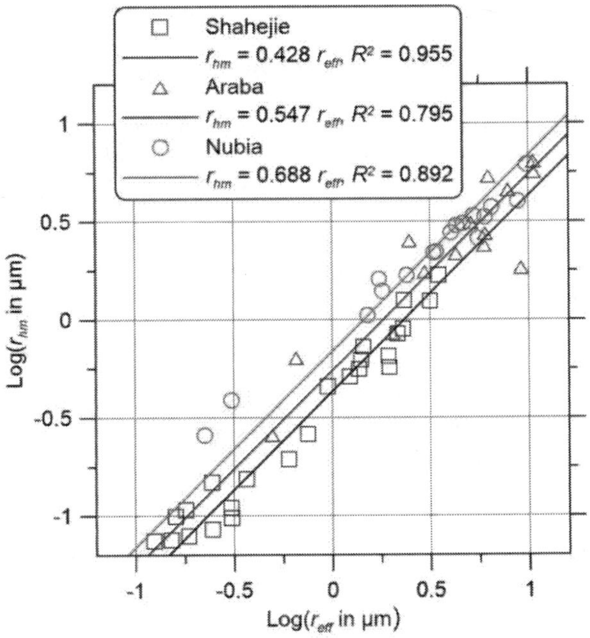

Fig. 4. Relationship between r_{eff} and characteristic pore radii r_{hm} for three sandstone formations.

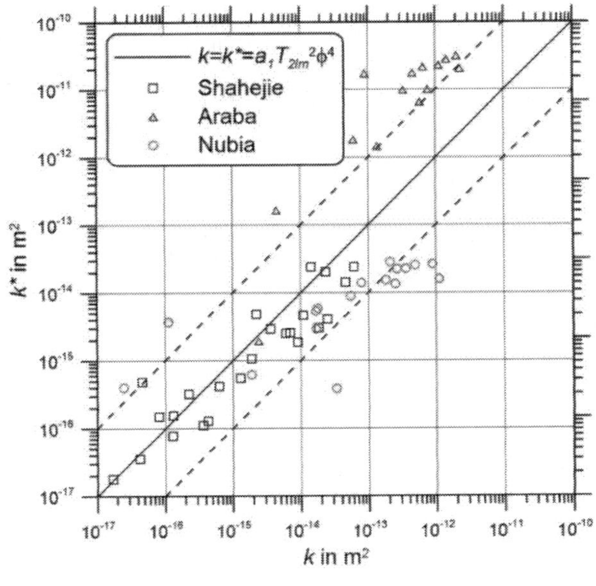

Fig. 5. Comparison of measured permeability k with $k^*=a_1T_{2lm}^{\ 2}\phi^4$ (original SDR equation 6) for three sandstone formations. A uniform prefactor $a_1 = 24.4{\cdot}10^{-9}$ m²/s² has been determined considering the samples of the three formations The two dashed lines on either side of the diagonal indicate a deviation of one order of magnitude from the measured permeability value.

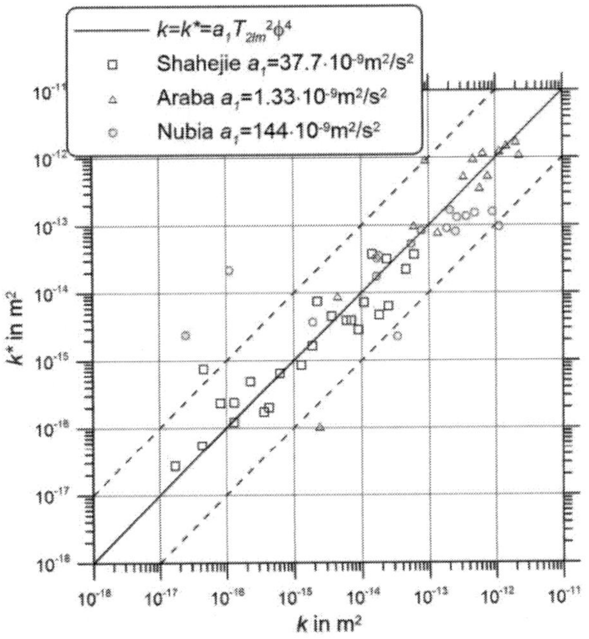

Fig. 6. Comparison of measured permeability k with $k^*=a_1T_{2lm}^{\ 2}\phi^4$ (original SDR equation 6) for three sandstone formations. Individual prefactors a_1 have been determined for the three sandstone formations. The two dashed lines on either side of the diagonal indicate a deviation of one order of magnitude from the measured permeability value.

In order to evaluate the predictive quality of the original or modified SDR equations, we determine the average absolute deviation (in log space) between predicted permeability k^* and the measured permeability k (e.g., [15]):

$$d = \frac{1}{n}\sum_{j=1}^{n}\left|\log_{10}(k_j) - \log_{10}(k_j^*)\right|. \tag{12}$$

A value of $d = 1$ denotes an average absolute deviation of one order of magnitude (or a factor 10). One order of magnitude above or below the measured value is rated as an acceptable estimation if samples of different formations are regarded [16]. In the case of samples originating from the same formation, we aim at a better predictive quality with values considerably smaller than $d = 0.5$.

Figure 5 compares the measured permeability k with the permeability k^* predicted by the original SDR equation 6 with $\overline{T_2} = T_{2lm}$. A prefactor $a_1 = 24.4{\cdot}10^{-9}$ m²/s² has been determined considering all samples of the three formations in the adjustment procedure. We get a moderate agreement between measured and predicted permeability with $d = 0.822$.

The use of a separate adjustment procedure for each sandstone formation yields a better agreement $d = 0.410$ between measured and predicted permeability as demonstrated in Figure 6. The legend indicates the individual prefactors for the three sandstone formations that vary over two orders of magnitude.

Figure 7 displays the comparison between measured permeability k and the permeability k^* predicted by the SDR equation 6 using T_{2hm} instead of T_{2lm}. A uniform prefactor $a_1 = 214.6{\cdot}10^{-9}$ m²/s² has been determined using all samples of this study. The deviation between measured and predicted permeability is $d = 0.936$.

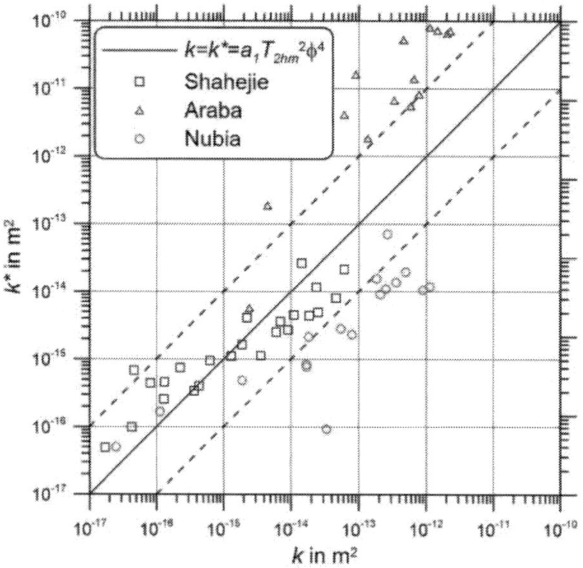

Fig. 7. Comparison of measured permeability k with $k^*=a_1 T_{2hm}^{\ 2}\phi^4$ (equation 6) for three sandstone formations. A uniform prefactor $a_1 = 214.6 \cdot 10^{-9}$ m²/s² has been determined considering the samples of the three formations The two dashed lines on either side of the diagonal indicate a deviation of one order of magnitude from the measured permeability value.

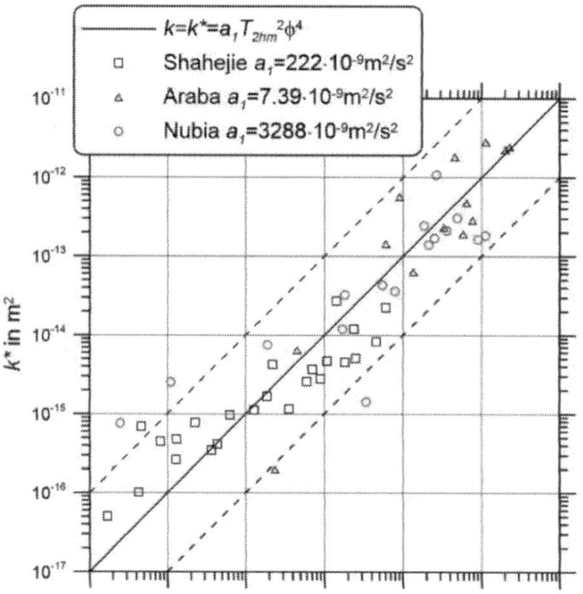

Fig. 8. Comparison of measured permeability k with $k^*=a_1 T_{2hm}^{\ 2}\phi^4$ (equation 6) for three sandstone formations. Individual prefactors a_1 have been determined for the three sandstone formations. The two dashed lines on either side of the diagonal indicate a deviation of one order of magnitude from the measured permeability value.

Figure 8 displays the comparison of measured permeability k and the permeability k^* predicted by the SDR equation using T_{2hm} instead of T_{2lm} with individual prefactors ($d = 0.435$).

The original SDR equation 6 considers a porosity exponent $b = 4$. Previous investigations have confirmed the suitability of this exponent. However, it is questionable whether the same exponent b should be used if characteristic pore radii replace the characteristic relaxations times. We investigate the predictive quality of equation 4 with different characteristic pore radii as proxies for r. The exponent b varies from 0 to 10. Figure 9 displays the average logarithmic deviation d as a function of the exponent b. The curves r_{am}, r_{lm} and r_{hm} indicate a minimum in the investigated range of the exponent b. We get the lowest average logarithmic deviation with $d = 0.211$ for the weighted harmonic mean r_{hm} and a porosity exponent $b = 2$.

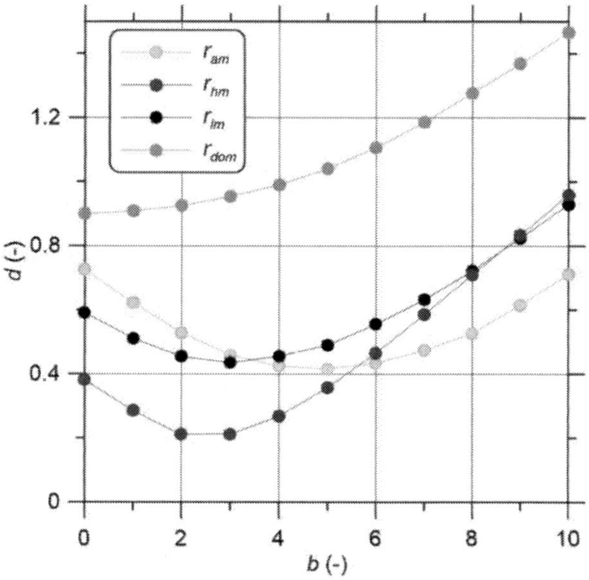

Fig. 9. Average logarithmic deviation d as function of the porosity exponent b in equation 4 using different characteristic pore radii $\bar{r}$ of the three sets of sandstone samples

Figure 10 shows the very good agreement between measured and predicted permeability for the model using the weighted harmonic mean of pore radius in equation 4.

The resulting prefactors a_1 of equation 6 (with $b = 4$) and the prefactors a of equation 4 (with $b = 2$) and the corresponding deviations d are compiled in Table 1.

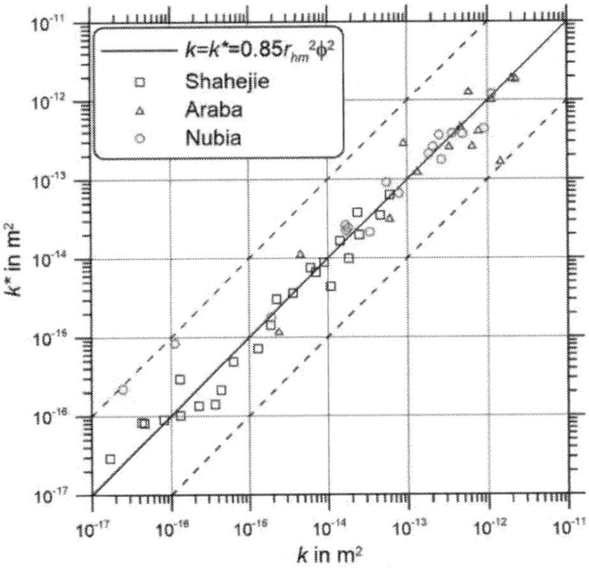

Fig. 10. Comparison of measured permeability k with $k^*=a\ r_{hm}^2\ \phi^2$ (equation 4) for three sandstone formations. A uniform prefactor $a = 0.85$ has been determined considering the samples of the three formations. The two dashed lines on either side of the diagonal indicate a deviation of one order of magnitude from the measured permeability value.

Table 1. The prefactors a_1 of equation 6 (in 10^{-9} m²/s²) and the prefactors a of equation 4 (dimensionless) with the resulting average deviations d (equation 12)

	Shahejie		Araba		Nubia		All samples	
Equation 6 ($b = 4$)	a_1	d	a_1	d	a_1	d	a_1	d
T_{2am}	3.11	0.443	0.54	0.323	17.3	0.769	3.38	0.745
T_{2hm}	221.56	0.411	7.39	0.373	3289	0.520	214.63	0.936
T_{2lm}	37.75	0.314	1.33	0.336	144.42	0.608	24.37	0.822
T_{2peak}	23.47	0.517	0.54	0.437	497.82	1.258	23.13	1.102
Equation 4 ($b = 2$)	a (-)	d	a (-)	d	a (-)	d	a (-)	d
r_{am}	0.01	0.222	0.08	0.445	0.003	0.675	0.01	0.528
r_{hm}	0.92	0.173	1.03	0.253	0.65	0.217	0.85	0.211
r_{lm}	0.16	0.280	0.18	0.401	0.03	0.488	0.10	0.454
r_{dom}	0.10	0.760	0.08	0.525	0.10	1.490	0.09	0.925

5 Discussion

Various models of permeability prediction require porosity and average pore radius as input parameters. The porosity can be derived either from NMR relaxation or by petrophysical laboratory experiments (e.g., triple weighing). The original SDR equation considers the log-mean relaxation time T_{2lm} as a proxy for the pore radius. A linear relationship between r_{eff} and the log-mean relaxation time is expected. We recognize from the graphs in Figure 1 that only the samples of the Araba Formation indicate a nearly linear relationship with a power law exponent 0.956. The slope in the double logarithmic plot is considerably smaller for the samples of the Shahejie Formation. The situation becomes even worse for the Nubia Formation with a negative slope and a coefficient of determination $R^2 = 0.169$.

The weighted harmonic mean T_{2hm} can be used as an alternative proxy for the pore radius. A look at Figure 2 indicates only slight changes. A nearly linear relationship between r_{eff} and T_{2hm} is found for the Araba Formation. The samples of the Nubia formation indicate a weak positive slope. The weighted arithmetic mean T_{2am} indicates similar relationships with weak correlations. The maxima T_{2peak} show a more or less linear relationship with r_{eff} for all three formations with $R^2 > 0.73$ for the Shahejie and Araba Formations but a low value for the Nubia Formation ($R^2 = 0.25$).

The transformation of characteristic relaxation time into a characteristic pore radius uses equation 10 with an individual value of surface relaxivity for each sample. The algorithm of ρ-determination requires an additional expense to get the specific surface area per unit pore volume S_{por}. Considering a wide variation in S_{por} for the samples of a single formation, we get a similar variation in ρ. The graphs in Figure 3 confirm linear relationships between r_{eff} and the characteristic pore radii r_{lm} for the samples of Shahejie and Araba Formations. The situation improves if r_{lm} is replaced by r_{hm}. The resulting graphs in Figure 4 present linear relationships with coefficients of determination $R^2 > 0.79$ for all sandstone formations. The factors of the linear relationship vary in a narrow range between 0.428 and 0.688. Therefore, the characteristic pore radius r_{hm} proves to be a reliable proxy for r_{eff} with a linear relationship $r_{hm} = c\ r_{eff}$. The factor $c = 0.528$ has been determined considering all samples of the three formations. The harmonic mean attributes more weight to the smaller pores or to the pore throats that control the fluid flow. This might be a possible explanation why r_{hm} proves to be a better proxy of r_{eff} in comparison with r_{lm}.

The permeability prediction based on the original SDR equation 6 with $\overline{T}_2 = T_{2lm}$ provides only limited predictive quality if a constant prefactor a_1 is used for all sandstone formations. As shown in Figure 5, the permeability prediction works well for the samples of the Shahejie Formation, whereas the SDR equation overestimates k for most samples of the Araba Formation and underestimates k for some samples of the Nubia Formation. An improvement can be achieved if separate prefactors a_1 are determined for each formation (see Figure 6). The variation in the prefactors is related to the average values of surface relaxivity. We get the lowest prefactor with $a_1 = 1.33 \cdot 10^{-9}$ m²/s² for the Araba Formation, which is characterized by a low average surface relaxivity ($\rho = 9.7$ µm/s), and the highest prefactor with $a_1 = 144 \cdot 10^{-9}$ m²/s² for the Nubia Formation with a higher average surface relaxivity ($\rho = 140$ µm/s). The wide variation in the prefactor a_1 underlines the significance of ρ in permeability prediction with the original SDR equation.

We included other characteristic relaxation times in our study. As summarized in Table 1, the use of T_{peak} causes the largest values of the average deviation d and consequently the lowest predictive quality for all considered sandstone formations. The weighted harmonic mean T_{hm} proves to be advantageous for the Nubia Formation, whereas the weighted arithmetic mean T_{am} results in the lowest d for the Araba Formation.

The direct use of a characteristic pore radius in equation 4 improves considerably the permeability prediction for all sandstone formations. The weighted harmonic mean of pore radii r_{hm}, which proves to be the best proxy for the r_{eff}, enables a high quality permeability prediction with average deviations $d \leq 0.25$ for all considered formations. The porosity exponent $b = 2$ is in good agreement with the theoretical model of equation 4 with $m = 2$. The prefactors a that have been adjusted for the single formations vary only in a narrow range between 0.65 and 1.03. The application of a mean value of the prefactor a provides a reliable predictive quality without any adjustment procedure. However, the better predictive quality can only be achieved if reliable values of surface relaxivity are available for the individual samples. The results of this study confirm the suitability of the approach of ρ-determination as proposed in [2].

6 Conclusions

Three sets of sandstone samples from China and Egypt with a variation of permeability over four orders of magnitude have been selected for the evaluation of the original and modified SDR equations. Our study confirms the applicability of the SDR equation for permeability prediction with the porosity raised to the 4th power. The use of T_{2lm} in the SDR equation has proved to be an appropriate choice. The prefactor in the SDR equation largely depends on the surface relaxivity. Therefore, a calibration procedure has to be applied to adjust the prefactor for a certain lithology.

We modified the SDR equation by replacing the characteristic relaxation time by a characteristic pore radius. The modified equations enable an improved predictive quality. The best results are achieved with weighted harmonic mean r_{hm} that proves to be a reliable proxy for the effective hydraulic radius r_{eff}. However, the transformation of relaxation time into pore radius requires the knowledge of the specific surface area for each sample to determine an individual value of surface relaxivity.

We confirm the potential of the SDR equation in NMR applications. A careful calibration of the equation with suitable values for the prefactor and exponents will contribute to an improved permeability prediction.

Acknowledgements

The authors would like to thank M. Kassab (Egyptian Pertroleum Research Institute) for providing the sandstone samples of the Araba and Nubia Formations and contributing to the petrophysical measurements.

References

1. K. R. Brownstein, C. E. Tarr, Importance of classical diffusion in NMR studies of water in biological cells: Physical Review A, **19**, 2446–2453, doi: 10.1103/PhysRevA.19.2446 (1979).
2. Z. Zhang, A. Weller, Estimating the nuclear magnetic resonance surface relaxivity of Eocene sandstones: A comparison of different approaches: Geophysics, **86**, 2, JM11-JM22, doi: 10.1190/GEO2020-0501.1 (2021a).
3. H. Pape, L. Riepe, J. R. Schopper, A pigeon-hole model for relating permeability to specific surface: The Log Analyst, **23**, 5–13 (1982).
4. Z. Zhang, A. Weller, A comparative study of permeability prediction for Eocene sandstones —Part 1: Application of modified Swanson models to mercury injection capillary pressure and nuclear magnetic resonance data: Geophysics, 86, 6, M233-M243, doi: 10.1190/GEO2021-0194.1 (2021b).
5. G. E. Archie, The electrical resistivity log as an aid in determining some reservoir characteristics: Transactions of American Institute of Mineralogists Metallurgists and Petroleum Engineers, **146**, 54-62 (1942).
6. J. H. Schön, *Physical properties of rocks: A workbook* (Elsevier, 2011).
7. G. Towle, An analysis of the formation resistivity factor-porosity relationship of some assumed pore geometries: Paper C presented at third annual meeting of SPWLA, Houston (1962).
8. A. Weller, L. Slater, Permeability estimation from induced polarization: an evaluation of geophysical length scales using an effective hydraulic radius concept: Near Surface Geophysics, **17**, 581–594, doi: 10.1002/nsg.12071 (2019).
9. R. L. Kleinberg, Utility of NMR T2 distributions, connection with capillary pressure, clay effect, and determination of the surface relaxivity parameter, ρ2: Magnetic Resonance Imaging, **14**, no. 7–8, 761–7, doi: 10.1016/S0730-725X(96)00161-0 (1996).
10. D. O. Seevers, A nuclear magnetic method for determining the permeability of sandstones, in SPWLA 7th Annual Logging Symposium, edited, p. **14**, Soc. of Petrophys. and Well Log Anal (1966).
11. Z. Zhang, A. Weller, Fractal dimension of pore space geometry of an Eocene sandstone formation: Geophysics, **79**, no. 6, D377-387. doi: 10.1190/GEO2014-0143.1 (2014).
12. M. Kassab, A. Weller, H. Abuseda, Integrated petrographical and petrophysical studies of sandstones from the Araba Formation for reservoir characterization: Arabian Journal of Geosciences, 15:944, doi: 0.1007/s12517-022-10221-3 (2022).
13. A. Weller, M.A. Kassab, W. Debschütz, C.D. Sattler, Permeability prediction of four Egyptian sandstone formations, Arabian Journal of Geosciences, **7**, 5171-5183, doi: 10.1007/s12517-013-1188-7 (2014).
14. API (American Petroleum Institute), Recommended Practices for Core Analysis, *API Recommended Practice 40, chapter 6 – permeability determination, second edition*, (1998).

15. A. Weller, L. Slater, A. Binley, S. Nordsiek, S. Xu, Permeability prediction based on induced polarization: Insights from measurements on sandstone and unconsolidated samples spanning a wide permeability range: Geophysics, **80**, 2, D161-D173, doi:10.1190/GEO2014-0368.1 (2015).
16. J. Robinson, L. Slater, A. Weller, K. Keating, T. Robinson, C. Rose, and B. Parker, On permeability prediction from complex conductivity measurements using polarization magnitude and relaxation time: Water Resources Research, **54**, 3436-3452, doi: 10.1002/2017WR022034 (2018).

Towards Multiscale Digital Rocks: Application of a Sub-Resolution Production Model to a multiscale Sandstone

Rafael Salazar-Tio[1,*], *Andrew* Fager[1], *Guangyuan* Sun[1], *Bernd* Crouse[1], *Rui* Xu[1], *Brett* Wendt[2], and *Adam* Lewis[2]

[1]Dassault Systèmes, USA
[2]ConocoPhillips Alaska, USA

Abstract. Many digital rock methodologies use a direct simulation approach, where only resolved pores are accounted for. This approach limits the types of rocks that can be analyzed, excluding some types of carbonates, unconventionals, and complex sandstones from the digital rock analysis. This is due to the challenge for single scale imaging to capture the full range of relevant pore sizes present in multiscale rocks. In this paper, a physical model is presented, within the context of an established direct simulation approach, to predict the production of hydrocarbons including the contribution of sub-resolution pores. The direct simulation component of the model employs a multiphase lattice Boltzmann method to simulate multiphase fluid flow displacement in resolved pores. In the production model, the amount of hydrocarbons present in the sub-resolution pores is identified and a physical description of the production behavior is provided. This allows a relative permeability curve to be predicted for rocks where mobile hydrocarbons are present in pores smaller than the image resolution. This simplified model for the oil movement in the unresolved pore space is based on a physical interpretation of different regions marked by simulation resolution limits in a USBM wettability test curve. The proposed methodology is applied to high-resolution microCT images of a sandstone that contains pores at multiple scales, some resolved and some not resolved. To allow for benchmarking, experimental routine and special core analysis data was also obtained. Good agreement to experimental results is observed, specifically in absolute and relative permeability. The presented multiscale model has the potential to extend the classes of reservoir rocks eligible for digital rock analysis and paves the way for further advancements in the modelling of multiscale rocks, particularly unconventionals and carbonates.

1 Introduction

Reservoir characterization is a critical step in oil field exploration and production. Well logging analysis and petrophysical interpretation is usually supported by laboratory property measurements of physical core samples from the wells, such as porosity, saturations, capillary pressure (P_c), wettability, absolute permeability (K_0) and relative permeability (k_r). Laboratory core analysis can be a time consuming and costly process, and considering that most wells are not core sampled, the value extracted from the available core needs to be maximized, trying to make economically efficient use of these limited core samples [1]. Additionally, core analysis planning needs to consider also that conducting laboratory core analysis experiments sometimes leads to destructive impact on the rock samples. Relative permeability, in particular, is a difficult and time-consuming measurement that, depending on rock type, can take several weeks to a year to complete. With all these considerations, and with the advance of pore scale imaging techniques and computer power in the past decades, digital rock approaches have become more and more a promising technology to complement laboratory measurements [2-4].

Although digital rock technology is promising, there are still challenges when dealing with real applications involving less than ideal rocks, like heterogeneous lithologies or high clay content sandstones for instance. It is computationally costly to simulate a high-resolution large domain model that can resolve pore sizes over multiple orders of magnitude. On one hand, one would like to use a small field of view microCT imaging in order to capture pore-throat connections with a sufficient number of voxels. On the other hand, a large enough field of view is required in order to obtain a statistically representative elementary volume (REV) [5]. For instance in a clay-rich sandstone sample, we could find two different scale of pores sizes: inter granular pores with pore throat diameter around $\sim 10\mu m$scale and clay internal pores around $\sim 10nm$ scale. In order to achieve an REV model, microCT images need to be taken at a voxel resolution consistent with inter granular pores. At this scale clay internal pores fall below the voxel resolution and will be missed. In this case it is impossible to cover all pore scales in one digital rock analysis. There are several studies on the effect of unresolved pores and how to predict petrophysical properties, such as porosity, capillary pressure, absolute and relative permeabilities [6-10, 26-30] but none, to our knowledge, on how to predict relative permeability curves by using multiphase fluid flow from direct numerical simulation in high clay content sandstones, such as the case just described, leading to a model with significant amount of unresolved porosity.

This work investigate this case, a high clay content sandstone, by using an extension of current numerical digital rock workflows [11] to simulate fluid flow in rocks where a connected pore-space is resolved and unresolved pores do not provide a significant contribution to single phase flow,

* Corresponding author: Rafael.SalazarTio@3DS.com

however for multiphase flow can provide critical connectivity paths for the wetting phase and impact the overall behavior of k_r and P_c curves. The two samples used for this work are clay-rich reservoir sandstones; having nearly 50% porosity not resolved at typical microCT resolutions. In the remainder of this paper, we present the methodology and results of this application: A multi-mineral image segmentation is used to identify pores, grains and clay. A capillary pressure based physical model is presented to emulate oil movement in unresolved pore volumes. Porosity, k_0 and k_r are calculated in the total pore space and are compared well with SCAL laboratory measures available for these samples.

2 Method

The workflow contains three parts: pre-simulation work, simulation, and post-simulation work, which we describe in this section.

2.1. Pre-sim work

The main goal of the pre-sim workflow is to ensure that the microCT images for the selected mini-plug are suitable for digital rock analysis with a good resolution in a representative region. In the mini-plug sample preparation step, we consider sample location, sample size and microCT image resolution. Once the microCT image is done, it follows the image processing to obtain a 3D segmented image. The pore space is analyzed for connectivity and sufficient statistics. Finally, an estimation of total porosity is done.

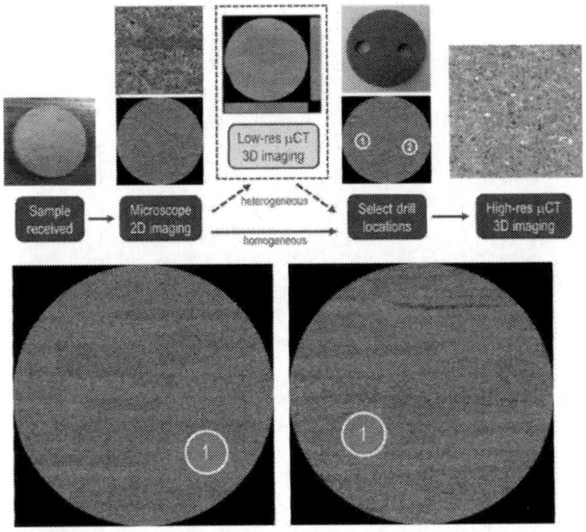

Fig. 1. Schematics of the pre-sim workflow (top). In the bottom, microscopy image of Rock A (left) and Rock B (right) with selected mini-plug locations.

2.1.1 Sample preparation

A schematic of the workflow for sample selection is shown in Figure 1. First, a large panoramic microscopy image is obtained of a polished surface of the available core sample, typically a few inches square at 4 $\mu m/voxel$ resolution. If heterogeneities are observed on the sample, an optional 3D microCT image can be taken of the core material at low resolution, typically 20 $\mu m/voxel$. These images are used in two ways: to select a suitable location for the mini-plug, and to get a rough estimation of grain size. A rule of thumb of 8-10 grains along the diameter of the mini-plug is used for the sample size [5], and the corresponding voxel-size is estimated by dividing the diameter by the number of voxels on the micro-CT camera. For the two samples in this paper, a 2.5 mm mini-plug diameter was estimated, which implies 1.0 $\mu m/voxel$ resolution given the 2500 $voxels$ in the microCT camera. We also show in Figure 1 the two-sample mini-plug locations used in this paper, Rock A, and Rock B.

2.1.2 Image processing

MicroCT images from mini-plugs typically require some image processing before they can be used successfully for segmentation. In the workflow, we first remove artefacts like beam hardening, rings, and intensity biases along the z-axis. The image noise level is reduced to our standard operational range by using a total variation filter, and grey-scale intensity is set to some default reference values. The main objective is to capture an accurate grey-scale interphase between pore-solid, in order to obtain the sub-resolution accuracy of the surface elements that are a defining feature of the LBM flow simulator utilized here.

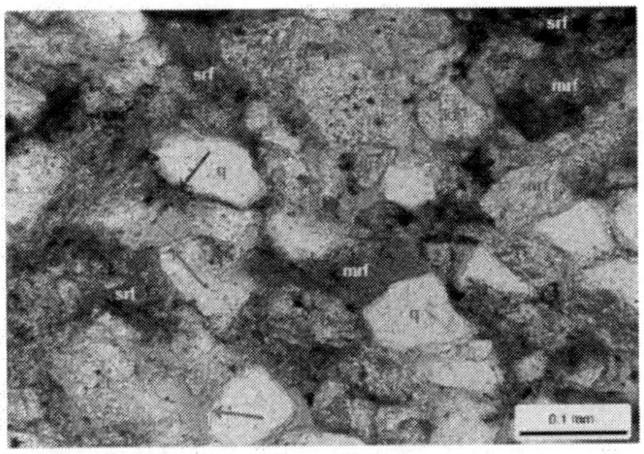

Fig. 2. Different mineral grains in Rock A, showing high clay content and metamorphic rock fragments. Quartz (q), metamorphic (mrf), shale (srf), chert (ct), dolostone (drf), siltstone (ss), volcanic (vrf), K-feldspar (kf), intergranular pores (blue).

In many cases, when complex mineralogy is present on the rock sample (see example in Figure 2), a multi-phase segmentation into different minerals is required. This is particularly important in cases with high clay content, with some partially dissolved grains of feldspars, or in cases with other unresolved porous structures, like in some carbonates. Some of these additional phases can contribute to the total porosity with pores of sizes smaller than the microCT resolution; therefore, an estimation of these fractions is necessary. A multi-mineral segmentation can also be used to model local wettability differences, in mix-wet multi-phase flow simulation scenarios.

An example of a multi-mineral segmentation is shown in Figure 3. This segmentation uses differences in grey values and intensity gradients to identify voxels along the solid-pore boundary and produce the resolved pore structure to be used in the flow simulation. An additional segmentation for just the solid phase was done, mainly to estimate the porosity contribution from the clay regions, by using the differences in texture to separate minerals [25]. The mineral segmentation is then merged back to the solid-pore segmentation as shown in Figure 3 for Rock A, where we can identify the following volumetric fractions: 10.8% resolvable porosity (black), 12.2% microporous clay-type (yellow), grains 65% (light blue), heavy grains 9% (orange), bright grains 3% (dark blue).

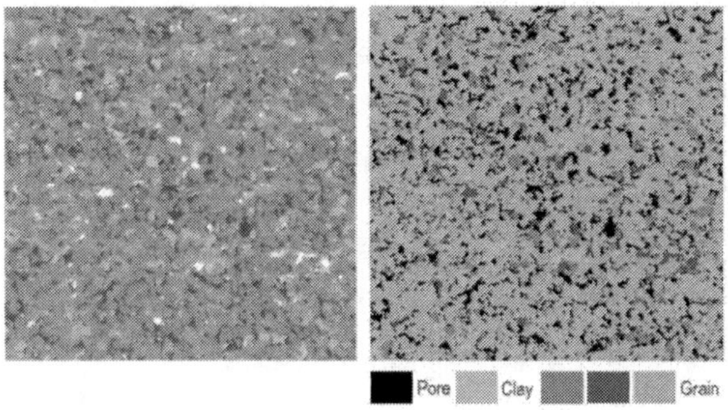

Fig. 3. Sample microCT image for Rock A and corresponding multi-mineral segmented image.

2.1.3 Pore Size Analysis

A pore space analysis (PSA) of the segmented resolvable pore phase is done as part of the pre-sim workflow in order to determine the voxel resolution and REV size for the final simulation model. Starting from calculating a coverable map for maximal inscribed sphere, a pore-size distribution can be computed as the distribution of pore volumes for each sphere radius. In this spherical approximation, the diameter of pores with a large aspect ratio can be underestimated. We also compute another similar volume distribution, but constrained by connectivity to an inlet surface for the model, and for decreasing values of the sphere radius. This connected accessible volume is used to estimate mercury intrusion capillary pressure (MICP) laboratory measurements by using the Young-Laplace relationship between capillary tube radius and capillary pressure, and to compute the critical radius for which the connectivity between inlet and outlet is established. A minimum of 3 $voxels$ is used to ensure accuracy in our multiphase flow simulation results, which corresponds to 3 μm for the microCT resolution used in this study. Finally, we estimate the convergence of the coefficient of variation (COV) of porosity and estimated permeability for different model sizes. COV is defined as the ratio of the standard deviation to the mean value. From the COV convergence analysis we determine what model size is statistically REV.

2.1.4 Estimation of total porosity

Given the physical limit imposed by the voxel-size in the microCT image, the fraction of the total porosity corresponding to pores smaller than the voxel-size cannot be directly measured from the image. However, using some considerations, an estimation of this unresolved porosity can be done. The first consideration follows the method described in [12] to estimate the non-clay unresolved porosity by fitting a Thomeer model for capillary pressure versus saturation as calculated using the PSA-MICP method described in the previous section. Using the resolvable porosity, voxel-size, critical radius and Thomeer G parameter, the non-clay unresolved porosity is estimated [12]. The second consideration involves including porosity contributions of some segmented phases from the image that could potentially have sub-resolution pores, like the clay phase shown in Figure 3 in yellow. Based on some clay mineral porosity studies [13, 14], we consider here an average of 50% contribution of the volume fraction corresponding to this segmented clay phase to complement the porosity estimation.

2.2 Simulation setup

The fluid flow numerical simulation approach used in this study is the Lattice Boltzmann method (LBM), which has been extensively used in digital rock applications, as a direct simulation approach for resolvable pore-scale events and capturing physical properties, such as residual oil on wettability dependence scenarios, capillary number effects and viscous fingering phenomena [11, 15-18]. The LBM is based on the kinetic equation, and represents a statistical description of the molecular behavior of fluid particles. The LBM can be used to simulate the dynamic behavior of fluid flow without directly solving the continuum fluid mechanics equations. Moreover, the LBM based fluid solvers are considered competitive alternatives to traditional Navier-Stokes PDE-based numerical methods [19, 20], particularly in applications involving complex geometries, like porous media flow in digital rock. The decoupling of streaming and collision steps also makes the LBM faster and more efficient in terms of computational efforts, at least compared to classical FE/FD-solvers.

The LBM solver used for this work is based on an extension of the multiphase Shan-Chen model (SC-LBM) [21]. Some of the recent extensions improve numerical stability and accuracy under the operating conditions required for digital rock workflows, such as low-resolution models and it includes the ability to deal with high viscosity ratios of the immiscible fluid phases [22-24]. In this multiphase LBM model, the fluid phase separation is the result of interaction forces between fluid components, while wettability is determined by interaction forces between fluid and pore walls [22]. Regarding the model geometry used in this LBM solver, in addition to the 3D volume elements corresponding to the voxels from the microCT scanned images, pore/solid surface elements, called surfels, are also used, which allow high fidelity representation of the pore/solid geometry interface with an effective sub voxel resolution accuracy [25, 26]. No slip boundary condition is applied between fluid and surfels.

2.2.1 Absolute permeability setup

Single-phase fluid flow can be simulated with the LBM solver in the resolved pore space. As per definition in Darcy's law, permeability is the linear proportionality ratio between flow rate and pressure gradient, which in this study is realized by a driving force applied in the flow direction, another option is to use pressure boundary conditions. The setup for this simulation is used to simulate absolute permeability individually in each direction: x, y, z. A periodic boundary condition is applied in the flow direction, while no-flow boundary conditions are used in the directions transversal to the flow. The permeability is calculated using Darcy's Law when flow rate have converged.

2.2.2 Relative permeability setup

A similar setup as in single phase absolute permeability is used here, regarding the driving force applied in the flow direction, and no-flow boundary conditions in the directions transverse to flow, but using a multiphase LBM solver within a steady-state relative permeability displacement setup. In this setup, the water fractional flow starts from zero and keeps increasing until a predefined water saturation is reached. After both water and oil flow become stable, effective permeabilities are calculated and fractional flow increases again until the next water saturation.
Additional details for this method can be found in [11]. In multiphase fluid flow simulation, another parameter that needs to be defined is surface wettability by assigning a contact angle value to each surfel element. In our setup we consider a distribution of contact angles specified throughout the surface based on local contact by oil, as illustrated in Figure 4, where some pores/solid surface regions are considered more water wet and others more oil/neutral wet.

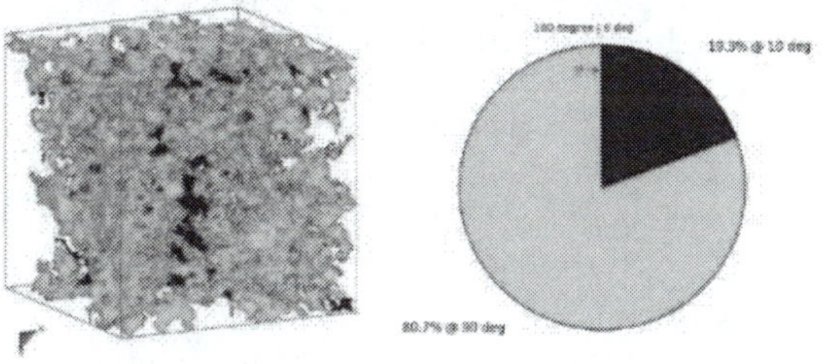

Fig. 4. Wettability setting for Steady-State Kr simulation.

The steady-state relative permeability displacement is conducted at constant capillary number N_c. The definition of capillary number used in this work is:

$$N_c = \frac{U_w * \mu_w + U_o * \mu_o}{IFT} \quad (1)$$

Where $U_{w,o}$ are the superficial velocities of water and oil, and $\mu_{w,o}$ are the dynamic viscosities of water and oil, while IFT is the interfacial tension between water and oil. In this work we use $N_c = 5 \times 10^{-7}$.

2.3 Post-sim and multi-scale model

2.3.1 Absolute permeability post-processing

Absolute permeability is calculated based on Darcy's law

$$K_{eff} = \frac{\nu * U}{g} \quad (2)$$

Where, ν is the kinematic viscosity, U is the volumetric superficial velocity from simulation, and g is the body force applied in the flow direction. The resulting absolute permeability value reflects only the flow contribution from the resolved pore space, while the contribution from the unresolved pore space is considered negligible in comparison. For the high clay content (above 10%) sandstones we are considering in this study, up to half of the pore space can be unresolved in the microCT images. However, as pointed out earlier, pores within clay can be several orders of magnitude smaller than inter granular resolved pores. Since permeability is proportional to pore-throat radius square, the flow contribution from clay pores is negligible when compared to the contribution from the resolvable pores, therefore the need for a multiscale procedure for absolute permeability in this case is less important [9].

2.3.2 Multi-scale post-processing for relative permeability

The resolved pore space relative permeability values for water and oil are calculated by Darcy's law at each converged saturation [11]. In this multiphase flow scenario, the fractional flow contribution from clay pores is also negligible when compared to the contribution from the resolvable pores, therefore we also consider here negligible the contributions of unresolved clay pores to the total flow, and use as a good approximation to the total flow the simulated flow only in the resolved pores, and the corresponding calculated relative permeability values K_r. However, the representation of $K_r(Sw_{sim})$ curves only considering saturations in the resolved pore space Sw_{sim} is incomplete and unsuitable for comparison with experimental results. The correct unknown curves are $K_r(Sw_{total})$ that consider saturations in the total pore space Sw_{total} instead. In this section, we explain how we can use a simplified model for the oil movement in the unresolved pore space and how we can rescale Sw_{sim} to Sw_{total} at each permeability calculation. This simplified model is based on a physical interpretation (capillary, wettability) of different regions marked by simulation resolution limits in a USBM wettability test curve.

One of the challenges in computing relative permeability curves in high clay content sandstones is that the unresolved pore space percentage can be larger than the initial water saturation percentage in experimental measurements. This indicates that there is oil in the unresolved pore space too. Such an amount of oil can be calculated from a USBM wettability test result. Figure 5 shows a schematic plot of a wettability test, illustrating the relationship between capillary pressure and water saturation. Let us consider the Young-Laplace equation to interpret the simulation resolution limits in this curve

$$\Delta P = \frac{IFT * 2\cos\theta}{R} \quad (3)$$

Assuming a contact angle $\theta = 0$ near the end of USBM test and a minimum value for the pore radius R given by the microCT resolution, we can estimate a maximum limit for the capillary pressure ΔP that can be simulated in the resolved pore space of $IFT \times 2/resolution$, where IFT is the interfacial surface tension between oil and water. Finer resolution values could resolve smaller pores and simulate a larger range of pressures, but at a higher computational cost.

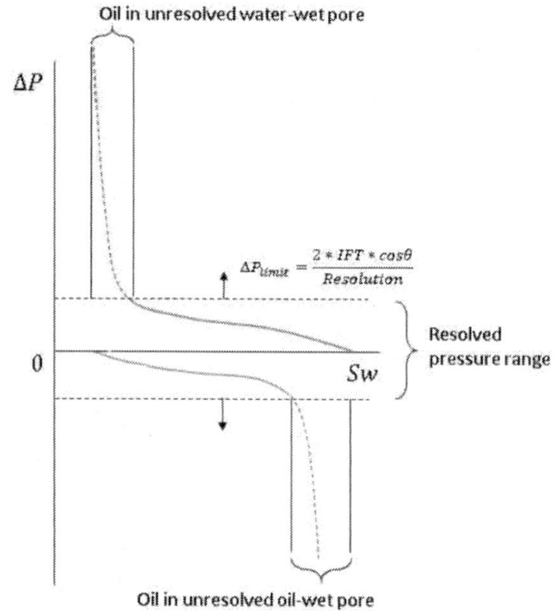

Fig. 5. Schematic plot illustrating the relationship between resolution, capillary pressure and amount of oil in unresolved water-wet and oil wet pore space.

The direct numerical simulation for relative permeability K_r can only be performed within the resolved pressure range shown in Figure 5. The red dash line in Figure 5 shows the amount of oil in the unresolved water-wet pore, which will be spontaneously imbibed before the beginning of the K_r simulation. The blue dash line shows the amount of oil in the unresolved oil-wet pore, which will not be displaced by water in the K_r simulation. For the rest of unresolved oil, we can assume a mix wet condition and model the oil displacement linearly during the K_r simulation. Figure 6 shows a schematic plot of how and when each part of the unresolved oil changes during the K_r simulation in the proposed model.

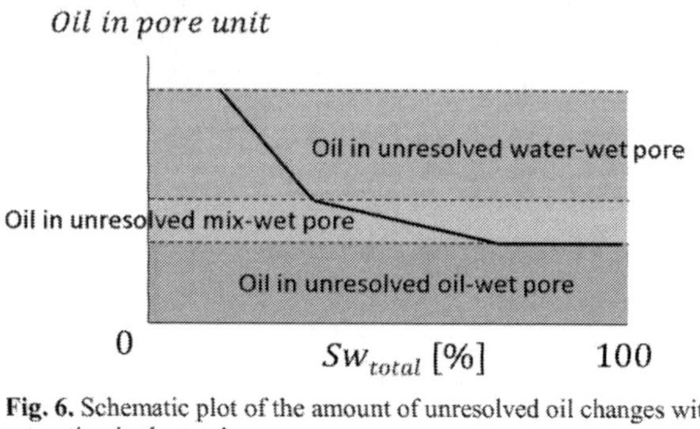

Fig. 6. Schematic plot of the amount of unresolved oil changes with saturation in the total pore space.

Since the proposed simplified model for the oil movement in the unresolved pore space is the main contribution in this paper, it is worthy to present next a summary of the steps to implement the model:

1. Provide an estimate of the total porosity and decompose it in resolvable and unresolvable porosities. Provide an estimate of the initial water saturation with respect to the total porosity.
2. Compute the amount of water/oil in the unresolvable pore space using the previous quantities.
3. Compute resolvable pressure simulation range using Equation (3); use it in the USBM wettability test curve to identify the regions marked in Figure 5, and estimate the allocation of oil in the unresolved pore space in 3 categories:
   a. water-wet: un-accessible pore volume above the max simulation pressure ΔP
   b. oil-wet: un-accessible pore volume below the min simulation pressure $-\Delta P$
   c. mix-wet: remaining pore volume to complete the total amount of oil in the unresolved pore space
4. Use a simple oil production model in the unresolved pore space (Figure 6):
   a. Oil in unresolved water-wet pore (blue): displaced before the simulation starts.
   b. Oil in unresolved oil-wet pore (red): never displaced by water.
   c. Oil in unresolved mix-wet pore (green): displaced from beginning to end of simulation using a linear rate.

The physical interpretation behind this model is based on simple relations between capillary displacements for different wettability and pore-size cases.

3 Results

We present in this section the results of the application of the methodology detailed in the previous sections to two reservoir rock samples: Rock A, and Rock B, both being tight sandstones with more than 10% clay content, some organic content, and other non-quartz minerals. As indicated in section 2.1.1 the microCT voxel-size used for both samples is 1 $\mu m/voxel$, and the REV estimated from the COV analysis described in section 2.1.3, about 800 voxels with a COV value around 25%. We select 1000^3 $voxels$ model size for porosity analysis and K_0 simulations, which are not too computationally expensive, but we just use 800^3 voxels for K_r simulations that are computationally more demanding.

3.1 Porosity

We apply the method described in section 2.1.4 to the segmented models of Rock A and Rock B. It is worth to notice that these rock samples have about half porosity unresolved below 1 μm, being from a digital rock perspective very challenging to characterize, however the resolved porosity is connected and the critical radius is properly resolved for flow simulations. The different values of porosity contributions are presented in Table 1.

 https://doi.org/10.1051/e3sconf/202336701015

Table 1. Porosity Calculation of Rock A, B

	Rock A	Rock B
Experimental	20.7%	19.6
Resolved	10.8%	10.2%
Resolved connected	10.2%	9.5%
Correct resolved connected	13.78%	12.37%
Clay image fraction	12.2%	13.9%
Clay porosity (assuming 50%)	6.1%	6.95%
Total porosity (calculated)	19.88%	19.32%

Experimental values of porosity are available for these two samples, and are compared with the image based total porosity estimation in Figure 7. The differences between experimental values and image estimations are within 5%, which is quite remarkable for samples with such significant unresolved porosity. Given the availability of the experimental porosities, for the sake of accuracy in the following simulations, we will be using the experimental values of total porosity; however, image porosity estimation can be used if experimental values were not available.

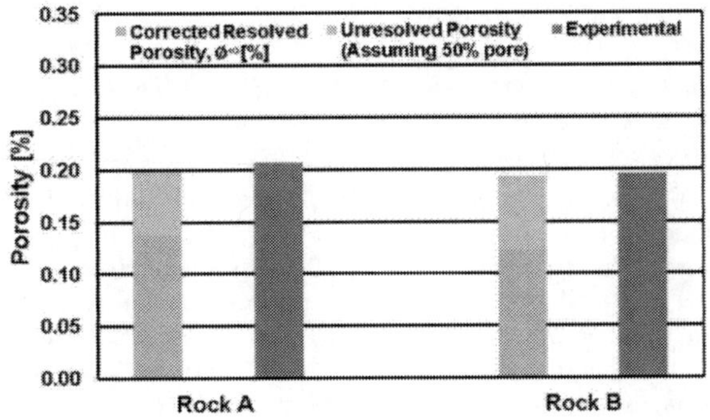

Fig. 7. Porosity Calculation of Rock A, and Rock B

3.2 Absolute Permeability

The results of the single-phase flow simulation setup described in section 2.2.1 in the resolved pore space are shown in Figure 8 where the flow velocity magnitude is represented in a color log scale. Flow in large connected channels contributes the most to the average flow speed and total permeability. Usually once the main flow paths in the pore space backbone are properly captured by the resolved porosity, the unresolved pore space contribution to flow seems to not be relevant anymore.

In Figure 9 we present the resolved pore space absolute permeability simulation results of Rock A and Rock B in x, y, z directions compared with the experimental values. The average simulated absolute permeability for Rock A is $21.2\,mD$ versus the experimental value $19.7\,mD$. The average simulated absolute permeability for Rock B is $16.0\,mD$ versus the experimental value $14.5\,mD$.

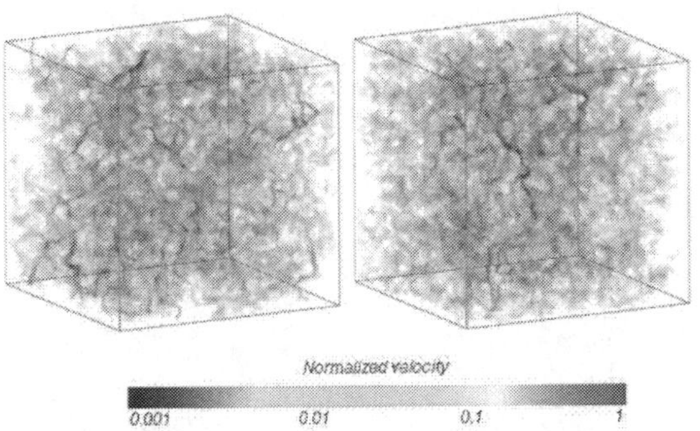

Fig. 8. Resolved pore space 1000^3 absolute permeability simulation in z-direction result of rock A (left) and rock B (right)

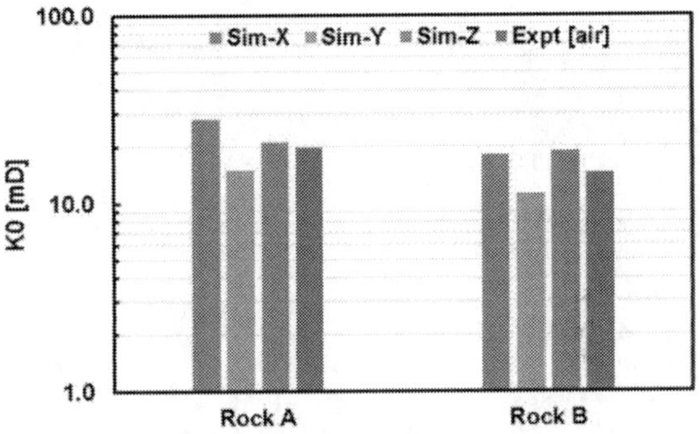

Fig. 9. Absolute permeability K_0 in x, y, z direction vs. experimental data

3.3 Relative Permeability

Following the methodology described in section 2.2.2, we setup the system wettability based on previous experience, knowledge of reservoir properties and USBM wettability lab measurements, as weakly water-wet to neutral-wet condition, by assigning a combination of 10 and 90 $degrees$ contact angles as shown in Figure 4. The initial water saturation in the simulation for the resolved pore space must be considered together with the unresolved initial water saturation, to be consistent with the experimental expected initial water saturation. In the simulation for the resolved pore space, we use an initial water saturation close to zero by assuming all water is in the unresolved pore space. This setup is consistent with the experimental initial water saturation for Rock A, which is 32.3%, considering that from the microCT images we can only resolve 52.2% of the total pore space, being 47.8% unresolved pore space missing in simulation. We use the same wettability and initial water setup for Rocks A and B.

In Figure 10 we illustrate the resolved pore space K_r simulation results for Rock A with two snapshots of oil and water distribution at early and late stages of displacement. At the bottom of Figure 10, we show the oil blobs color coded by blob size in log scale. Dark red represents the largest blob. Blue represents the smallest blob. At initial stages oil is well connected. At late stages when water saturation increases, oil blobs are more disconnected as more oil blobs are snapped off. The corresponding steady-state relative permeability curve calculated from simulation in the resolved pore space $K_r(Sw_{sim})$ is shown in Figure 12 using green lines.

Fig. 10. Resolved pore space 800^3 voxels model for Rock A and steady-state relative permeability simulation results. Water (green)-oil (yellow) iso-surface visualisation (top). Oil blob iso-surface visualisation by blob size (dark red is largest blobs, blue is smallest blobs) (bottom).

We introduced an unresolved pore space oil production model in section 2.3.2 to describe how oil moves in the unresolved pore in order to calculate the correct $K_r(Sw_{total})$ curves for the total pore space by using USBM lab test results. In this study, the microCT resolution is 1.0 $\mu m/voxel$, IFT is 26 mN/m, so using Equation (3), the pressure limit is 7.54 psi. In Figure 11 we show the USBM test results for Rock A together with the estimated resolved pore space simulation pressure limits.

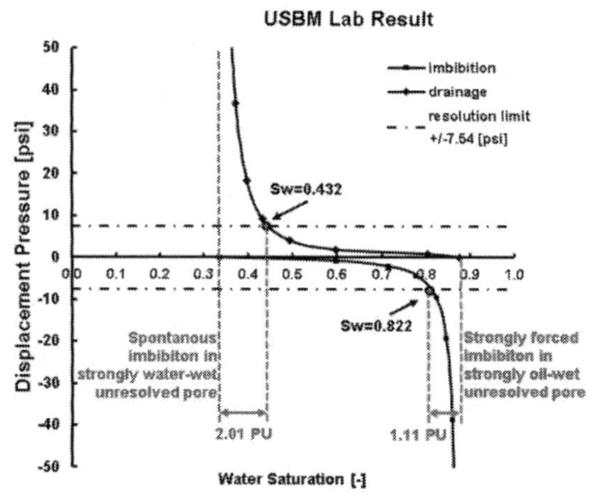

Fig. 11. USBM wettability test results for Rock A

The total porosity is 20.7% for Rock A, which corresponds to 20.7 Pore Unit (PU). Figure 11 shows that there is 2.01 PU of oil in unresolved water-wet pores and 1.11 PU of oil in unresolved oil-wet pores. Table 2 shows how the percentage of oil in neutral-wet unresolved pores is calculated, assuming $Sw_i = 0$ in the resolvable pore space simulation.

Table 2. Pore Unit calculation of initial oil in the unresolved pore space for Rock A: water-wet, neutral-wet and oil-wet.

	Percentage	Pore Unit
Total porosity (Lab)	20.7 %	20.7 PU
Resolved pore (Simulation pore space)	10.2 %	10.2 PU
Unresolved pore	10.5 %	10.5 PU
Connate water saturation in total pore space	32.2 %	6.67 PU
Assumed initial water in unresolved pore		6.67 PU
Derived initial oil in unresolved pore		3.83 PU
From USBM Result:		
Oil in water-wet unresolved pore		2.01 PU
Oil in oil-wet unresolved pore		1.11 PU
Oil in neutral-wet unresolved pore		0.71 PU

Following the linear production oil model illustrated in Figure 6, and using the input values from Table 2, we compute the red curves in Figure 12, that shows $K_r(Sw_{total})$ curves in total pore space after rescaling water saturation from resolved pore to total pore in sample Rock A, the black dots are lab test results. Overall, the simulation results and the experimental values match well in most of the K_r curves, which is remarkable considering that the resolved pore space in the simulation is only half of the total pore space. The mismatch of the first K_r point could be because the simulation computes the oil permeability at a lower oil saturation and a lower capillary pressure compared with lab, when oil has already been fully drained from the unresolved water-wet pore space. The direct numerical simulation reaches S_{OR} at the end of the solid line. An additional step is used to model the oil displacement through the unresolved pore space. The mismatch on the S_{OR} point could be due to such an effect. Another baseline for comparison can be obtained by compressing the green curve in the range 50%-100%, which corresponds to considering the half-unresolved porosity filled with unmovable water; this will overestimate the saturation crossing point.

Figure 13 shows $K_r(Sw_{sim})$ curves for Rock B simulated in the resolved pore space and $K_r(Sw_{total})$ in the total pore space by applying the same sub-resolution oil production model as per Rock A. Due to lack of lab results, a possible S_{OR} point from extended run was not done here for comparison. The simulated prediction for K_r is promising since Rock A and Rock B are similar rock types. Both have similar average grain size, belong to the same formation, have similar mineralogy components, and clay content, the use of a similar oil production model is therefore justified.

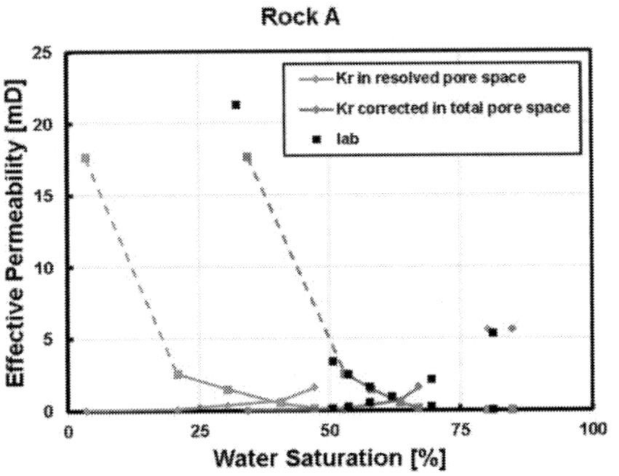

Fig. 12. $K_r(Sw_{sim})$ curves in the resolved pore space from direct simulation, $K_r(Sw_{total})$ curves in the total pore space using a production oil model, lab Kr results for Rock A.

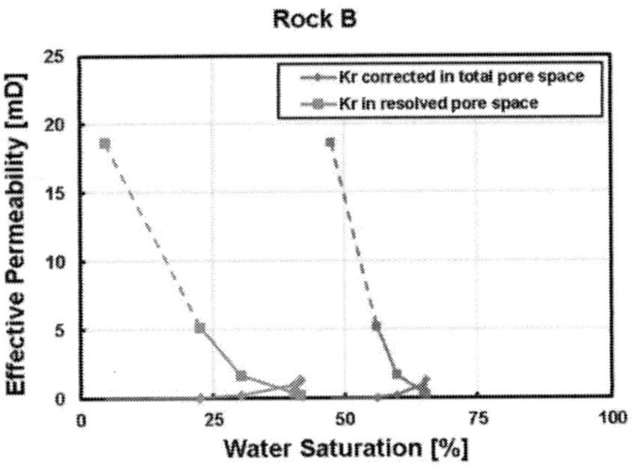

Fig. 13. $K_r(Sw_{sim})$ curves in the resolved pore space from direct simulation, $K_r(Sw_{total})$ curves in the total pore space using a production oil model, for Rock B.

4 Conclusions

A novel physics based sub-resolution oil production model is presented, which together with a digital rock workflow for multiphase fluid flow is applied to two clay rich sandstone reservoir rocks containing pores at two different scales. This simplified model is based on a physical interpretation (capillary, wettability) of different regions marked by simulation resolution limits in a USBM wettability test curve. For these high clay content sandstones, with half of the pore space below the microCT image resolution, the current workflow predicts porosities within 5%, absolute permeability within 10% and relative permeability K_r curves consistent with existing lab results. The current workflow demonstrate that is possible to use simplified models to extend the scope of application of digital rock technology to some challenging rock types, like high clay content sandstones, without having to explicitly simulate fluids movement in the unresolved pore regions using other multiscale multiphase flow methods such as [26-30]. In cases where the main percolating path is capture, but still a significant pore space fraction is unresolved, the method proposed in this paper could be computationally less expensive than full multiscale multiphase fluid flow methods [26-30].

References

1. J.J. Buckles, R. D. Hazlett, S. Chen, K. G. Eggert, D. W. Grunau, W. E. Soll, *Los Alamos Science*, **22**, 112-121, 1994.
2. S. Bakke, P. E. Øren, SPE 35479, *SPE Journal*, **2**, 136-149, 1997.
3. H. Andrä, N. Combaret, J. Dvorkin, E. Glatt, J. Han, M. Kabel, Y. Keehm, et al., *Computers & Geosciences*, **50**, 25–32, 2013.
4. H. Andrä, N. Combaret, J. Dvorkin, E. Glatt, J. Han, M. Kabel, Y. Keehm, et al., *Computers & Geosciences*, **50**, 33–43, 2013.
5. J. Bear, *Dover Publications Inc*., 1988.
6. H. Long, C. Nardi, N. Idowu, A. Carnerup, M. A. Knackstedt, T. Varslot. SCA 2013-013, *Int. Symp. Soc. Core Analysts*, 2013.
7. S. Zhang, A. Byrnes, J. Howard, URTEC 2019-416, *Unconventional Resources Technology Conference*, Denver, Colorado, USA, 2019.
8. S. N. Apourvari, C. H. Arns, *Advances in Water Resources*, **95**, 161-175, 2016.
9. M. S. Jouini, S. Vega, A. A. Ratrout, *Geophysical Prospecting*, **63**, 405-421, 2014.
10. H. Zhang, P. Yuan, J. Wu, A. Mezzatesta, G. Jin, R. Satti, N. Koliha, J. Bautista, B. Crouse, and D. Freed, SPE 183883, *Middle East Oil and Gas Show & Conf.*, 2017
11. G. R. Jerauld, J. Fredrich, N. Lane, Q. Sheng, B. Crouse, D. M. Freed, A. Fager, and R. Xu, SPE 188688, *2017 SPE Abu Dhabi Int. Pet. Exhib. & Conf.,* Abu Dhabi, U.A.E., Nov. 13-16, 2017.
12. N. Saxena, A. Hows, R. Hofmann, J. Freeman, M. Appel, *Transp Porous Med* **129**, 403–412, May, 2019
13. P.H. Nadeau, A. Hurst, *Journal of Sedimentary Research*, **61**, 921-925, 1991.
14. A. Alansari, A. M. A. Salim, H. T. Janjuhah, A. H. B. A. Rahman, N. M. Fello, *Journal of Natural Gas Geoscience*, **4**, 139-150, 2019
15. B. Crouse, D. M. Freed, N. Koliha, G. Balasubramanian, R. Satti, D. Bale, and S. Zuklic, SCA 2016-058, *Int. Symp. Soc. Core Analysts*, Snow Mass, CO, USA. Aug. 2016.
16. R. Xu, B. Crouse, D. M. Freed, A. Fager, G. R. Jerauld, N. Lane, Q. Sheng, SCA 2018-066, *Int. Symp. Soc. Core Analysts,* Trondheim, Norway, Aug. 2018.
17. G. Sun, B. Crouse, D. M. Freed, R. Xu, J. Bautista, R. Zhang, H. Otomo, Y. Li, H. Chen, H. Fan, M. Dressler, *Fuentes, el reventón energético*. **16**, no. 2, 2018.
18. A. Fager, B. Crouse, G. Sun, R. Xu, D. M. Freed, SPE 195734, *2019 SPE Offshore Europe Conf. & Exhib.*, Aberdeen, UK, Sep. 2019

19. S. Chen, H. Chen, D. Martinez, and W. Mattheus, *Phys. Rev. Lett.*, **67**, no. 27, 3776-3779, Dec. 1991.
20. H. Chen, S. Chen, W. Matthaeus, *Phys. Rev. A*, **45**, no. 8, R5339-R5342, Apr. 1992.
21. S. Chen and G. Doolen, *Annu. Rev. Fluid Mech.,* **30**, no. 1, 329-264, 1998.
22. H. Otomo, H. Fan, R. Hazlett, Y. Li, I. Staroselsky, R. Zhang, R., and H. Chen, H., *Comptes Rendus Mécanique*, **343**, 559 (2015).
23. H. Otomo, H. Fan, Y. Li, M. Dressler, I. Staroselsky, R. Zhang, and H. Chen, Jour. Comp. Sci., 17, 334 (2016).
24. H. Otomo, B. Crouse, M. Dressler, D.M. Freed, I. Staroselsky, R. Zhang, H. Chen, Computer Fluids, 172, 674 (2018)
25. I. Aranda-Carreras, V. Kaynig, C. Rueden, K.W. Eliceiri, J. Schindelin, A. Cardona, and H. Sebastian Seung, Bioinformatics, 33, 2424 (2017)
26. S. K. Masalmeh, X. Jing, S. Roth, C. Wang, H. Dong, M. Blunt, Society of Petroleum engineers SPE-177572-MS (2015)
27. L. C. Ruspini, G. Lindkvist, S. Bakke, L. Alberts, A. M. Carnerup, and P. E. Øren, Society of Petroleum Engineers SPE-180268-MS (2016)
28. O. Dinariev, N. Evseev, D. Klemin, SCA 2019-002, *Int. Symp. Soc. Core Analysts, 2019.*
29. M. Suhrer, X. Nie, J. Toelke, Proceedings of international petroleum technology conference, IPTC 20035 (2020)
30. A. Fager, H. Otomo, R. Salazar-Tio, G. Balasubramanian, B. Crouse, R. Zhang, H. Chen, J. Schembre-McCabe, *Int. Symp. Soc. Core Analysts, 2021*

AUTHOR INDEX